Mutation Breeding in Coffee with Special Reference to Leaf Rust

Ivan L. W. Ingelbrecht ·
Maria do Céu Lavado da Silva ·
Joanna Jankowicz-Cieslak
Editors

Mutation Breeding in Coffee with Special Reference to Leaf Rust

Protocols

IAEA

Joint FAO/IAEA Centre
Nuclear Techniques in Food and Agriculture

Editors
Ivan L. W. Ingelbrecht
Plant Breeding and Genetics Laboratory,
Joint FAO/IAEA Centre of Nuclear
Techniques in Food and Agriculture
IAEA Laboratories Seibersdorf
International Atomic Energy Agency
Vienna International Centre
Vienna, Austria

Joanna Jankowicz-Cieslak ⓘ
Plant Breeding and Genetics Laboratory,
Joint FAO/IAEA Centre of Nuclear
Techniques in Food and Agriculture
IAEA Laboratories Seibersdorf
International Atomic Energy Agency
Vienna International Centre
Vienna, Austria

Maria do Céu Lavado da Silva ⓘ
Associated Laboratory TERRA
LEAF—Linking Landscape, Environment
Agriculture, and Food Research Center
CIFC—Centro de Investigação das
Ferrugens do Cafeeiro, Instituto Superior de
Agronomia
Universidade de Lisboa
Lisbon, Portugal

This Springer imprint is published by the registered company Springer-Verlag GmbH, DE, part of Springer Nature.
The registered company address is: Heidelberger Platz 3, 14197 Berlin, Germany

Preface

Since its establishment in 1964, the Joint FAO/IAEA Centre of Nuclear Techniques in Food and Agriculture has played a significant role in fostering the use of induced mutagenesis for crop improvement in FAO and IAEA Member States to help tackle transboundary plant pests and diseases, enhance food and nutrition security and adapt to climate change. This is being done primarily by coordinating and supporting demand-driven R&D and technology transfer, by providing crop irradiation services, and by collecting and disseminating information on plant mutation breeding.

Mutation induction in plants aims to generate novel genetic diversity for plant breeders targeting increased yield, improved quality and enhanced resistance to biotic and abiotic stresses. Plant mutation breeding has a track record of success with global impact on agricultural productivity. To date, over 3300 mutant varieties have been released in more than 220 crop species as listed in the Mutant Variety Database of the Joint FAO/IAEA Centre (https://mvd.iaea.org/). Since the turn of the twentieth century, mutation breeding has become fully integrated with advanced biotechnology, genomics and informatics tools adding precision, speed and efficiency to the mutation breeding process.

A Coordinated Research Project (CRP) titled "Efficient Screening Techniques to Identify Mutants with Disease Resistance for Coffee and Banana" (D22005; December 2015–November 2020) was launched by the Joint FAO/IAEA Centre to develop innovative R&D tools and protocols and investigate whether induced mutagenesis could generate useful genetic variation in banana and coffee leading to resistance to Fusarium Wilt Tropical Race 4 and Coffee Leaf Rust, respectively. A first protocol book ensuing from this CRP on mutation induction and screening techniques in banana for resistance to the devastating Fusarium Wilt Tropical Race 4 strain was published in 2022 (Efficient Screening Techniques to Identify Mutants with TR4 Resistance in Banana: Protocols | SpringerLink). This second CRP protocol book is focused on mutation induction and screening techniques of Arabica coffee with special reference to Leaf Rust. Arabica coffee provides a source of income to nearly 125 million people worldwide. Over 90% of the production takes place in developing countries. The first chapter introduces general principles and practices for mutation-assisted breeding along with current breeding limitations of Arabica coffee. A second introductory chapter provides an overview of the Coffee Leaf Rust disease, a major threat to Arabica coffee cultivation, especially in Latin America (Climate Change and Coffee: Combatting Coffee Rust I IAEA).

Since the 1920s, natural spontaneous mutant traits of significant economic value have been found in Arabica coffee plantations or collections such as dwarfism, fruit size and colour, and reduced caffeine content. However, to our knowledge, no Arabica coffee variety has been released following induced mutagenesis and studies on mutation-assisted breeding of Arabica coffee are scarce. Hence, a major objective of the CRP was to establish robust protocols and conditions for mutation induction using physical and chemical mutagens. Within the genus Coffea, Arabica coffee is unique because it is a self-pollinating, amphidiploid species unlike the other species such as *Coffea canephora* (aka Robusta coffee) which is a cross-pollinating diploid. Thus, mutagenesis techniques and methods for population advancement applicable to annual, diploid, seed-propagated crops can be equally followed for Arabica coffee. In vitro cell culture of Arabica coffee started in the 1970s, primarily as an alternative method for multiplication besides seeding. Since then, in vitro methods for mass propagation of coffee from single cells have been published. This opens exciting opportunities to integrate induced mutagenesis with advanced cell culture techniques to produce chimera-free mutant plants, a major bottleneck for mutation breeding of perennial crops with a long juvenile phase such as coffee. Hence, different propagules were used as targets for mutagenesis studies of Arabica coffee under this CRP. In this book, protocols for mutation induction and dose optimization of seed, seedlings, cuttings and in vitro cells are presented by Costa Rica, Nigeria, China and Austria. Screening techniques for Leaf Rust resistance are presented by Portugal, Costa Rica and P. R. China. Towards the end of this CRP, several mutant populations were under development that can be further advanced and screened for Leaf Rust resistance using these protocols. Finally, molecular methods for mutation detection are described, including the use of a coffee Exome Capture kit and High Resolution Melt analysis which can aid the selection process.

It is our hope that this book will serve as a timely resource for breeders and researchers interested in broadening the genetic base or improving Arabica coffee for enhanced Leaf Rust resistance or other targeted traits through induced mutagenesis.

We also hope that this book will stimulate the integrated use of single-cell mutagenesis with advanced molecular techniques for accelerated breeding of perennial crops and trees which so far have lagged, behind the annual seed crops.

Stephan Nielen
EMBRAPA
Recursos Genéticos e Biotecnologia
Brasília, Federal District, Brazil

Ivan L. W. Ingelbrecht
Plant Breeding and Genetics
Laboratory, Joint FAO/IAEA Centre
of Nuclear Techniques in Food
and Agriculture
IAEA Laboratories
Seibersdorf, Austria

Acknowledgements

We would like to thank all staff of the Plant Breeding and Genetics Laboratory and Section, past and present, who were involved in the development and implementation of the IAEA Coordinated Research Project (CRP): "Efficient Screening Techniques to Identify Mutants with Disease Resistance for Coffee and Banana" (code: D22005). Special thanks to Mr. Till Brad and Mr. Nielen Stephan who led the CRP during its first three years as IAEA Scientific Secretary. We further thank the CRP contract and agreement holders Mr. Yi Kexian (P. R. China), Mr. Wu Weihuai (P. R. China), Mr. Dada Keji Emmanuel (Nigeria), Mr. Melgarejo Gutierrez Tomas Adan (Peru), Ms. do Céu Lavado da Silva Maria (Portugal), Mr. Varzea Vitor (Portugal), Ms. Schwarzacher Trude (UK) and Ms. Laimer Margit (Austria) for their valuable contributions throughout this CRP including the protocols presented in this book. We further acknowledge Mr. Gatica-Arias Andrés (Costa Rica) for stimulating discussions on coffee and his contributions to this protocol book. Thanks also to the consultants who were involved at the inception of this project for their guidance and inputs into the final CRP proposal. Funding for this CRP and supporting research at the FAO/IAEA PBG Laboratory work was provided by the FAO and the IAEA through the Joint FAO/IAEA Centre of Nuclear Techniques in Food and Agriculture, with additional support from Belgium through the Peaceful Use Initiative: "Enhancing Climate Change Adaptation and Disease Resilience in Banana-Coffee Cropping Systems in East Africa" (code: EBR-BEL01-18-03).

Contents

Contributors

Emanuel Araya-Valverde Centro Nacional de Innovaciones Biotecnológicas, San José, Costa Rica

Helena Gil Azinheira CIFC, Centro de Investigação das Ferrugens do Cafeeiro, Instituto Superior de Agronomia, Universidade de Lisboa, Oeiras, Portugal;
LEAF, Linking Landscape, Environment, Agriculture and Food Research Center, Associate Laboratory TERRA, Instituto Superior de Agronomia, Universidade de Lisboa, Lisboa, Portugal

Souleymane Bado Plant Breeding and Genetics Laboratory, Joint FAO/IAEA Centre of Nuclear Techniques in Food and Agriculture, IAEA Laboratories Seibersdorf, International Atomic Energy Agency, Vienna International Centre, Vienna, Austria

Xuehui Bai Dehong Institute of Tropical Agriculture, Ruili, Yunnan, P. R. China

Miguel Barquero-Miranda Phytoprotection Laboratory, Costa Rican Coffee Institute, Coffee Center Research, San Pedro, Heredia, Costa Rica

Dora Batista CIFC, Centro de Investigação das Ferrugens do Cafeeiro, Instituto Superior de Agronomia, Universidade de Lisboa, Oeiras, Portugal;
LEAF, Linking Landscape, Environment, Agriculture and Food Research Center, Associate Laboratory TERRA, Instituto Superior de Agronomia, Universidade de Lisboa, Lisboa, Portugal

Alejandro Bolívar-González Laboratorio Biotecnología de Plantas, Escuela de Biología, Universidad de Costa Rica, San Pedro, Costa Rica

Rashmi Boro Plant Biotechnology Unit, Department of Biotechnology, University of Natural Resources and Life Sciences (BOKU), Vienna, Austria

Eduviges G. Borroto Fernandez Plant Biotechnology Unit, Department of Biotechnology, University of Natural Resources and Life Sciences (BOKU), Vienna, Austria

María José Cordero-Vega Phytoprotection Laboratory, Costa Rican Coffee Institute, Coffee Center Research, Heredia, Costa Rica

Reina Céspedes Coffee Research Centre, Costa Rican Coffee Institute, San Pedro, Barva, Heredia, Costa Rica

Keji Dada Plant Breeding Unit, Cocoa Research Institute of Nigeria (CRIN), Ibadan, Oyo State, Nigeria

Inês Diniz CIFC, Centro de Investigação das Ferrugens do Cafeeiro, Instituto Superior de Agronomia, Universidade de Lisboa, Oeiras, Portugal;
LEAF, Linking Landscape, Environment, Agriculture and Food Research Center, Associate Laboratory TERRA, Instituto Superior de Agronomia, Universidade de Lisboa, Lisboa, Portugal

Fabián Echeverría-Beirute Escuela de Agronomía, Instituto Tecnológico de Costa Rica-San Carlos, San Carlos, Costa Rica

Noel Arrieta Espinoza San Francisco Bay Gourmet Coffee, Lincoln, CA, USA

Andrés Gatica-Arias Laboratorio Biotecnología de Plantas, Escuela de Biología, Universidad de Costa Rica, San Pedro, Costa Rica

Thomas Gbokie Jr College of Plant Protection, Nanjing Agricultural University, Nanjing, P. R. China

Abdelbagi Mukhtar Ali Ghanim Plant Breeding and Genetics Laboratory, Joint FAO/IAEA Centre of Nuclear Techniques in Food and Agriculture, IAEA Laboratories Seibersdorf, International Atomic Energy Agency, Vienna International Centre, Vienna, Austria

Florian Goessnitzer Plant Breeding and Genetics Laboratory, Joint FAO/IAEA Centre of Nuclear Techniques in Food and Agriculture, IAEA Laboratories Seibersdorf, International Atomic Energy Agency, Vienna International Centre, Vienna, Austria

Leonor Guerra-Guimarães CIFC, Centro de Investigação das Ferrugens do Cafeeiro, Instituto Superior de Agronomia, Universidade de Lisboa, Oeiras, Portugal; LEAF, Linking Landscape, Environment, Agriculture and Food Research Center, Associate Laboratory TERRA, Instituto Superior de Agronomia, Universidade de Lisboa, Lisboa, Portugal

Tieying Guo Dehong Institute of Tropical Agriculture, Ruili, Yunnan, P. R. China

Veronika Hanzer Plant Biotechnology Unit, Department of Biotechnology, University of Natural Resources and Life Sciences (BOKU), Vienna, Austria

Chunping He Hainan Key Laboratory for Monitoring and Control of Tropical Agricultural Pests, Environment and Plant Protection Institute, Chinese Academy of Tropical Agricultural Sciences, Haikou, Hainan, P. R. China

Xing Huang Hainan Key Laboratory for Monitoring and Control of Tropical Agricultural Pests, Environment and Plant Protection Institute, Chinese Academy of Tropical Agricultural Sciences, Haikou, Hainan, P. R. China

Ivan L. W. Ingelbrecht Plant Breeding and Genetics Laboratory, Joint FAO/IAEA Centre of Nuclear Techniques in Food and Agriculture, IAEA Laboratories Seibersdorf, International Atomic Energy Agency, Vienna International Centre, Vienna, Austria

Joanna Jankowicz-Cieslak Plant Breeding and Genetics Laboratory, Joint FAO/IAEA Centre of Nuclear Techniques in Food and Agriculture, IAEA Laboratories Seibersdorf, International Atomic Energy Agency, Vienna International Centre, Vienna, Austria

Margit Laimer Plant Biotechnology Unit, Department of Biotechnology, University of Natural Resources and Life Sciences (BOKU), Vienna, Austria

Yanqiong Liang Environment and Plant Protection Institute, Chinese Academy of Tropical Agricultural Sciences, Haikou, Hainan, P. R. China

Le Li Environment and Plant Protection Institute, Chinese Academy of Tropical Agricultural Sciences, Haikou, Hainan, P. R. China;
NHC Key Laboratory of Tropical Disease Control/Key Laboratory of Tropical Translational Medicine of Ministry of Education, School of Tropical Medicine, Hainan Medical University, Haikou, Hainan, P. R. China

Jinhong Li Dehong Institute of Tropical Agriculture, Ruili, Yunnan, P. R. China

Rui Li Environment and Plant Protection Institute, Chinese Academy of Tropical Agricultural Sciences, Haikou, Hainan, P. R. China

Xiaoyu Zoe Li Key Laboratory of Plant Resources Conservation and Sustainable Utilization/Guangdong Provincial Key Laboratory of Applied Botany, South China Botanical Garden, Chinese Academy of Sciences, Tianhe District, Guangzhou, P. R. China

Qing Liu Key Laboratory of Plant Resources Conservation and Sustainable Utilization/Guangdong Provincial Key Laboratory of Applied Botany, South China Botanical Garden, Chinese Academy of Sciences, Tianhe District, Guangzhou, P. R. China

Andreia Loureiro CIFC, Centro de Investigação das Ferrugens do Cafeeiro, Instituto Superior de Agronomia, Universidade de Lisboa, Oeiras, Portugal;
LEAF, Linking Landscape, Environment, Agriculture and Food Research Center,

Associate Laboratory TERRA, Instituto Superior de Agronomia, Universidade de Lisboa, Lisboa, Portugal

Ying Lu Environment and Plant Protection Institute, Chinese Academy of Tropical Agricultural Sciences, Haikou, Hainan, P. R. China

Ramón Molina-Bravo Laboratorio de Cultivo de Tejidos y Células Vegetales, y Laboratorio de Biología Molecular, Universidad Nacional, Heredia, Costa Rica

Stephan Nielen EMBRAPA, Recursos Genéticos e Biotecnologia, Brasília, DF, Brazil

Radisras Nkurunziza Plant Breeding and Genetics Laboratory, Joint FAO/IAEA Centre of Nuclear Techniques in Food and Agriculture, IAEA Laboratories Seibersdorf, International Atomic Energy Agency, Vienna International Centre, Vienna, Austria;
Laboratory for Applied in Vitro Plant Biotechnology, Department of Plants and Crops, Faculty of Bioscience Engineering, Ghent University, Ghent, Belgium

Emmanuel Ogwok Department of Science and Vocational Education, Faculty of Education, Lira University, Lira, Uganda

J. S. Pat Heslop-Harrison Department of Genetics and Genome Biology, University of Leicester, Leicester, UK;
Key Laboratory of Plant Resources Conservation and Sustainable Utilization/ Guangdong Provincial Key Laboratory of Applied Botany, South China Botanical Garden, Chinese Academy of Sciences, Tianhe District, Guangzhou, P. R. China

Ana Paula Pereira CIFC, Centro de Investigação das Ferrugens do Cafeeiro, Instituto Superior de Agronomia, Universidade de Lisboa, Quinta do Marquês, Oeiras, Portugal;
LEAF, Linking Landscape, Environment, Agriculture and Food Research Center, Associated Laboratory TERRA, Instituto Superior de Agronomia, Universidade de Lisboa, Lisbon, Portugal

Jorge Rodríguez-Matamoros Laboratorio Biotecnología de Plantas, Escuela de Biología, Universidad de Costa Rica, San Pedro, Costa Rica

José Andrés Rojas-Chacón Escuela de Agronomía, Instituto Tecnológico de Costa Rica-San Carlos, San Carlos, Costa Rica

Trude Schwarzacher Department of Genetics and Genome Biology, University of Leicester, Leicester, UK;
Key Laboratory of Plant Resources Conservation and Sustainable Utilization/ Guangdong Provincial Key Laboratory of Applied Botany, South China Botanical Garden, Chinese Academy of Sciences, Tianhe District, Guangzhou, P. R. China

Maria do Céu Lavado da Silva CIFC, Centro de Investigação das Ferrugens do Cafeeiro, Instituto Superior de Agronomia, Universidade de Lisboa, Oeiras, Portugal;
LEAF, Linking Landscape, Environment, Agriculture and Food Research Center,

Associate Laboratory TERRA, Instituto Superior de Agronomia, Universidade de Lisboa, Lisboa, Portugal

Elodia Sánchez-Barrantes Laboratorio Biotecnología de Plantas, Escuela de Biología, Universidad de Costa Rica, San Pedro, Costa Rica

Samira Tajedini Plant Breeding and Genetics Laboratory, Joint FAO/IAEA Centre of Nuclear Techniques in Food and Agriculture, IAEA Laboratories Seibersdorf, International Atomic Energy Agency, Vienna International Centre, Vienna, Austria

Shibei Tan Environment and Plant Protection Institute, Chinese Academy of Tropical Agricultural Sciences, Haikou, Hainan, P. R. China

Sílvia Tavares CIFC, Centro de Investigação das Ferrugens do Cafeeiro, Instituto Superior de Agronomia, Universidade de Lisboa, Oeiras, Portugal;
Department of Plant and Environmental Sciences, Copenhagen Plant Science Center, University of Copenhagen, Frederiksberg, Denmark

Paulina Tomaszewska Department of Genetics and Genome Biology, University of Leicester, Leicester, UK;
Department of Genetics and Cell Physiology, Faculty of Biological Sciences, University of Wrocław, Wrocław, Poland

Kimberly Ureña-Ureña Phytoprotection Laboratory, Costa Rican Coffee Institute, Coffee Center Research, Heredia, Costa Rica

César Vargas-Segura Laboratorio Biotecnología de Plantas, Escuela de Biología, Universidad de Costa Rica, San Pedro, Costa Rica

Vítor Várzea CIFC, Centro de Investigação das Ferrugens do Cafeeiro, Instituto Superior de Agronomia, Universidade de Lisboa, Quinta do Marquês, Oeiras, Portugal;
LEAF, Linking Landscape, Environment, Agriculture and Food Research Center, Associated Laboratory TERRA, Instituto Superior de Agronomia, Universidade de Lisboa, Lisbon, Portugal

Norman Warthmann Plant Breeding and Genetics Laboratory, Joint FAO/IAEA Centre of Nuclear Techniques in Food and Agriculture, IAEA Laboratories Seibersdorf, International Atomic Energy Agency, Vienna, Austria

Stefaan P. O. Werbrouck Laboratory for Applied in Vitro Plant Biotechnology, Department of Plants and Crops, Faculty of Bioscience Engineering, Ghent University, Ghent, Belgium

Weihuai Wu Hainan Key Laboratory for Monitoring and Control of Tropical Agricultural Pests, Environment and Plant Protection Institute, Chinese Academy of Tropical Agricultural Sciences, Haikou, Hainan, P. R. China

Jingen Xi Hainan Key Laboratory for Monitoring and Control of Tropical Agricultural Pests, Environment and Plant Protection Institute, Chinese Academy of Tropical Agricultural Sciences, Haikou, Hainan, P. R. China

Kexian Yi Hainan Key Laboratory for Monitoring and Control of Tropical Agricultural Pests, Environment and Plant Protection Institute, Chinese Academy of Tropical Agricultural Sciences, Haikou, Hainan, P. R. China

Hongbo Zhang Dehong Institute of Tropical Agriculture, Ruili, Yunnan, P. R. China

Jinlong Zheng Environment and Plant Protection Institute, Chinese Academy of Tropical Agricultural Sciences, Haikou, Hainan, P. R. China

Hua Zhou Dehong Institute of Tropical Agriculture, Ruili, Yunnan, P. R. China

Introduction

Mutation Breeding in Arabica Coffee

Ivan L. W. Ingelbrecht, Noel Arrieta Espinoza, Stephan Nielen, and Joanna Jankowicz-Cieslak

Abstract Coffee is a perennial (sub)tropical crop and one of the most valuable commodities globally. Coffee is grown by an estimated 25 million farmers, mostly smallholders, and provides livelihoods to about 125 million people. The Coffea genus comprises over 120 species. Two species account for nearly the entire world coffee production: *C. arabica* L. (Arabica coffee) and *C. canephora* Pierre ex A. Froehner (Canephora coffee) with the former supplying about 65% of the world's consumption. Arabica coffee is a self-pollinated, amphidiploid species (2n = 4x = 44) whereas other *Coffea* species are diploid (2n = 2x = 22) and generally cross-pollinated. Induced mutagenesis using physical and chemical mutagens has been a successful strategy in producing over 3,300 mutant varieties in over 220 crop species with global impact. Spontaneous Arabica coffee mutants of significant economic importance have been found since the early 1900s, following the spread of Arabica coffee cultivation across the globe. However, Arabica coffee has so far not been improved through induced mutagenesis and studies on coffee mutagenesis are scarce. In this chapter, principles and practices of mutation-assisted breeding along with current breeding limitations of Arabica coffee are briefly reviewed, as an introduction to subsequent protocol chapters on mutation induction, advanced cell and tissue culture, Leaf Rust resistance screening and the application of novel molecular/genomics tools supporting mutation-assisted improvement and genetics research of Arabica coffee.

I. L. W. Ingelbrecht (✉) · J. Jankowicz-Cieslak
Plant Breeding and Genetics Laboratory, Joint FAO/IAEA Centre of Nuclear Techniques in Food and Agriculture, IAEA Laboratories Seibersdorf, International Atomic Energy Agency, Vienna International Centre, Vienna, Austria
e-mail: i.ingelbrecht@iaea.org

N. A. Espinoza
San Francisco Bay Gourmet Coffee, 173 Aviation Blvd, Lincoln, CA 95648, USA

S. Nielen
EMBRAPA, Recursos Genéticos e Biotecnologia, Brasília, DF, Brazil

I. L. W. Ingelbrecht et al. (eds.), *Mutation Breeding in Coffee with Special Reference to Leaf Rust*, https://doi.org/10.1007/978-3-662-67273-0_1

1 General Principles of Plant Mutation Breeding

Mutations can be defined as sudden heritable changes in the DNA of living organisms, not caused by genetic recombination or segregation. Mutational events can be easily produced in the laboratory with two principal types of mutagens: physical or chemical. Among the physical methods gamma and X-rays are the most frequently used (reviewed by Spencer-Lopes et al. 2018). Alkylating agents, especially EMS, and sodium azide are the most frequently used chemical mutagens (reviewed by Ingelbrecht et al. 2018). Mutations induced through physical or chemical mutagens occur randomly throughout the genome. Induced mutagenesis has generated a vast amount of genetic variability with a significant role in plant breeding, genetics, and functional genomics. Applied mutagenesis has been particularly successful for the genetic improvement of annual seed crops such as barley, rice, wheat, and sorghum amongst many others. Records from the Joint FAO/IAEA Centre for Nuclear Applications in Food and Agriculture, Austria, maintained in the Mutant Variety Database (MVD), show that over 3,300 crop varieties with one or more traits resulting from induced mutagenesis have been released since the 1960s (IAEA 2023).

The history of plant mutation breeding has been reviewed in several publications (van Harten 1998; Forster and Shu 2012; Bado et al. 2015) and thus will only be briefly described here. In 1901, Hugo de Vries, a Dutch botanist and one of the first plant geneticists, coined the term "mutation" to describe seemingly new forms that suddenly arose in his experiments on the evening primrose. Proof of the mutation theory of de Vries was firmly established by the pioneering work of Lewis John Stadler who induced mutations in barley and maize using X-rays (Stadler 1928). Up to that time, only natural spontaneous mutations were selected to generate novel genetic diversity in plants.

Following the spread of Arabica coffee from Africa to other continents, natural spontaneous Arabica coffee mutants appeared within the widely grown plantations. Several of these mutants attracted the attention of breeders and were described as new varieties in the different regions where they were grown. The most important spontaneous mutants are those affecting plant height, fruit shape and -colour, and leaf colour. Examples include Caturra Vermelho and Caturra Amarelo, dwarf, high-yielding mutants of Bourbon Amarelo observed in the 1930s in Brazil and officially registered as varieties in 1999 (Guimarães Mendes et al. 2007); Pacas, a dwarf mutant of Bourbon found in 1949 in El Salvador; Villa Sarchi, a dwarf mutation of Bourbon found in Costa Rica and released in 1957; and, Maragogype, a large bean size mutant within the Typica variety discovered in Brazil. In addition, a few male sterile plants have been found in Brazil and in Ethiopian accessions in the CATIE collection in Costa Rica (Wintgens 2012; Arabica Coffee Varieties | Variety Catalog; (worldcoffeeresearch.org) (https://varieties.worldcoffeeresearch.org/varieties)). More recently, natural mutations conferring very low caffeine content were discovered at the Instituto Agronômico de Campinas, Brazil. Out of 3,000 coffee trees representing 300 *C. arabica* accessions from Ethiopia, three plants contained only 0.07% caffeine, in contrast to the normal caffeine content of 1.2% in

C. arabica (Silvarolla et al. 2004). These plants have the potential of being the basis for the development of a new coffee varieties giving rise to "naturally decaffeinated" coffee.

The Joint FAO/IAEA Centre of Nuclear Applications in Food and Agriculture has been promoting and disseminating the efficient use of mutation techniques as a tool for crop improvement since the 1960s. Several authors have documented the global impact of induced mutant varieties (Ahloowalia et al. 2004; Kharkwal and Shu 2009). From the 1980s onwards plant mutagenesis has become increasingly integrated into a range of enabling biotechnology and genomics/bioinformatics tools to fast-track the breeding process, mutant selection or mutant trait discovery (Mokry et al. 2011; Schneeberger and Weigel 2011; Ghosh et al. 2018; Knudsen et al. 2022).

Merits of induced mutagenesis as a complementary tool for crop improvement are:

- Different plant propagules can be subjected to mutagenesis treatments, including seed, entire plants and vegetative propagules, regenerable tissues or single cells
- It can be (much) faster than conventional breeding
- Mutation induction is simple and can rapidly produce novel variation (e.g., gamma and X-ray mutagenesis, EMS treatments)
- Desired traits are introduced directly into well adapted, elite lines/varieties
- Track record of success for numerous crops and traits, including morphological and physiological traits, yield, disease resistance
- A public domain, non-GMO technology greatly facilitating adoption of end products.

Limitations of plant mutation breeding include:

- Not all types of useful variation can be induced as mutations
- Mutations are random and the desired mutation is rare thus large mutant populations are required to recover the desired mutation or trait
- Mutant selection mostly relies on phenotyping starting at the M_2 generation though recent developments could provide more efficient, early selection systems
- The space and labour required to grow out large mutant populations, particularly in case of horticultural crops and trees
- The length of time required to develop mutant varieties in crops with a long reproductive cycle as in perennial crops.

Mutation breeding has been especially successful with annual, inbreeding, diploid crops that are seed-propagated, because it is relatively quick to advance populations from the initial mutant population (M_1) to advanced mutant lines. However, vegetatively propagated crops and perennial species—including Arabica coffee—have lagged, due to the limitations listed above. Recent advances in in vitro cell culture offer new opportunities and strategies for vegetative crops and trees through single-cell mutagenesis as described further in this protocol book for Arabica coffee. Likewise, new genomic, bioinformatics and genotyping tools enable screening mutant

populations in early generations and can provide a means for short-cutting generations and fast-tracking mutation breeding. This is especially relevant for perennial
crops and trees with long juvenile periods.

2   Breeding Limitations in Arabica Coffee

Arabica coffee production is facing multiple threats including the interrelated challenges of climate change and transboundary pests and diseases such as Coffee Leaf
Rust (CRL) and Coffee Berry Disease (CBD) (Bunn et al. 2015; Läderach et al. 2017;
Solymosi and Techel 2019). Pests and diseases affecting Arabica coffee cultivation
with special reference to CLR are reviewed in Chapter "Coffee Leaf Rust Resistance:
An Overview". Protocols for Leaf Rust screening and molecular diagnostics are
presented in Chapters "Screening for Resistance to Coffee Leaf Rust", "Inoculation
and Evaluation of *Hemileia vastatrix* Under Laboratory Conditions", "Evaluation of
Coffee (*Coffea arabica* L. var. Catuaí) Tolerance to Leaf Rust (*Hemileia vastatrix*)
Using Inoculation of Leaf Discs Under Controlled Conditions" and "A PCR-Based
Assay for Early Diagnosis of the Coffee Leaf Rust Pathogen *Hemileia vastatrix*".

Nearly all coffee is grown between the Tropics of Cancer and Capricorn where
conditions for coffee cultivation are ideal. This band of latitudes is known as the
coffee belt. Arabica coffee is usually cultivated in relatively cool mountain climates
at 400–2800 m asl. Arabica coffee is sensitive to environmental factors such as
exposure to direct sunlight, temperature, and rainfall (Muschler 2001). It is within
the coffee belt that the most drastic changes in climate have occurred in recent years.
The implications of these changes in coffee production can range from physiological
and phenological disorders of plants, to the reduced adaptability of plants to areas
with limiting conditions. In addition, pests and diseases also see their physiology and
phenology altered, sometimes promoted favorably which implies greater pressure on
the production systems. Overall, climate change is impacting coffee production both
through changes in weather patterns, viz. rising temperatures, excessive rainfall, or
longer droughts, and through changed/expanded habitats of important coffee diseases
such as CLR (Avelino et al. 2015; van der Vossen et al. 2015). Coffee Berry Disease,
still limited to the African continent, is a latent threat for the Americas in view of
the favorable agroecological conditions offered by Latin America for this fungus.

Plant breeding requires genetic variation of useful traits to improve crops.
However, the genetic diversity within the primary gene pool of *C. arabica* is very
narrow (Scalabrin et al. 2020) so the required genetic variability to address above-
mentioned constraints is lacking. Most of the genetic diversity is found in Ethiopia
and South Sudan, the centres of origin of *C. arabica* (Sylvain 1958; Thomas 1942).
Since the 1960s coffee yields have stagnated in all coffee producing countries except
Brazil, Colombia and Vietnam (Montagnon et al. 2019). Other challenges of Arabica
coffee breeding are inherent to the perennial nature of this crop. The generation time

factor—3–5 years from seed to seed—remains a major issue for coffee breeding programs.

Currently, two main approaches are followed for the genetic improvement of Arabica coffee. Since the 1950s the traditional varieties formed the basis for pedigree breeding mainly with the 'Timor Hybrid' (an interspecific hybrid of *C. arabica* and *C. canephora* resulting from a natural cross) that has resistance to CLR (Bettencourt 1973; Silva et al. 2018). However, pedigree breeding is a long process requiring 30 years or more to release a stable, homogeneous and distinguishable variety. To date, most Arabica coffee plantations around the world are established with the varieties resulting from breeding efforts initiated some 50 years ago. However, these varieties are susceptible to disease outbreaks, especially CLR, and are poorly adapted to the changing climatic conditions observed in many coffee growing regions during the past decade. Obtaining CLR resistant varieties will allow to produce coffee with reduced pesticide use or in organic farming systems (Arrieta 2014). Since the 1990s, F1 hybrids are being developed as an alternative breeding strategy in view of their improved performance over traditional varieties in terms of yield and disease resistance, and because of the reduced timeframe of 10–20 years from breeding to commercial release (Frédérick et al. 2019). However, unlike the traditional varieties, F1 hybrids are not true breeding and thus require a different mechanism for mass production, typically via clonal propagation (World Coffee Research | F1 Hybrid Trials (https://worldcoffeeresearch.org/programs/next-generation-f1-hybrid-varieties)). More recently, male sterility has been used for F1 hybrid seed production in Arabica coffee (Frédérick et al. 2019).

3 Mutation Breeding in Arabica Coffee

3.1 Background

Arabica coffee production is threatened by disease outbreaks and climate change while conventional breeding is hampered by the very narrow genetic base within its primary gene pool, as summarized above. Induced mutagenesis may have significant value for Arabica coffee by increasing genetic variability for genetic studies and breeding purposes. The plant Mutant Variety Database (MVD) lists over 3300 released mutant varieties in a wide range of crop plants. Over 80% of these resulted from exposure to physical mutagens. Ionizing radiation such as gamma rays and X-rays have been the most widely used techniques for mutation induction. Example successes in mutation breeding of woody plants reported in the MVD are shown in Table 1. Note that Arabica coffee is not listed in the MVD. To our knowledge no Arabica mutant variety has been released following induced mutagenesis. Thus, Arabica coffee remains a major crop that has not been improved by mutation breeding, though Arabica coffee varieties resulting from natural, spontaneous mutations are being grown commercially.

Table 1 Examples of perennial crops and trees improved using radiation breeding (IAEA 2023)

Crop species	No. of mutants	Main improved traits	Mutagen	Plant propagule
Apple	13	Early maturity, more coloured and bigger fruits, plant structure, smooth sheen fruits, extended cold storability, wide adaptation, well-balanced taste, altered acids to sugars ratio and high yield, resistance to *Podosphera leucotricha* and *Venturia inaequalis*	EMS, gamma rays, X-rays	Growing shoot, dormant trees, buds, scions, seed
Cherry	3	Growth habit, fruit quality, flower colour, high yield, big seed fruit, early maturity, self-fertile, resistance to rain splinting, seed set without pollination	Gamma rays, ion beam, X-rays, colchicine	Buds, cuttings, scions, pollen, in vitro culture
Citrus	3	Almost-seedless, high yield and sweet taste, improved seed production traits and good cooking quality	Gamma rays	Buds
Loquat	1	Large fruit size and good taste	Gamma rays	Seed
Indian jujube	2	Early maturity, 2 crops/year, increased fruit size and better taste (peach flavour), round fruits, pink rose, sweeter taste, more stable fruits yield	MNH	Seed
Fig	1	Not specified	Gamma rays	Seed
Pomegranate	2	Dwarfism	Gamma irradiation	Seed
Plum	1	Early maturity, self-compatible, better branching and fruit setting	Gamma rays	Cuttings
Pineapple	1	Altered leaf colour	Irradiation with chronic gamma rays (gamma greenhouse)	Not specified
Pear	8	Plant structure, maturity, good cooking quality, bacterial diseases, resistance to black spot disease	Gamma rays	Suckers
Peach	7	Early maturity, large fruit size, high yield, good fruit quality, resistance to black spot disease	Gamma rays	Buds, pollen

According to the MVD, induced mutagenesis has been successful for inducing resistance to fungal diseases in 334 cases (Fig. 1). These include tree crops such as pear (Sanada et al. 1993; Saito 2016) and crops with polyploid genomes such as wheat (Sigurbjörnsson and Micke 1974) and sugarcane (IAEA 2023).

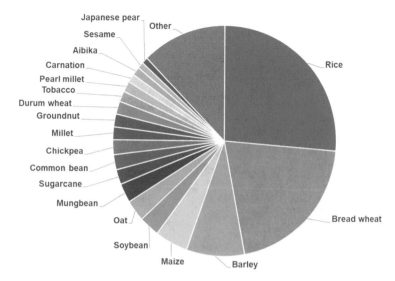

Fig. 1 Released mutant varieties with induced resistance to fungal diseases according to the FAO/IAEA Mutant Variety Database (IAEA 2023). Only plant species with at least three released varieties are listed with rice, wheat, barley and maize having the highest counts; 'other' includes plant species with one or two released mutant varieties

So far, there has been limited research on induced mutagenesis of Arabica coffee. The first attempt to induce new mutations in *C. arabica* was reported by Carvalho et al. (1954) using X-ray irradiation of seeds with doses up to 1,500 Gray (Gy, see below Sect.3.2). Main effects established were early termination of seedling growth if treatment was higher than 125 Gy and a general slow growth of mutagenized seedlings as compared to controls. Among the surviving seedlings, variation in the form of abnormal leaves was also observed. Moh and Orbegosos (1960) used thermal neutrons, X-rays and gamma-rays for induced mutagenesis in *C. arabica* and frequently obtained angustifolia (*ag*) mutants characterized by long and narrow leaves. Appearance of this phenotype already in the M_1 generation was explained by possible chromosomal aberrations. Interestingly, the mutant leaf type was similar for the entire plant and not sectorial, which excludes the presence of chimerism. Recently, similar observations were made in M_1 mutant coffee plants derived from gamma-ray irradiation at the FAO/IAEA Plant Breeding and Genetics Laboratory, Seibersdorf, Austria (Fig. 2). Moh (1961) speculates that the lack of chimerism in the M_1 plants indicates that the coffee plant originates from only one initial cell in the embryo shoot apex. This would, however, be one of very rare cases among the angiosperms. It is also conceivable that the uniform mutant leaf phenotype is the result of diplontic selection between cells of the meristem or that only one initial cell survived after irradiation. A final answer to the question of M_1 uniformity is yet to be given. In later induced mutation breeding experiments, analysis of traits of economic importance such as yield were put forward and monitored over several

Fig. 2 Arabica coffee M_1 mutants obtained from gamma-ray irradiation of seed at the FAO/IAEA Plant Breeding and Genetics Laboratory, Seibersdorf, Austria, 29 months after irradiation. **a** Wild type; **b** dwarf and leaf morphology mutant; **c**, **d** leaf morphology mutants. Note that the mutant leaf morphology characteristic is not sectorial but is similar for the entire plant

mutant generations. However, apart from the occurrence of leaf mutations, no correlation between varying yield and radiation dose could be established (Carvalho et al. 1984). These early experiments in coffee mutation breeding however, do suffer from the relatively small number of plants analyzed. A protocol for phenotypic characterization of an M_1 Arabica coffee greenhouse-based mutant population is presented in Chapter "Use of Open-Source Tools for Imaging and Recording Phenotypic Traits of a Coffee (*Coffea arabica* L.) Mutant Population".

3.2   The Need for Radiosensitivity Testing

Treatment of a plant or plant part with a mutagen affects its vigour, growth rate, germination, and fertility. Mutation rates vary with mutagen dosage. The higher the dosage of a mutagen, the more frequent the mutations and hence also, the greater the chance of undesired damage and lethality. The optimal dose is the one that, on the one hand, limits adverse effects that prevent the creation of a sufficiently large and vigorous mutant population, and, on the other hand, produces sufficient mutations to have a reasonable chance of recovering the desired mutation or mutant trait in the population, while preserving the (elite) genetic background. Hence, dose optimization is typically the first step in experimental or applied mutagenesis. Here, key principles and considerations for optimizing physical mutagenesis relevant to Arabica coffee will be briefly described. General principles and protocols for dose optimization using physical (Spencer-Lopes et al. 2018) and chemical mutagens

(Ingelbrecht et al. 2018; Jankowicz-Cieslak and Till 2016) and subsequent mutant population development (Ghanim et al. 2018) have been published.

Common units in physical mutagenesis include Gray (Gy), used to quantify the dose of radiation absorbed by the plant material and Gy/s or Gy/min, which is the unit for absorbed dose rate, a characteristic of the radiation source and irradiator used for mutagenesis. Radiosensitivity is a property of the target material, e.g., seed versus vegetative tissues, and of the species/variety. In addition, radiosensitivity is subject to external factors, such as, for example, the water content of the target material. Depending on the explant, the water content can be regulated. For example, the water content of seed can be adjusted to ca 12–14% through equilibration in a desiccator containing a 60% glycerol solution prior to irradiation, which is standard procedure at the FAO/IAEA PBG Laboratory, Seibersdorf, Austria.

Radiosensitivity testing refers to the determination of the optimum dose(s) of radiation of a particular plant propagule to be used as a basis for selecting the dose levels for bulk irradiation. In practice, radiosensitivity testing is performed across a series of mutagen doses in the lab or greenhouse over a short period of time. Growth responses or lethality is measured compared to a non-radiated control, to determine the GR_{30} (30% Growth Reduction) and GR_{50} (50% Growth Reduction) or LD_{30} (30% Lethal Dose) and LD_{50} (50% Lethal Dose) values respectively. These ranges have been observed to preserve the fitness of the M_1 plants (first mutant generation) while inducing sufficient stable, genetic variability for genetic studies or breeding purposes. In case of radiosensitivity testing of seed, the GR value is usually determined from the reduction of seedling height or leaf growth of the M_1 plants compared to untreated M_0 controls. In case of radiosensitivity testing of lethality of seed, seedling survival is measured over a range of doses compared to untreated controls. Importantly, the biological effects observed at the M_1 stage are the result of transient physiological effects and from genetic effects that are passed on to the next M_2 generation. Mutations are single cell events and thus mutagenic treatment of seed or other multicellular tissues may carry one or several mutations, each occupying a small part of the resulting M_1 plant. Such M_1 plants are therefore chimeras. Plant scientists or breeders need to be aware of the complications caused by chimerism and apply techniques to resolve them. Mutagenic treatment of single cells followed by plant regeneration does in principle result in chimera-free plants.

The doses chosen for bulk irradiation and development of the M_1 population, depend on different factors, such as the available resources to grow out and screen the mutant populations. The breeding system of the species under study plays a key role (van Harten 1998). For example, for annual diploid, self-fertile crops such as barley or sorghum, background mutations can be relatively easily removed through backcrossing. This is much more challenging or impossible for (obligately) vegetatively propagated crops or trees as in the case of Arabica coffee due to their long juvenile phase. Ideally up to three different doses are applied for bulk irradiation, including doses lower than LD_{50} or $GR_{50,}$ to ensure that at least one level will yield a sufficient number of the required mutant types. The frequency of induced mutations depends on the type of mutagen, the applied dose and the target materials. The plant species or variety, ploidy level, developmental stage, physiological state, etc. may all result in differences in response to radiation. Therefore, standardization of the target material and keeping records of all relevant information about the radiation source and treatment conditions is critical.

3.3   Choice of Material for Mutation Induction

Since Arabica coffee is self-fertilizing, the cheapest and most appropriate propagation system, especially in a commercial setting, is through seeding. However, for research and experimentation purposes, other plant propagules such as seedlings, cuttings or grafts can also be applied. In vitro cells and tissues are another attractive target in case of Arabica coffee given the availability of methods for de novo regeneration through somatic embryogenesis. Different plant propagules that can be used as targets for mutation induction in Arabica coffee with their advantages and limitations are summarized in Fig. 3.

In choosing the target material for dose optimization and mutagenesis treatments, it is important to consider the life cycle of a coffee tree, the seed and germination process. The coffee plant takes approximately three years from seed germination to produce the first fruit. It takes 6–9 months from flowering to mature cherries ready for harvest. The coffee cherry is the whole fruit, and has a skin, pulp, and parchment that cover the seed of the coffee. Inside the fruit are usually two seeds. Protocols for the establishment of in vitro tissue culture systems for Arabica coffee and methods

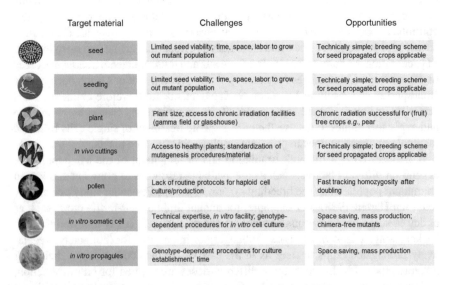

Target material	Challenges	Opportunities
seed	Limited seed viability; time, space, labor to grow out mutant population	Technically simple; breeding scheme for seed propagated crops applicable
seedling	Limited seed viability; time, space, labor to grow out mutant population	Technically simple; breeding scheme for seed propagated crops applicable
plant	Plant size; access to chronic irradiation facilities (gamma field or glasshouse)	Chronic radiation successful for (fruit) tree crops e.g., pear
in vivo cuttings	Access to healthy plants; standardization of mutagenesis procedures/material	Technically simple; breeding scheme for seed propagated crops applicable
pollen	Lack of routine protocols for haploid cell culture/production	Fast tracking homozygosity after doubling
in vitro somatic cell	Technical expertise, in vitro facility; genotype-dependent procedures for in vitro cell culture	Space saving, mass production; chimera-free mutants
in vitro propagules	Genotype-dependent procedures for culture establishment; time	Space saving, mass production

Fig. 3 Target materials for mutagenesis treatment of Arabica coffee with limitations and advantages

for mutation induction using in vitro tissues and cells are described in Chapters "In Vitro Plantlet Establishment of *Coffea arabica* from Cut Seed Explants", "Somatic Embryogenesis and Temporary Immersion for Mass Propagation of Chimera-Free Mutant Arabica Coffee Plantlets", "Protocol on Mutation Induction in Coffee Using In Vitro Tissue Cultures", "Mutation Induction Using Gamma-Ray Irradiation and High Frequency Embryogenic Callus from Coffee (*Coffea arabica* L.)", "Chemical Mutagenesis of Embryogenic Cell Suspensions of *Coffea arabica* L. var. Catuaí Using EMS and NaN$_3$", "Chemical Mutagenesis of *Coffea arabica* L. var. Venecia Cell Suspensions Using EMS" and "Chemical Mutagenesis of Zygotic Embryos of *Coffea arabica* L. var. Catuaí Using EMS and NaN$_3$". Protocols for mutation induction of seed and ex vitro vegetative propagules are described in Chapters "Physical Mutagenesis of Arabica Coffee Seeds and Seedlings", "Mutation Induction in *Coffea arabica* L. Using In Vivo Grafting and Cuttings", "Chemical Mutagenesis of Mature Seed of *Coffea arabica* L. var. Venecia Using EMS" and "Chemical Mutagenesis of Coffee Seeds (*Coffea arabica* L. var. Catuaí) Using NaN$_3$".

3.4 Enabling Biotechnology and Genomics Tools

In vitro plant cell and tissue culture techniques offer the possibility of rapid true-to-type multiplication. Somatic embryogenesis (SE) is an in vitro vegetative propagation technique that can produce clones of plants in large quantities. Research on in vitro tissue culture of Arabica coffee began in the 1970s. Since then, protocols for in vitro regeneration of Arabica coffee through SE have been developed, both direct and indirect methods have been reported (Barry-Etienne et al. 2002; Murvanidze et al. 2021). Somatic embryos can be produced from leaves of trees as starting material. Two innovations aimed at developing commercial scale multiplication systems include the use of bioreactors and direct sowing of somatic embryos in nurseries (Barry-Etienne et al. 1999; de Rezende Maciel et al. 2016; Etienne and Berthouly 2002; Etienne et al. 2013, 2018; Menendez-Yuffa et al. 2010). These micropropagation techniques are intended to enable mass propagation of elite Arabica coffee materials, such as F1 hybrids which cannot be propagated by seeding. The availability of protocols for in vitro regeneration of Arabica coffee offers exciting opportunities to integrate advanced cell culture techniques with induced mutagenesis to produce chimera-free mutant plants, a major bottleneck in induced mutagenesis of perennial crops with a long juvenile phase such as coffee. Some horticultural techniques, such as cuttings, can also enable cloning and multiplication of coffee plants.

The analysis of segregating molecular markers has confirmed earlier genetic and cytogenetic evidence that *C. arabica* is a functional diploid (Lashermes et al. 2008). Alkimim et al. (2017) and Saavedra et al. (2023) reported the use of marker-assisted selection for pyramiding multiple CLR and CBD resistance alleles. Genomic tools and large-scale sequencing enable a better understanding and characterization of the diversity and function of the *Coffea* genetic resources. This knowledge can then be utilised by breeders to select the best parental materials for incorporation into breeding programmes. Genomic selection (GS) allows breeders to select traits that are influenced by large numbers of small-effect alleles in a wide range of genotypes. Using GS in the context of resistance breeding for perennial crops increases the efficiency of breeding programs by shortening the breeding cycles (Alves et al. 2015). The release of reference genomes of *C. canephora* and *C. arabica* broadened the possibilities and facilitated significant progress for *C. arabica* genomic analysis (Denoeud et al. 2014; Dereeper et al. 2015; https://worldcoffeeresearch.org/resour ces/coffea-arabica-genome; Scalabrin et al. 2020). Sant'Ana et al. (2018) used the *C. canephora* reference genome to find SNP markers in the *C. arabica* genome associated with lipids and di-terpenes composition in a GWAS study of 107 diverse *C. arabica* genotypes. Knowledge of the molecular genetic structure of genes of interest to coffee breeders can then be applied for molecular breeding of Arabica coffee, including for example, for gene-based selection in mutation breeding programs. Protocols on the development and use of a Coffee Exome Capture kit, the application of High-Resolution Melt analysis, and the use of molecular cytogenetics for the detection of induced mutations in coffee are described in Chapters "Targeted Sequencing in Coffee with the Daicel Arbor Biosciences Exome Capture Kit", "High Resolution Melt (HRM) Genotyping for Detection of Induced Mutations in Coffee (*Coffea arabica* L var. Catuaí)" and "Protocols for Chromosome Preparations: Molecular Cytogenetics and Studying Genome Organization in Coffee", respectively.

References

Ahloowalia BS, Maluszynski M, Nichterlein K (2004) Global impact of mutation-derived varieties. Euphytica 135:187–204

Alkimim ER, Caixeta ET, Sousa TV, Pereira AA, Baião de Oliveira AC, Zambolim L, Sakiyama NS (2017) Marker-assisted selection provides arabica coffee with genes from other *Coffea* species targeting on multiple resistance to rust and coffee berry disease. Mol Breed 37:6. https://doi.org/10.1007/s11032-016-0609-1

Alves AA, Laviola BG, Formighieri EF, Carels N (2015) Perennial plants for biofuel production: bridging genomics and field research. Biotechnol J 10:505–507

Arrieta EN (2014) Resistencia genética en café: Estrategia de manejo no quimíco de la Roya del Cafeto. In: FAO Memorias del Seminario Científico Internacional Manejo Agroecológico de la Roya del Café. Panamá

Avelino J, Cristancho M, Georgiou S, Imbach P, Aguilar L, Bornemann G, Läderach P, Anzueto F, Hruska AJ, Morales C (2015) The coffee rust crises in Colombia and Central America (2008–2013): impacts, plausible causes and proposed solutions. Food Secur 7:303–321. https://doi.org/10.1007/s12571-015-0446-9

Bado S, Forster BP, Nielen S, Ali AM, Lagoda PJL, Till BJ, Laimer M (2015) Plant mutation breeding: current progress and future assessment, chap 2. In: Janick J (ed) Plant breeding reviews. Wiley-Blackwell, Hoboken, NJ, pp 23–88

Barry-Etienne D, Bertrand B, Vásquez N & Etienne H (1999) Direct sowing of Coffea arabica somatic embryos mass-produced in a bioreactor and regeneration of plants. Plant Cell Rep. 19:111–117

Barry-Etienne D, Bertrand B, Vasquez N, Etienne H (2002) Comparison of somatic embryogenesis-derived coffee (*Coffea arabica* L.) plantlets regenerated in vitro or ex vitro conditions: morphological, mineral and water characteristics. Ann Bot 90:77–85

Bettencourt AJ (1973) Considerações gerais sobre o 'Híbrido de Timor'. Instituto Agronômico de Campinas, Circular nº 23, p 20

Bunn C, Läderach P, Jimenez JGP, Montagnon C, Schilling T (2015) Multiclass classification of agro-ecological zones for Arabica coffee: an improved understanding of the impacts of climate change. PLoS ONE. https://doi.org/10.1371/journal.pone.0140490

Carvalho A, Antunes Filho H, Nogueira RK (1954) Genética de coffea. XX. Resultados preliminares do tratamento de sementes de café com raios-X. Bragantia, Campinas, 13 XVII–XX (Nota, 7)

Carvalho A, Fazuoli LC, Medina Filho HP (1984) Efeitos de raios X na indução de mutações em Coffea arábica. Bragantia, Campinas 43:553–567

Denoeud F, Carretero-Paulet L, Dereeper A, Droc G, Guyot R, Pietrella M, Zheng C, Alberti A, Anthony F, Aprea G, Aury JM, Bento P, Bernard M, Bocs S, Campa C, Cenci A, Combes MC, Crouzillat D, Da Silva C, Daddiego L, De Bellis F, Dussert S, Garsmeur O, Gayraud T, Guignon V, Jahn K, Jamilloux V, Joët T, Labadie K, Lan T, Leclercq J, Lepelley M, Leroy T, Li LT, Librado P, Lopez L, Muñoz A, Noel B, Pallavicini A, Perrotta G, Poncet V, Pot D, Priyono, Rigoreau M, Rouard M, Rozas J, Tranchant-Dubreuil C, VanBuren R, Zhang Q, Andrade AC, Argout X, Bertrand B, de Kochko A, Graziosi G, Henry RJ, Jayarama, Ming R, Nagai C, Rounsley S, Sankoff D, Giuliano G, Albert VA, Wincker P, Lashermes P (2014) The coffee genome provides insight into the convergent evolution of caffeine biosynthesis. Science 345:1181–1184

Dereeper A, Bocs S, Rouard M, Guignon V, Ravel S, Tranchant-Dubreuil C, Poncet V, Garsmeur O, Lashermes P, Droc G (2015) The coffee genome hub: a resource for coffee genomes. Nucleic Acids Res 43:D1028–D1035

de Rezende Maciel AL, Almendagna Rodrigues F, Pasqual M, Siqueira de Carvalho CH (2016) Large-scale, high-efficiency production of coffee somatic embryos. Crop Breed Appl Biotechnol 16:102–107

Etienne H, Berthouly M (2002) Temporary immersion systems in plant micropropagation. Plant Cell Tissue Organ Cult 69:215–231. https://doi.org/10.1023/A:1015668610465

Etienne H, Bertrand B, Georget F, Lartaud M, Montes F, Dechamp E, Verdeil JL, Barry-Etienne D (2013) Development of coffee somatic and zygotic embryos to plants differs in the morphological, histochemical and hydration aspects. Tree Physiol 33(6):640–653. https://doi.org/10.1093/treephys/tpt034

Etienne H, Breton D, Breitler J-C, Bertrand B, Déchamp E, Awada R, Marraccini P, Léran S, Alpizar E, Campa C, Courtel P, Georget F, Ducos J-P (2018) Coffee somatic embryogenesis: how did research, experience gained and innovations promote the commercial propagation of elite clones from the two cultivated species? Front Plant Sci 9:1630

Forster BP, Shu QY (2012) Plant mutagenesis in crop improvement basic terms and applications, chap 1. In: Shu QY, Kakagawa H, Forster BP (eds) Plant mutation breeding and biotechnology. CAB International and FAO, Wallingford, UK, pp 9–20

Frédérick G, Lison M, Alpizar E, Courtel P, Bordeaux M, Hidalgo JM, Marraccini P, Breitler J-C, Déchamp E, Poncon C, Etienne H, Bertrand B (2019) Starmaya: the first Arabica F1 coffee hybrid produced using genetic male sterility. Front Plant Sci 10. https://doi.org/10.3389/fpls.2019.01344

Ghosh S, Watson A, Gonzalez-Navarro OE et al (2018) Speed breeding in growth chambers and glasshouses for crop breeding and model plant research. Nat Protoc 13:2944–2963. https://doi.org/10.1038/s41596-018-0072-z

Guimarães Mendes NA, Carvalho GR, Botelho CE, Fazuoli LC, Silvarolla MB (2007) História das primeiras cultivares de café plantadas no Brasil. In: Carvalho CHS (ed) Cultivares de café. EMBRAPA, Brasília

IAEA (2023) The joint FAO/IAEA mutant variety database. Available at: https://nucleus.iaea.org/sites/mvd. Accessed 27 Jan 2023

Ingelbrecht I, Jankowicz-Cieslak, J, Szurman M, Till BJ, Szarejko I (2018) Chemical mutagenesis. In: Spencer-Lopes MM, Forster BP, Jankuloski L (eds) Manual on mutation breeding, 3rd edn. FAO/IAEA Publication, FAO, Rome, Italy

Jankowicz-Cieslak J, Till BJ (2016) Chemical mutagenesis of seed and vegetatively propagated plants using EMS. Curr Protoc Plant Biol 1:617–635. https://doi.org/10.1002/cppb.20040

Kharkwal MC, Shu QY (2009) The role of induced mutations in world food security. In: Shu QY (ed) Induced plant mutations in the genomics era. Food and Agriculture Organization of the United Nations, Rome, pp 33–38

Knudsen S, Wendt T, Dockter C, Thomsen HC, Rasmussen M, Egevang Jørgensen M, Lu Q, Voss C, Murozuka E, Østerberg JT, Harholt J, Braumann I, Cuesta-Seijo JA, Kale SM, Bodevin S, Tang Petersen L, Carciofi M, Pedas PR, Opstrup Husum J, Nielsen MTS, Nielsen K, Jensen MK, Møller LA, Gojkovic Z, Striebeck A, Lengeler K, Fennessy RT, Katz M, Garcia Sanchez R, Solodovnikova N, Förster J, Olsen O, Møller BL, Fincher GB, Skadhauge B (2022) FIND-IT: accelerated trait development for a green evolution. Sci Adv 8(34):26. https://doi.org/10.1126/sciadv.abq2266

Läderach P, Ramirez–Villegas J, Navarro-Racines C, Zelaya C, Martinez–Valle A, Jarvis A (2017) Climate change adaptation of coffee production in space and time. Clim Change 141:47–62. https://doi.org/10.1007/s10584-016-1788-9

Lashermes P, Carvalho Andrade A, Etienne H (2008) Genomics of coffee, one of the world's largest traded commodities. In: Moore PH, Ming R (eds) Genomics of tropical crop plants. Springer

Menendez-Yuffa A, Barry-Etienne D, Bertrand B, Georget F, Etienne H (2010) A comparative analysis of the development and quality of nursery plants derived from somatic embryogenesis and from seedlings for large-scale propagation of coffee (*Coffea arabica* L.). Plant Cell Tissue Organ Cult 102(3):297–307

Moh CC (1961) Does a coffee plant develop from one initial cell in the shoot apex of an embryo? Radiat Bot 1:97–99

Moh CC, Orbegoso G (1960) The induction of angustifolia mutants in coffee in the R1 generation by ionizing radiations. Genetics 45:1000 (Abstract)

Mokry M, Nijman IJ, van Dijken A, Benjamins R, Heidstra R, Sheres B, Cuppen E (2011) Identification of factors required for meristem function in Arabidopsis using a novel next generation sequencing fast forward genetics approach. BMC Genomics 12:256. https://doi.org/10.1186/1471-2164-12-256

Montagnon C, Marraccini P, Bertrand B (2019) Breeding for coffee quality. In: Oberthur et al (eds) Specialty coffee managing quality. Cropster Innsbruck, Austria, pp 109–143

Mukhtar Ali Ghanim A, Spencer-Lopes MM, Thomas W (2018) In: Spencer-Lopes MM, Forster BP, Jankuloski L (eds) Manual on mutation breeding, 3rd edn. FAO/IAEA Publication, FAO, Rome, Italy

Murvanidze N, Nisler J, Lerou O, Werbrouck SPO (2021) Cytokinin oxidase/dehydrogenase inhibitors stimulate 2iP to induce direct somatic embryogenesis in *Coffea arabica*. Plant Growth Regul 94:195–200. https://doi.org/10.1007/s10725-021-00708-6

Muschler RG (2001) Shade improves coffee quality in a sub-optimal coffee-zone of Costa Rica. Agrofor Syst 85:131–139

Saavedra LM, Caixeta ET, Barka GD, Borém A, Zambolim L, Nascimento M, Cruz CD, de Oliveira ACB, Pereira AA (2023) Marker-assisted recurrent selection for pyramiding leaf rust and coffee berry disease resistance alleles in *Coffea arabica* L. Genes 14:189. https://doi.org/10.3390/genes14010189

Saito T (2016) Advances in Japanese pear breeding in Japan. Breeding Science 66: 46–59. https://doi.org/10.1270/jsbbs.66.46

Sanada T, Kotobuki K, Nishida T, Fujita H, Ikeda F (1993) A new Japanese pear cultivar 'Gold Nijisseiki' resistant mutant to black spot disease of Japanese pear. Jpn J Breed 43(3):455–461. https://doi.org/10.1270/jsbbs1951.43.455

Sant'Ana GC, Pereira LFP, Pot D, Ivamoto ST, Domingues DS, Ferreira RV, Pagiatto NF, da Silva BSR, Nogueira LM, Kitzberger CSG, Scholz MBS, de Oliveira FF, Sera GH, Padilha L, Labouisse J-P, Guyot R, Charmetant P, Leroy T (2018) Genome-wide association study reveals candidate genes influencing lipids and diterpenes contents in *Coffea arabica* L. Sci Rep 8:465

Scalabrin S, Toniutti L, Di Gaspero G et al (2020) A single polyploidization event at the origin of the tetraploid genome of *Coffea arabica* is responsible for the extremely low genetic variation in wild and cultivated germplasm. Sci Rep 10:4642. https://doi.org/10.1038/s41598-020-61216-7

Schneeberger K, Weigel D (2011) Fast-forward genetics enabled by new sequencing technologies. Trends Plant Sci 16(5):282–288. https://doi.org/10.1016/j.tplants.2011.02.006

Sigurbjörnsson B, Micke A (1974) Philosophy and accomplishments of mutation breeding. In: Polyploidy and induced mutations in plant breeding. IAEA, Vienna, pp 303–343

Silva RA, Zambolim L, Castro ISL et al (2018) The Híbrido de Timor germplasm: identification of molecular diversity and resistance sources to coffee berry disease and leaf rust. Euphytica 214:153. https://doi.org/10.1007/s10681-018-2231-2

Silvarolla MB, Mazzafera P, Fazuoli LC (2004) A naturally decaffeinated arabica coffee. Nature 429:826

Solymosi K, Techel G (2019) Brewing up climate resilience in the coffee sector: adaptation strategies for farmers, plantations, and producers. Available at: https://www.idhsustainabletrade.com/upl oaded/2019/06/Brewing-up-climate-resilience-in-the-coffee-sector-1.pdf

Spencer-Lopes MM, Jankuloski L, Mukhtar Ali Ghanim A, Matijevic M, Kodym A (2018) Physical mutagenesis. In: Spencer-Lopes MM, Forster BP, Jankuloski L (eds) Manual on mutation breeding, 3rd edn. FAO/IAEA Publication, FAO, Rome, Italy

Stadler LJ (1928) Mutations in barley induced by X-rays and radium. Science 68:186. https://doi.org/10.1126/science.68.1756.186

Sylvain PG (1958) Ethiopian coffee—its significance for the world coffee problems. Econ Bot 12:111–139

Thomas AS (1942) The wild Arabica coffee on the Boma Plateau of Anglo-Egyptian Sudan. Emp J Exp Agric 10:207–212

van der Vossen H, Bertrand B, Charrier A (2015) Next generation variety development for sustainable production of arabica coffee (*Coffea arabica* L.): a review. Euphytica 204(2):244. https://doi.org/10.1007/s10681-015-1398-z. S2CID 17384126

van Harten AM (1998) Mutation breeding: theory and practical applications. Cambridge University Press

Wintgens JN (2012) Coffee: Growing, Processing, Sustainable Production. A Guidebook for Growers, Processors, Traders, and Researchers. (2nd ed.), Wiley-VCH Verlag GmbH & Co. KGaA, Weinheim, Germany

Coffee Leaf Rust Resistance: An Overview

Leonor Guerra-Guimarães, Inês Diniz, Helena Gil Azinheira, Andreia Loureiro, Ana Paula Pereira, Sílvia Tavares, Dora Batista, Vítor Várzea, and Maria do Céu Lavado da Silva

Abstract Coffee is one of the most important cash crops and beverages. Several diseases caused by fungi, bacteria, and viruses can affect coffee plantations and compromise production. Coffee leaf rust (CLR), caused by the biotrophic fungus *Hemileia vastatrix* is the top fungal disease, representing a permanent threat to sustainable Arabica coffee production for more than a century. This review provides a comprehensive survey of the most common coffee diseases, their importance, and geographic distribution, with an emphasis on coffee leaf rust. Summing up the progress obtained so far from different research fields on the coffee–*H. vastatrix* interaction, we revisited the pathogen genetic diversity and population dynamics, and the complex mechanisms underlying plant resistance/immunity. We also highlight how new advanced technologies can provide avenues for a deeper understanding of this pathosystem, which is crucial for devising more reliable and long-term strategies for disease control.

L. Guerra-Guimarães · I. Diniz (✉) · H. G. Azinheira (✉) · A. Loureiro · A. P. Pereira ·
S. Tavares · D. Batista · V. Várzea · M. do C. L. da Silva
CIFC, Centro de Investigação das Ferrugens do Cafeeiro, Instituto Superior de Agronomia, Universidade de Lisboa, Quinta do Marquês, 2784-505 Oeiras, Portugal
e-mail: inesdiniz@isa.ulisboa.pt

H. G. Azinheira
e-mail: hmga@edu.ulisboa.pt

L. Guerra-Guimarães · I. Diniz · H. G. Azinheira · A. Loureiro · A. P. Pereira · D. Batista ·
V. Várzea · M. do C. L. da Silva
LEAF, Linking Landscape, Environment, Agriculture and Food Research Center, Associate Laboratory TERRA, Instituto Superior de Agronomia, Universidade de Lisboa, Tapada da Ajuda, 1349-017 Lisboa, Portugal

S. Tavares
Department of Plant and Environmental Sciences, Copenhagen Plant Science Center, University of Copenhagen, 1871 Frederiksberg, Denmark

© The Author(s) 2023
I. L. W. Ingelbrecht et al. (eds.), *Mutation Breeding in Coffee with Special Reference to Leaf Rust*, https://doi.org/10.1007/978-3-662-67273-0_2

1 Introduction

Coffee is one of the most widely consumed beverages in the world and one of the most traded commodities globally. The main coffee-producing countries are Brazil, Vietnam, and Colombia, while the European Union and the United States of America are the largest consuming and importing markets globally (FAO 2022).

The two cultivated species, *Coffea arabica* L. (Arabica) and *Coffea canephora* Pierre ex A. Froehner (Robusta) accounted in 2020, on average, for about 60% and 40% of the world's coffee production, respectively (ICO 2020). *C. arabica* is predominantly cultivated in the highlands and preferred by consumers due to its low bitterness, its aromatic characteristics, and its low caffeine content. *C. canephora* is more suitable for intertropical lowlands and characterized by a stronger bitterness and higher caffeine content (Lécolier et al. 2009).

Coffee, like other crops, is affected by several factors, including diseases, which may cause considerable yield losses. Moreover, there is clear evidence that the geographical distribution of several pathogens is expanding due to climate change and increasing global trade (Nnadi and Carter 2021). There are several ways to control diseases, ranging from chemical and biological control to good cropping practices. However, breeding for disease resistance is considered the most efficient and sustainable disease control strategy (Silva et al. 2022 and references therein).

Following a brief description of the most common coffee diseases, this review focuses on the advances in coffee leaf rust (CLR) research, mainly regarding pathogen infection, pathogen genetic diversity and population dynamics, and plant defense mechanisms. This knowledge is of utmost importance as an informed base to breed efficiently for durable resistance and devise innovative crop protection approaches.

2 Coffee's Main Diseases

A plant disease results from the interaction between a susceptible host plant, a virulent pathogen, and favorable environmental conditions (Agrios 2005). Diseases caused by fungi, bacteria, and viruses (Table 1) are the major limiting factors in coffee production. According to Maghuly et al. (2020), approximately 26% of the global annual coffee production is lost due to diseases, threatening the income of nearly 125 million people worldwide.

Coffee leaf rust (CLR), caused by the biotrophic fungus *Hemileia vastatrix* Berkeley & Br. (phylum Basidiomycota, class Pucciniomycetes, order Pucciniales), is the major disease affecting Arabica coffee (Talhinhas et al. 2017; Silva et al. 2022 and references therein) inducing losses of over $1 billion annually (Kahn 2019). CLR was first recorded in 1861 near Lake Victoria (East Africa), but its first major outbreak was in 1869 in Ceylon (now Sri Lanka), leading to the eradication of coffee cultivation in this country, with devastating social and economic consequences. Nowadays, CLR is present in all the coffee-growing regions (Wellman 1952; Rodrigues Jr. et al.

Table 1 List of the main diseases of *Coffea* spp.

Class	Disease common name	Pathogen	Country/region of occurrence
Fungus	Coffee leaf rust (CLR)	*Hemileia vastatrix*	Worldwide
	Powdery rust or grey rust of coffee	*Hemileia coffeicola*	West Africa
	Coffee berry disease (CBD)	*Colletotrichum kahawae*	Africa
	Brown eye spot of coffee (BES)	*Cercospora coffeicola*	Brazil
	Coffee wilt disease (CWD)	*Fusarium xylarioides* (teleomorph = *Gibberella xylarioides*)	Central and East African countries
	American leaf spot of coffee	*Mycena citricolor*	Central America, Colombia
Bacterium	Bacterial halo blight (BHB) of coffee	*Pseudomonas syringae* pv. *garcae*	Brazil, Kenya, Ethiopia, Uganda and China
	Coffee leaf scorch (CLS)	*Xylella fastidiosa*	Brazil, Costa Rica, Porto Rico
Virus	Coffee ringspot (CoRSV)	Coffee ringspot virus	Brazil, Costa Rica

1975; McCook and Peterson 2020; Keith et al. 2021). In the last decade, the epidemic resurgence of CLR, known as "the big rust", had strong economic and social impacts on several countries across Latin America and the Caribbean (Baker 2014; Avelino et al. 2015). Moreover, CLR has been reported to have expanded its area of distribution to regions of higher altitude where previously it was not detected, namely above 1000–1100 m in Central America and above 1600 m in Colombia (Rozo et al. 2012; Avelino et al. 2015). It has led to food security issues as a result of the high dependence on coffee production by most coffee farmers and laborers (Avelino et al. 2015). There is little evidence that the big rust was caused by the evolution of new virulence in *H. vastatrix*. Rather, a combination of more conducive weather patterns, changing climatic conditions, and recurring economic shocks appear to be responsible (Rhiney et al. 2021 and references therein).

H. vastatrix infects the lower surface of the leaves, where it produces chlorotic spots preceding the differentiation of suprastomatal, bouquet-shaped, orange-coloured uredinia (Fig. 1a). Infected leaves fall off, leaving long expanses of twigs without any foliage (Fig. 1b). Another coffee rust [powdery rust (or grey rust) of coffee] is caused by the fungus *Hemileia coffeicola* Maublanc and Roger. Both *H. vastatrix* and *H. coffeicola* have *C. arabica* and other *Coffea* species as hosts, but *H. coffeicola* is only of local importance in some West African countries (Rodrigues Jr. 1990; Ritschel 2005). The symptoms of the disease are characterised by a dusty or powdery coating of yellow uredosori covering the underside of the coffee leaves (Rodrigues Jr. 1990).

Coffee berry disease (CBD) caused by the hemibiotrophic fungus *Colletotrichum kahawae* J.M. Waller & P.D. Bridge (phylum Ascomycota, class Sordariomycetes,

Fig. 1 Coffee leaf rust (CLR) symptoms and urediniospore infection. **a** Chlorotic spots and uredosporic sori on the lower leaves surface; **b** severe defoliation associated with CLR in one coffee plant contrasting with resistant ones in the background; **c** *Hemileia vastatrix* infection process. Photos taken by the authors. Created with Biorender.com

order Glomerellales) is the most devastating disease affecting Arabica coffee production in Africa at high altitudes (Silva et al. 2006; van der Vossen and Walyaro 2009; Loureiro et al. 2012; Maghuly et al. 2020). It was first reported in 1922 in Kenya (McDonald 1926) and is still restricted to Africa, but represents a threat to high-altitude coffee areas of Latin America and Asia (Silva et al. 2006 and references therein; van der Vossen and Walyaro 2009). *C. kahawae* infects all stages of the crop, but maximum crop losses occur following infection of green berries with the formation of dark sunken lesions with sporulation causing their mummification and premature dropping. Under adequate climatic conditions and if no control measures are applied, this disease can destroy 50–80% of the developing green berries (Firman and Waller 1977; Silva et al. 2006; Hindorf and Omondi 2011).

Cercosporiosis, or brown eye spot (BES), is currently one of the main diseases of coffee in Brazil. It is caused by *Cercospora coffeicola* Berk. and Cooke (phylum Ascomycota, class Dothideomycetes, order Mycosphaerellales). The pathogen causes lesions on leaves and fruits, resulting in defoliation, decreased productivity, diminished coffee quality, and yield loss. In the nursery, this defoliation reduces the seedling's growth rate, which becomes inappropriate for planting and marketing. Also, in field conditions, the disease could be harmful to young trees (Botelho et al. 2017 and references therein). BES can appear as two distinct symptoms on leaves and in field conditions, the 'brown eye spot' and the 'black spot' (Andrade et al. 2021). The first one is the typical symptom caused by *C. coffeicola* on coffee leaves, which can be described as small necrotic spots consisting of a light-colored center and sometimes surrounded by a purple, brown ring with yellow edges, giving rise to the name brown eye spot. The atypical symptom is characterized by a black spot with the lesions being black, without the yellow halo.

Coffee wilt disease (CWD), or tracheomycosis, is caused by the vascular wilt pathogen *Fusarium xylarioides* Steyaert (teleomorph = *Gibberella xylarioides* R. Heim & Saccas) (phylum Ascomycota, class Sordariomycetes, order Hypocreales). CWD spreads across Africa, destroying coffee trees, reducing yields, and significantly impacting producer livelihoods. (Pinard et al. 2016; Maghuly et al. 2020; Flood 2021). It is frequently found in older and densely planted coffee trees (Assefaa et al. 2022). Through systematic sanitation and the establishment of breeding programs in affected countries, CWD appeared to have declined. However, in the 1990s, the disease re-emerged and increased to epidemic proportions affecting Robusta coffee in the Democratic Republic of the Congo, Uganda, and Tanzania and Arabica coffee in Ethiopia (Flood 2021). The first symptoms of CWD are yellowing of the leaves, which then wilt and develop brown necrotic lesions. The leaves then curl, dry up and fall off. This process may start on one part of the tree but eventually spreads to the rest of the plant. Once a tree is infected, there is no remedy other than to uproot the tree and burn it in situ to reduce the chances of spreading the infection. No new trees should be planted in the same place for at least six months to prevent remnants of the root system in the soil, which retain viable spores of the disease, from reinfecting new plants (Phiri and Baker 2009).

American leaf spot disease, also known as "Ojo de Gallo" is caused by the fungus *Mycena citricolor* (Berk. & M.A. Curtis) Sacc. (Phylum Basidiomycota, class Agaricomycetes, order Agaricales) and has been reported in Latin America. This fungus can grow on all plant organs, including leaves, stems, and fruits. Subcircular spots, initially brown, becoming pale-brown to straw-colored, are produced mainly on leaves. The spots have a distinct margin and are 6–13 mm in diameter but with no halo. The centers of older leaf spots may disintegrate, giving a shot-hole appearance. The main effect of the disease is leaf fall, with a consequent reduction in the growth and yield of the coffee plants (Wang and Avelino 1999; Krishnan 2017).

The bacterial halo blight (BHB) of coffee caused by the Gram-negative bacterium *Pseudomonas syringae* pv. *garcae* (Psgc) Young, Dye & Wilkie of the family *Pseudomonadaceae*, was first described in 1955 by Amaral et al. (1956) in the municipality of Garça in the Brazilian state of São Paulo, and later in Paraná and Minas Gerais states (Badel and Zambolim 2019 and references therein). Bacterial halo blight has been reported in the African Continent, in Kenya, Ethiopia, and Uganda (Ramos and Shavdia 1976; Korobko and Wondinagegne 1997) and in China (Bai et al. 2013). It has been estimated that BHB can cause losses up to 70% in nurseries and in the field, predominantly in regions above 1000 m in the presence of severe wind (Zoccoli et al. 2011). Necrotic spots surrounded or not by chlorotic haloes in leaf borders are the most common symptoms of BHB disease. However, flowers, fruits, and branches can also be affected (Badel and Zambolim 2019). The bacterium survives mainly as an epiphyte associated with plant debris. It penetrates the host tissue through natural openings (stomata) or wounds and is disseminated by water and wind-driven aerosol particles (Zoccoli et al. 2011). A recent study suggests that seeds may also be a source of inoculum (Belan et al. 2016).

Coffee leaf scorch (CLS), also referred to as atrophy of branches, is caused by *Xylella fastidiosa* subsp. *pauca (Xfp)*, a Gram-negative bacterium belonging to the family *Xanthmonadaceae*. CLS has been reported in Brazil, Costa Rica, and Porto Rico (Beretta et al. 1996; Rodriguez et al. 2001; Bolaños et al. 2015). Strains of the bacterium isolated from coffee and citrus are closely related, and both share the sharpshooter insect of the *Cicadellidae* family as a dissemination vector. The bacterium colonizes the xylem vessels of host plants, as well as the foregut of its insect vectors (Badel and Zambolim 2019). CLS symptoms include apical and marginal leaf scorch, defoliation, reduction of the internode length, the leaf size, the plant height, fruit size and quantity, terminal clusters of small chlorotic and deformed leaves, lateral shoot dieback, and overall stunting (Li et al. 2001). CLS disease is widespread and often occurs if coffee is adjacent to citrus orchards affected by *X. fastidiosa* (citrus variegated chlorosis disease). Although there appears to be some degree of host specialization within the subspecies of *X. fastidiosa*, cross-infection has been reported in commercial grapevine cultivars and olive trees (Badel and Zambolim 2019).

Coffee ringspot virus (CoRSV), currently classified as Coffee ringspot dichorhavirus by the International Committee on Taxonomy of Viruses (ICTV), belong to the genus *Dichorhavirus*, of *Rhabdoviridae* family (Genus: Dichorhavirus, ICTV). CoRSV has been reported in the main coffee-growing states of Minas Gerais (Brazil) (Ramalho et al. 2014, 2016) and some regions of Costa Rica (Rodrigues et al. 2002). The virus is transmitted by *Brevipalpus phoenicis* (Geijskes) (Acari: *Tenuipalpidae*) and can infect coffee leaves and fruits. Symptoms on leaves are typical concentric chlorotic rings, sometimes forming bands on the veins. On coffee berries, CoRSV develops chlorotic or necrotic lesions that are frequently invaded by secondary fungal or bacterial opportunists. In severely affected trees, leaves fall and fruit drops, which can affect coffee production and quality (Ramalho et al. 2014). The severity of the disease has been attributed to ecological disturbances associated with expanding coffee areas and to chemical pest control favoring the vector.

3 CLR Causal Agent: *Hemileia vastatrix*

3.1 Life Cycle and Infection Process

Hemileia vastatrix is a hemicyclic fungus producing urediniospores, teliospores, and basidiospores. Urediniospores and teliospores are produced in the same sorus but at different times. Urediniospores are dikaryotic and represent the asexual cycle, reinfecting the leaves whenever environmental conditions are favorable. Teliospores rarely occur and germinate in situ, producing a promycelium from which four basidiospores are formed. Basidiospores cannot infect coffee, but no other host plant has been identified (Talhinhas et al. 2017 and references therein).

As an obligate biotroph, *H. vastatrix* can only feed, grow and reproduce on living coffee leaves by the differentiation of a specific hypha called haustorium. This organ forms after penetration of the wall of a live host cell, expanding on the inner side of the cell wall while invaginating the surrounding host plasma membrane (Garnica et al. 2014). In addition to their role in nutrient uptake, haustoria are actively involved in establishing and maintaining the biotrophic relationship through the secretion of effector proteins (Voegele and Mendgen 2011; Kemen et al. 2005; Bozkurt and Kamoun 2020).

The *H. vastatrix* infection process on coffee leaves (Fig. 1c), like other rust fungi (Heath 1997; Voegele and Mendgen 2011), involves specific events, including urediniospore germination, appressorium formation over stomata, penetration, and inter- and intracellular colonization (Rodrigues Jr. et al. 1975; Silva et al. 1999, 2006 and references therein). Urediniospore germination requires water and is optimal at about 24 °C. After appressorium formation, the fungus penetrates, forming a penetration hypha that grows into the substomatal chamber. This hypha produces at the advancing tip two thick lateral branches; each hypha and its branches resemble an anchor. From each lateral branch of the anchor is borne a hypha, the haustorial mother cell (HMC), that gives rise to a haustorium, which primarily infects the stomatal subsidiary cells. The fungus pursues its growth with the formation of more intercellular hyphae, including HMCs, and many haustoria in the spongy and palisade parenchyma cells and even of the upper epidermis. A dense mycelium is observed below the penetration area, and a uredosporic sorus protrudes like a "bouquet" through the stomata about 21 days after inoculation.

3.2 Genetic and Physiological Diversity

The first experimental evidence for the physiological specialization of *H. vastatrix* was identified in India by Mayne (1932), who differentiated the local rust samples into four physiologic races. The world surveys of coffee rust races initiated in the 1950s at the Coffee Rusts Research Center (Centro de Investigação das Ferrugens do Cafeeiro—CIFC) in Portugal enabled the characterization of more than 55 rust races from about 4000 rust samples received from different coffee-growing countries. These races have been identified according to their spectra of virulence on a set of 27 coffee plant differentials (Silva et al. 2022 and references therein).

Coffee rust race identification relies on applying Flor's gene-to-gene theory to the *Coffea* sp.–*H. vastatrix* interaction (Noronha-Wagner and Bettencourt 1967). The resistance genes characterized on coffee plants were designated by "S_H genes". Major genes S_H1, S_H2, S_H4, and S_H5 were found in *C. arabica*, the gene S_H3 is considered to derive from *C. liberica*, and S_H6, S_H7, S_H8, and S_H9 were only found in "Timor hybrid"—HDT (natural *Coffea arabica* × *canephora* hybrid) derivatives, therefore coming from the Robusta side of the hybrid (Rodrigues Jr. et al. 1975; Bettencourt and Rodrigues Jr. 1988). Concomitantly, it was possible to infer nine virulence genes (v_1–v_9) on *H. vastatrix*.

The race genotypes comprise v_1–v_9 virulence genes from isolates derived from *C. arabica* and tetraploid interspecific hybrids, whereas the virulence genes from some isolates that attack diploid coffee species are not known due to the unavailability of genetic studies in these plants (Rodrigues Jr. et al. 1975; Bettencourt and Rodrigues Jr. 1988; Talhinhas et al. 2017). However, virulence profiling, particularly of isolates infecting tetraploid interspecific hybrids, like HDT derivatives, can only go as far as the available collection of coffee differential genotypes allows, leaving many virulence profiles incomplete or entirely unidentified (Talhinhas et al. 2017; Silva et al. 2022).

Work carried out at CIFC allowed to find rust races with the ability to infect all known coffee genotypes of *C. arabica*, as well as some genotypes of HDT, like CIFC HDT 832/1 and CIFC HDT 832/2 used as sources of resistance in the coffee populations of Catimor and Sarchimor that were spread to the coffee world (Talhinhas et al. 2017; Silva et al. 2022; CIFC records).

The evolution of virulence in *H. vastatrix* is parallel to the existent resistance genotypes. For example, in a coffee country that grows only pure Arabica like Geisha (S_H1), Kent's (S_H2), Agaro (S_H4), and Bourbon and Typica (S_H5), the probability of finding different associations of v_1, v_2, v_4, v_5 is 100%. Contrarily, the probability of finding the virulence genes v_6, v_7, v_8, and v_9 associated with the interspecific tetraploid hybrids, like HDT, is very low or inexistent (CIFC records).

3.3 Implications of Rust Molecular Diversity and Population Dynamics in Coffee Breeding

The global dissemination of *H. vastatrix* across continents seems to be intimately linked to the historical evolution of the global coffee industry. It is no surprise that coffee hosts act as a major selective pressure shaping the geographical distribution and local prevalence of rust races, leading to the recorded large spectrum of virulence profiles and the recurrent emergence of new ones. However, it is puzzling how *H. vastatrix* can so rapidly evolve to overcome resistance in coffee cultivars, considering that no sexual phase in the pathogen's life cycle has yet been identified. Numerous studies have been engaged to characterize population genetic variability in coffee rust, failing to detect any clear genetic structuring pattern or a direct link between phenotypic diversity and molecular diversity (Talhinhas et al. 2017 and references therein; Kosaraju et al. 2017; Santana et al. 2018; Quispe-Apaza et al. 2021; Bekele et al. 2022). Those studies using several kinds of DNA markers (AFLP, RFLP, rDNA-ITS, SRAP, SSRs) and mainly addressing local populations reported different levels of genetic variability, from low to high, but consistently no evidence of population structure could be found, either relating to race, host or geographical origin. The first insights on the population evolutionary history of this pathogen arouse only recently from efforts on genomic research with the analysis of genome-wide SNP data. For the first time, *H. vastatrix* populations were found to be clearly structured into three divergent genetic lineages with marked host specialization, differentiating rusts infecting diploid coffee species from those infecting tetraploid coffee species (Silva et al. 2018). Moreover, evidences of recombination and footprints of introgression were also found, alerting to the possibility of virulence factors exchange between rust lineages. Episodes of introgression by hybridization have probably been rare in time, but the risk of having different rust lineages within cruising range should be taken into account to guide disease control and breeding strategies. While recombination has been detected both from DNA marker and genomic data (Maia et al. 2013; Cabral et al. 2016; Silva et al. 2018), albeit through unknown mechanisms, probably of a parasexual nature. Both these events may, however, be contributing to virulence increase and could explain the high pace of pathotype emergence. More recently, additional genomic data allowed to further discover a population genetic subdivision in a worldwide sampling of rusts from *C. arabica* and interspecific hybrids, revealing three divergent genetic subgroups: a low differentiated and globally distributed rust lineage, as well as two other highly differentiated rust lineages, one occurring specifically in Africa and the other in East Timor (Rodrigues et al. 2022). Most interestingly, the divergence of these lineages could be explained by only a small set of SNP loci putatively under the effect of positive selection (Rodrigues et al. 2022). These results suggest that genetic variation underlying host adaptation relies on a small portion of the genome and that the genes associated with these *loci* may be critically important for the species to survive in novel environments/coffee hosts. Important genome reference assemblies for *H. vastatrix* have been created in the last years (Cristancho et al. 2014; Porto et al. 2019). The recent availability of a chromosome-level genome resource for *H. vastatrix* (Tobias et al. 2022) offers renewed prospects for characterizing virulence loci, envisioning the future development of candidate diagnostic markers associated with rust pathotypes and alternative strategies for selective breeding.

4 Coffee Rust Resistance

4.1 Concepts of Plant Disease Resistance

Plants' perception and response to pathogen invasion evolved alongside the pathogen infection strategy as a sophisticated multilayer surveillance system leading to complete host immunity, partial resistance, or susceptibility (Bettgenhaeuser et al. 2014; Li et al. 2020). Plant defenses are structured in constitutive/passive and induced responses. The first comprises preformed barriers, such as waxy epidermal cuticles, rigid cell walls, and antimicrobial secondary metabolites that offer protection against pathogens. The second one results from a suite of plasma membrane receptors that detect the pathogens or their molecules and is based on the recognition of conserved microbe-associated or pathogen-associated molecular patterns (MAMPs/PAMPs) and host danger-associated molecular patterns (DAMPs) by pattern recognition receptors (PRRs) that activate PAMP-triggered immunity (PTI) (Delplace et al. 2022). Alongside this broad-range defense system, plants developed intracellular resistance proteins to detect specific effector proteins secreted by pathogens activating the effector-trigger immunity (ETI) (Andersen et al. 2018). This two-branch immunity system (PTI and ETI) leads to the induction of a continuous and overlapping downstream response, such as mitogen-activated protein kinase (MAPK) cascades, G-proteins, calcium flux, reactive oxygen species (ROS), transcriptional reprogramming, phytohormones, pathogenesis-related (PR) genes/ proteins and epigenetic modifications (Meng and Zhang 2013; Zhang et al. 2012; Lecourieux et al. 2006; Robert-Seilaniantz et al. 2011; Amorim et al. 2017; Zhu et al. 2016). These responses seem stronger and more prolonged during ETI when compared to PTI. ETI is typically associated with the hypersensitive reaction (HR), a form of programmed plant cell death localized at the infection sites (Torres et al. 2006; Jones and Dangl 2006). HR is one of the most important factors in the restriction of pathogen growth, particularly of obligate biotrophs (Heath 2000; Andersen et al. 2018) such as rust fungi (Periyannan et al. 2017).

ROS can trigger HR (Torres et al. 2006; Martins et al. 2020) and may act as a damaging or signaling molecule depending on the balance between ROS production and scavenging mechanisms. When the ROS-scavenging mechanisms fail, this leads to an excess of ROS accumulation responsible for oxidative damage, promoting lipid peroxidation, and damaging macromolecules (e.g., proteins, lipids, and sugars) and DNA (Das and Roychoudhury 2014). Membrane damage by peroxidation of polyunsaturated fatty acids (e.g., linoleic acid) can be initiated not only by ROS but also by lipid radicals or by lipoxygenases (LOXs) during the HR. Also associated with the rapid loss of cell membrane integrity during HR is an increase in oxidizing enzymes, such as superoxide dismutase (SOD) and peroxidases (Heath 2000; Daudi et al. 2012).

Plant peroxidases (POD) function in plant defense against pathogens through the production of antimicrobial quantities of H_2O_2, as well as in cell wall lignification or cross-linking with cell wall proteins (Penel 2000; Torres et al. 2006).

Moreover, several studies suggest that plant phenolic compounds are strongly involved in plant–pathogen interaction and may restrict pathogen spread (Vermerris and Nicholson 2006). As well phenylalanine ammonia-lyase (PAL), a key enzyme of the phenylpropanoid pathway, catalyzes the deamination of phenylalanine to trans-cinnamic acid, a precursor for several phenolic compounds such as salicylic acid (SA), phenylpropanoids, flavonoid, and lignin seems to be involved in immunity responses (Bagal et al. 2012).

Other proteins also reported as being involved in plant resistance are hydrolases, particularly sugar hydrolases (GH) and peptidases/proteases. The proteases (together with phosphatases) can lead to a complex regulation of cell wall proteins through post-translational modification and GHs confer high plasticity to cell wall polysaccharides and/or are directly involved in antifungal activity (Guerra-Guimarães et al. 2015). Indeed, ß-1,3-glucanase and chitinase (considered as PR-proteins) hydrolyze ß-1,3-glucan and chitin, respectively, the cell wall components of most fungal pathogens (as revised by Silva et al. 1999). Additionally, Germins and germin-like proteins (GLPs) which are homohexamer glycoproteins, have also been implicated in biotic and abiotic stress responses (Liao et al. 2021). Studies related to GLP and plant immunity showed their association with jasmonic acid (JA)-mediated defense response (Liu et al. 2016; Pei et al. 2019) or the connection between GLP overexpression and the improved resistance to fungal pathogens and the accumulation of ROS (Beracochea et al. 2015; Sultana et al. 2016).

4.2 Mechanisms of Coffee Resistance to H. vastatrix

Over the years, there has been considerable progress in understanding the mechanisms of coffee's complete resistance to *H. vastatrix* at the cellular and biochemical levels, and more recently via analytical chemistry, gene expression analysis, and the use of omics approaches.

Coffee resistance to *H. vastatrix* is characterized by the arrest of fungal growth, which may occur at pre-haustorial stages (pre-haustorial resistance) or after the formation of at least one haustorium (post-haustorial resistance). In both types of resistance, the hypersensitive reaction (HR) is one of the first cytological responses induced by *H. vastatrix*. This response occurs initially in stomatal cells associated with pre-haustorial fungal stages and later in plant cells invaded by haustoria, and spreads to adjacent noninvaded cells (Silva et al. 2006, and references therein; Silva et al. 2008; Diniz et al. 2012; Guerra-Guimarães et al. 2015).

The early perception of the pathogen invasion by a repertoire of recognition kinases at the plasma membrane, such as RLK and *LRR-RLK2*, or intracellularly by nucleotide-binding site–leucine-rich repeat (*NBS-LRR*) are successful steps in triggering host defense (Fernandez et al. 2004; Guzzo et al. 2009; Diniz et al. 2012; Diola et al. 2013; McCook and Peterson 2020). During coffee resistance to *H. vastatrix*, the up-regulation of two MAPKs (*MAPK2—Mitogen-activated protein kinase 2* and *MEK2—Dual specificity mitogen-activated protein kinase kinase 2*) suggests that they are important signaling elements of the defense response during the infection process (Diola et al. 2013). In addition, two calcium-related genes, *calcium-dependent protein kinase 5* (*CDPK5*) and *calmodulin-binding protein* (*CaMBP*) have been associated as part of the Ca^{2+} signaling in coffee resistance response (Diola et al. 2013). In coffee plants, a small list of transcriptional factors has been associated with resistance to *H. vastatrix*: Ap2 (*AP2 type transcription factor*) (Fernandez et al. 2004), bHLH (*basic helix-loop-helix DNA-binding protein*) (Florez et al. 2017), and *bZIP56* (*bZIP transcription factor*) (Diola et al. 2013). However, the WRKY family is the most well-studied transcriptional regulators in coffee plants, with 17 out of 22 genes that seem to be linked to *H. vastatrix* resistance. Of all known coffee WRKYs, CaWRKY1 (*CaWRKY1a* and *CaWRKY1b*) is activated as early as *H. vastatrix* penetration into host tissues and deployment of HR (Ganesh et al. 2006; Petitot et al. 2008, 2013a; Diniz et al. 2012). The role of phytohormones, like JA and ethylene (ET), in coffee defense against *H. vastatrix*, remains unclear (Guzzo et al. 2009; Ramiro et al. 2010; Diniz et al. 2012; Diola et al. 2013; Florez et al. 2017). However, several studies point out that SA is a key hormone in coffee defense against *H. vastatrix*, by the increased expression of SA pathway-related genes, such as SA-biosynthesis gene *PAL*, by SA-induce PR genes, *CaPR1, CaPR2, CaPR5, CaPR10* (Couttolenc-Brenis et al. 2020; Diniz et al. 2012, 2021; Diola et al. 2013; Guzzo et al. 2009; Ramiro et al. 2009) and by SA-mediated protein–protein interactions gene *NPR1* (*Non-expressor of pathogenesis-related*) (Diniz et al. 2012; Petitot et al. 2013b; Couttolenc-Brenis et al. 2020). NPR1 is a transcription co-factor and a *bone fide* SA receptor and, consequently, a positive regulator of several SA downstream responses such as HR (Saleem et al. 2021).

Light and transmission electron microscopic observations further suggest the involvement of ROS, such as H_2O_2 and O_2^- in the HR of coffee-resistant genotypes. Additionally, deposition of phenolic-like compounds in cell walls and cytoplasmic contents, plant cell wall lignification, haustoria encasement with callose and β-1,4-glucans; accumulation of intercellular pectin-like material containing polysaccharides and phenols, and plant cell hypertrophy were also observed (Rodrigues Jr. et al. 1975; Silva et al. 2001, 2002, 2006, 2008; Ramiro et al. 2009; Diniz et al. 2012).

Biochemical studies with coffee-resistant genotypes, revealed the early increase in the activity of several oxidative enzymes associated with ROS homeostasis, namely, POD, LOX, and SOD, as well as the enzyme PAL (Rojas et al. 1993; Silva et al. 2002, 2008; Guerra-Guimarães et al. 2009a). The early increase in the SA levels, quantified by HPLC/ESI-MS/MS, reinforces the involvement of an SA-dependent pathway in coffee resistance to *H. vastatrix* (Sá et al. 2014). Additionally, the evaluation of chlorogenic acid, an abundant polyphenol in coffee, performed by HPLC-DAD and LC-MS also revealed an early and significant increase in its content associated with the resistance (Leitão et al. 2011). Chitin and β-1,3-glucan, are the main components of *H. vastatrix* cell walls, including those from intercellular hyphae (infection hyphae and HMCs) and haustoria (Silva et al. 1999). An early increase of β-1,3-glucanase (PR2) and chitinase (PR3) activity was observed in crude extracts and in the apoplast of resistant coffee leaves (Maxemiuc-Naccache et al. 1992; Guerra-Guimarães et al. 2009b). Furthermore, basic isoforms specific to class I chitinases were detected earlier and only in the resistance, suggesting its involvement in the defense response of the coffee plants (Guerra-Guimarães et al. 2009b).

Going deep into the study of the coffee leaf apoplast, a proteomic analysis was performed revealing the increase in abundance of several cell wall glycohydrolases (GH3, GH31, and GH38 family), PR proteins [PR1, PR2, PR3, thaumatin/osmotin (PR5), GPLs (PR15 and PR16)], proteases (serin, cysteine and aspartic peptidases) and other enzymes (e.g.; metallophosphatases) playing a role in the coffee defense response (Azinheira et al. 2013; Guerra-Guimarães et al. 2015; Possa et al. 2020; Silva et al. 2022).

The resistant and susceptible coffee genotypes share most of the described host responses when infected by *H. vastatrix*. However, they are observed earlier and with greater magnitude during the resistance response, particularly in pre-haustorial resistance.

5 Conclusions

This chapter reviews decades of scientific knowledge accumulated on coffee–*H. vastatrix* interactions. From the first studies back in the sixties until today, a wealth of data has been gathered about this binomial relationship. What we know today has evolved significantly from Flor's gene-for-gene model to rust races' candidate markers within the current "omics" era (genomics, transcriptomics, metabolomics, and proteomics). The income of new data continuously challenges what we know and raises further questions. What is the true nature of the coffee resistance genes, and how are they regulated? How are the recently discovered small non-coding RNA (sRNA, miRNAs) involved in gene regulation? How can miRNA be related to coffee resistance to *H. vastatrix*? The answer to these and more questions relies on our ability to continue to explore the coffee–*H. vastatrix* interaction and adventure ourselves to go deep into barely explored coffee resistance research fields such as functional characterization and epigenetics. Despite all the significant progress made to date, a thorough exploration of *Coffea–Hemileia vastatrix* interactions using advanced technologies remains critical for developing new and efficient disease control strategies.

Acknowledgements Funding for this work was provided by the Food and Agriculture Organization of the United Nations and the International Atomic Energy Agency through their Joint FAO/IAEA Research Contract nº 20902/R0 of the IAEA Coordinated Research Project D22005 and by Foundation for Science and Technology (FCT) and FEDER funds through PORNorte under the project CoffeeRES ref. PTDC/ASP-PLA/29779/2017 and by FCT UNIT (UID/AGR/04129/2020) of LEAF—Linking Landscape, Environment, Agriculture and Food, Research Unit.

References

Agrios GN (2005) Plant pathology. Academic Press, New York

Amaral JF, Teixeira C, Pinheiro ED (1956) A bactéria causadora da mancha aureolada do cafeeiro. Arq Inst Biol 23:151–155

Amorim LL, da Fonseca Dos Santos R, Neto JPB, Guida-Santos M, Crovella S, Benko-Iseppon AM (2017) Transcription factors involved in plant resistance to pathogens. Curr Protein Pept Sci 18(4):335–351

Andersen EJ, Ali S, Byamukama E, Yen Y, Nepal MP (2018) Disease resistance mechanisms in plants. Genes 9:339

Andrade CCL, Resende MLV, Moreira SI, Mathioni SM, Botelho DMS, Costa JR, Andrade ACM, Alves E (2021) Infection process and defense response of two distinct symptoms of *Cercospora* leaf spot in coffee leaves. Phytoparasitica 49:727–737

Assefa A, Balcha W, Kloos H (2022) Assessment of spatio-temporal patterns of coffee wilt disease, constraints on coffee production and its management practices in Berbere district, Bale Zone, southeastern Ethiopia. Arch Phytopathol Plant Prot 55(8):991–1013. https://doi.org/10.1080/03235408.2022.2081525

Avelino J, Cristancho M, Georgiou S, Imbach P, Aguilar L, Bornemann G, Läderach P, Anzueto F, Hruska AJ, Morales C (2015) The coffee rust crises in Colombia and Central America (2008–2013): impacts, plausible causes and proposed solutions. Food Secur 7:303–321. https://doi.org/10.1007/s12571-015-0446-9

Azinheira HG, Gorgulho R, Vieira A, Talhinhas P, Batista D, Silva MC, Guerra-Guimarães L (2013) An integrative analysis of proteomic and transcriptomic data to understand *Coffea arabica* responses to *Hemileia vastatrix*. In: Viera N, Saibo N, Oliveira MM (eds) Book of abstracts of the FV 2013—XIII Congresso Luso-Espanhol de Fisiologia Vegetal, Lisboa, Portugal, 24–28 July 2013, Abstract, 312 pp

Badel JL, Zambolim L (2019) Coffee bacterial diseases: a plethora of scientific opportunities. Plant Pathol 68(3):411–425

Bagal UR, Leebens-Mack JH, Lorenz WW, Dean JFD (2012) The phenylalanine ammonia lyase (PAL) gene family shows a gymnosperm-specific lineage. BMC Genom 13:S1

Bai X, Zhou L, Hu Y, Ji G, Li J, Zhang H (2013) Isolation and identification of the pathogen of coffee bacterial blight disease. Chin J Trop Crops 34:738–742

Baker P (2014) The 'big rust': an update on the coffee leaf rust situation. Coffee Cocoa Int 40:37–39

Bekele KB, Senbeta GA, Garedew W, Caixeta ET, Ramírez-Camejo LA, Aime MC (2022) Genetic diversity and population structure of *Hemileia vastatrix* from Ethiopian Arabica coffee. Arch Phytopathol Plant Prot 55(13):1483–1503. https://doi.org/10.1080/03235408.2021.1983385

Belan LL, Pozza EA, Freitas MLO, Raimundi MK, Souza RM, Machado JC (2016) Occurrence of *Pseudomonas syringae* pv. *garcae* in coffee seeds. Aust J Crop Sci 10:1015–1021

Beracochea VC, Almasia NI, Peluffo L, Nahirñak V, Hopp EH, Paniego N, Heinz RA, Vazquez-Rovere C, Lia VV (2015) Sunflower germin-like protein HaGLP1 promotes ROS accumulation and enhances protection against fungal pathogens in transgenic *Arabidopsis thaliana*. Plant Cell Rep 34:1717–1733

Beretta M, Harakava R, Chagas M, Derrick K, Barthe G, Ceccardi T, Lee R, Paradela O, Sugimori H, Ribeiro I (1996) First report of *Xylella fastidiosa* in coffee. Plant Dis 80:821

Bettencourt AJ, Rodrigues Jr CJ (1988) Principles and practice of coffee breeding for resistance to rust and other diseases. In: Clarke RJ, Macrae R (eds) Coffee agronomy, vol IV. Elsevier Applied Science Publishers Ltd, London and New York, pp 199–234

Bettgenhaeuser J, Gilbert B, Ayliffe M, Moscou MJ (2014) Nonhost resistance to rust pathogens—a continuation of continua. Front Plant Sci 5:664

Bolaños C, Zapata M, Brodbeck B, Andersen P, Wessel-Beaver L, Jensen CE (2015) Distribución espacial de cafetos (*Coffea arabica* L.) posiblemente enfermos con el encorchamiento de la hoja causado por *Xylella fastidiosa* en Puerto Rico. J Agric Univ PR 99(2):157–165

Botelho DMS, Resende MLV, Andrade VT, Pereira AA, Patricio FRA, Junior PMR, Ogoshi C, Rezende JC (2017) *Cercosporiosis* resistance in coffee germplasm collection. Euphytica 213:117. https://doi.org/10.1007/s10681-017-1901-9

Bozkurt TO, Kamoun S (2020) The plant-pathogen haustorial interface at a glance. J Cell Sci 4:133. https://doi.org/10.1242/jcs.237958. PMID: 32132107

Cabral PGC, Maciel-Zambolim E, Oliveira SAS, Caixeta ET, Zambolim L (2016) Genetic diversity and structure of *Hemileia vastatrix* populations on Coffea spp. Plant Pathol 65:196–204. https://doi.org/10.1111/ppa.12411

Couttolenc-Brenis E, Carrión GL, Villain L, Ortega-Escalona F, Ramírez-Martínez D, Mata-Rosas M, Méndez-Bravo A (2020) Prehaustorial local resistance to coffee leaf rust in a Mexican cultivar involves expression of salicylic acid-responsive genes. Peer J 1–21. https://doi.org/10.7717/peerj.8345

Cristancho MA, Botero-Rozo DO, Giraldo W, Tabima J, Riaño-Pachón DM, Escobar C, Rozo Y, Rivera LF, Durán A, Restrepo S, Eilam T, Anikster Y, Gaitán AL (2014) Annotation of a hybrid partial genome of the coffee rust (*Hemileia vastatrix*) contributes to the gene repertoire catalog of the Pucciniales. Front Plant Sci 5:594

Das K, Roychoudhury A (2014) Reactive oxygen species (ROS) and response of antioxidants as ROS-scavengers during environmental stress in plants. Front Environ Sci 2:53

Daudi A, Cheng Z, O'Brien JA, Mammarella N, Khan S, Ausubel FM, Paul Bolwell G (2012) The apoplastic oxidative burst peroxidase in *Arabidopsis* is a major component of pattern-triggered immunity. Plant Cell 24:275–287

Delplace F, Huard-Chauveau C, Berthomé R, Roby D (2022) Network organization of the plant immune system: from pathogen perception to robust defense induction. Plant J 109:447–470

Diniz I, Talhinhas P, Azinheira HG, Várzea V, Medeira C, Maia I, Petitot A-S, Nicole M, Fernandez D, Silva MC (2012) Cellular and molecular analyses of coffee resistance to *Hemileia vastatrix* and nonhost resistance to *Uromyces vignae* in the resistance-donor genotype HDT832/2. Eur J Plant Pathol 133(1):141–157. https://doi.org/10.1007/s10658-011-9925-9

Diniz I, Figueiredo A, Sebastiana M, Munoz-Pajarez AJ, Valverde J, Azevedo H, Rodrigues AS, Prakash RS, Pereira AP, Guerra-Guimarães L, Azinheira HG, Várzea V, Batista D, Silva MC (2021) First steps on the resistance profiling of Kawisari coffee hybrid through cytological and gene expression analyses. In: Book of abstracts of the 28th international conference on coffee science (ASIC), Montpellier, France, 28 June–1 July 2021, Abstract, 60 pp

Diola V, Brito GG, Caixeta ET, Pereira LFP, Loureiro ME (2013) A new set of differentially expressed signaling genes is early expressed in coffee leaf rust race II incompatible interaction. Funct Integr Genomics 13(3):379–389. https://doi.org/10.1007/s10142-013-0330-7

Fernandez D, Santos P, Agostini C, Bon MC, Petitot AS, Silva MC, Guerra-Guimarães L, Ribeiro A, Argout X, Nicole M (2004) Coffee (*Coffea arabica* L.) genes early expressed during infection by the rust fungus (*Hemileia vastatrix*). Mol Plant Pathol 5(6):527–536. https://doi.org/10.1111/J.1364-3703.2004.00250.X

Firman ID, Waller JM (1977) Coffee berry disease and other *Colletotrichum* diseases of coffee, 4th edn. Commonwealth Mycological Institute, Surrey, England

Flood J (2021) Coffee wilt diseases. In: Climate-smart production of coffee: achieving sustainability and ecosystem services. CABI, UK. https://doi.org/10.19103/AS.2021.0096.25

Florez JC, Mofatto LS, do Livramento Freitas-Lopes RL, Ferreira SS, Zambolim EM, Carazzolle MF, Zambolim L, Caixeta ET (2017) High throughput transcriptome analysis of coffee reveals prehaustorial resistance in response to *Hemileia vastatrix* infection. Plant Mol Biol 95(6):607–623. https://doi.org/10.1007/s11103-017-0676-7

Food and Agriculture Organization of the United Nations (FAO) (2022) Markets and trade 2022. Available online: https://www.fao.org/markets-and-trade/commodities/coffee/en/. Accessed 13 Dec 2022

Ganesh D, Petitot AS, Silva MC, Alary R, Lecouls AC, Fernandez D (2006) Monitoring of the early molecular resistance responses of coffee (*Coffea arabica* L.) to the rust fungus (*Hemileia vastatrix*) using real-time quantitative RT-PCR. Plant Sci 170(6):1045–1051. https://doi.org/10.1016/j.plantsci.2005.12.009

Garnica DP, Nemri A, Upadhyaya NM, Rathjen JP, Dodds PN (2014) The ins and outs of rust haustoria. PLoS Pathog 10(9):10–13. https://doi.org/10.1371/journal.ppat.1004329

Guerra-Guimarães L, Cardoso S, Martins I, Loureiro A, Bernardes SA, Várzea V, Silva MC (2009a) Differential induction of superoxide dismutase in *Coffea arabica–Hemileia vastatrix* interactions. In: Proceedings of the 22th international conference on coffee science (ASIC), Campinas, Brazil, 14–19 Sept 2008, pp 1036–1039

Guerra-Guimarães L, Silva MC, Struck C, Loureiro A, Nicole M, Rodrigues CJ, Ricardo CPP (2009b) Chitinases of *Coffea arabica* genotypes resistant to orange rust *Hemileia vastatrix*. Biol Plant 53:702–706. https://doi.org/10.1007/s10535-009-0126-8

Guerra-Guimarães L, Tenente R, Pinheiro C, Chaves I, Silva MD, Cardoso FMH, Planchon S, Barros DR, Renaut J, Ricardo CP (2015) Proteomic analysis of apoplastic fluid of *Coffea arabica* leaves highlights novel biomarkers for resistance against *Hemileia vastatrix*. Front Plant Sci 6:478. https://doi.org/10.3389/fpls.2015.00478

Guzzo SD, Harakava R, Tsai SM (2009) Identification of coffee genes expressed during systemic acquired resistance and incompatible interaction with *Hemileia vastatrix*. J Phytopathol 157(10):625–638. https://doi.org/10.1111/j.1439-0434.2008.01538.x

Heath MC (1997) Signalling between pathogenic rust fungi and resistant or susceptible host plants. Ann Bot 80:713–720

Heath MC (2000) Hypersensitive response-related death. Plant Mol Biol 44:321–334

Hindorf H, Omondi CO (2011) A review of three major fungal diseases of *Coffea arabica* L. in the rainforests of Ethiopia and progress in breeding for resistance in Kenya. J Adv Res 2:109–120. https://doi.org/10.1016/j.jare.2010.08.006

International Coffee Organization (ICO) (2020) ICO coffee production 2020. Available online: https://www.ico.org/prices/poproduction.pdf. Accessed 13 Dec 2022

Jones JDG, Dangl JL (2006) The plant immune system. Nature 444:323–329

Kahn LH (2019) Quantitative framework for coffee leaf rust (*Hemileia vastatrix*), production and futures. Int J Agric Ext 7:77–87. https://doi.org/10.33687/ijae.007.01.2744

Keith L, Sugiyama L, Brill E, Adams B-L, Fukada M, Hoffman K, Ocenar J, Kawabata A, Kong A, McKemy J, Olmedo-Velarde A, Melzer M (2021) First report of coffee leaf rust caused by *Hemileia vastatrix* on coffee (*Coffea arabica*) in Hawaii. Plant Dis 34:2–4. https://doi.org/10.1094/pdis-05-21-1072-pdn

Kemen E, Kemen AC, Rafiqi M, Hempel U, Mendgen K, Hahn M, Voegele RT (2005) Identification of a protein from rust fungi transferred from haustoria into infected plant cells. Mol Plant Microb Interact MPMI 18(11):1130–1139. https://doi.org/10.1094/MPMI-18-113

Korobko A, Wondinagegne E (1997) Bacterial blight of coffee (*Pseudomonas syringae* pv. *garcae*) in Ethiopia. In: Rudoldh K, Burr TJ, Mansfield JW, Stead D, Vivian A, Von Kietzele J (eds) *Pseudomonas syringae* and related pathogens. Springer, Dordrecht, Netherlands, pp 538–541

Kosaraju B, Sannasi S, Mishra MK, Subramani D, Bychappa M (2017) Assessment of genetic diversity of coffee leaf rust pathogen *Hemileia vastatrix* using SRAP markers. J Phytopathol 165:486–493. https://doi.org/10.1111/jph.12583

Krishnan S (2017) Sustainable coffee production. Denver Botanic Gardens. https://doi.org/10.1093/acrefore/9780199389414.013.224

Lécolier A, Besse P, Charrier A, Thierry-Nicolas T, Noirot M (2009) Unraveling the origin of *Coffea arabica* 'Bourbon pointu' from La Réunion: a historical and scientific perspective. Euphytica 168:1–10. https://doi.org/10.1007/s10681-009-9886-7

Lecourieux D, Ranjeva R, Pugin A (2006) Calcium in plant defence-signalling pathways. New Phytol 171:249–269

Leitão S, Guerra-Guimarães L, Bronze MR, Vilas Boas L, Sá M, Almeida MHG, Silva MC (2011) Chlorogenic acid content in coffee leaves: possible role in coffee leaf rust resistance. In: Proceedings of the 23rd international conference on coffee science (ASIC), Bali, Indonesia, 3–8 Oct 2010, pp 743–747

Li W-B, Pria WD Jr, Teixeira DC, Miranda VS, Ayres A, Franco CF, Costa MG, He C-X, Costa P, Hartung JS (2001) Coffee leaf scorch caused by a strain of *Xylella fastidiosa* from citrus. Plant Dis 85:501–505

Li P, Lu YJ, Chen H, Day B (2020) The lifecycle of the plant immune system. Crit Rev Plant Sci 39:72–100

Liao L, Hu Z, Liu S, Yang Y, Zhou Y (2021) Characterization of germin-like proteins (GLPs) and their expression in response to abiotic and biotic stresses in cucumber. Horticulturae 7:412. https://doi.org/10.3390/horticulturae7100412

Liu Q, Yang J, Yan SS, Zhao J, Wang W, Yang T, Wang X, Mao X, Dong J et al (2016) The germin-like protein OsGLP2-1 enhances resistance to fungal blast and bacterial blight in rice. Plant Mol Biol 92:411–423

Loureiro A, Nicole MR, Várzea V, Moncada P, Bertrand B, Silva MC (2012) Coffee resistance to *Colletotrichum kahawae* is associated with lignification, accumulation of phenols and cell death at infection sites. Physiol Mol Plant Pathol 77(1):23–32. https://doi.org/10.1016/j.pmpp.2011.11.002

Maghuly F, Jankowicz-Cieslak J, Bado S (2020) Improving coffee species for pathogen resistance. CAB Rev 15(009):1–18. https://doi.org/10.1079/PAVSNNR202015009

Maia TA, Maciel-Zambolim E, Caixeta ET, Mizubuti ESG, Zambolim L (2013) The population structure of *Hemileia vastatrix* in Brazil inferred from AFLP. Australas Plant Pathol 42:533–542. https://doi.org/10.1007/s13313-013-0213-3

Martins D, Araújo SD, Rubiales D, Patto MCV (2020) Legume crops and biotrophic pathogen interactions: a continuous cross-talk of a multilayered array of defense mechanisms. Plants 9:1460

Maxemiuc-Naccache V, Braga MR, Dietrich SMC (1992) Chitinase and beta-1,3-glucanase changes in compatible and incompatible combinations between coffee leaf disks and coffee rust (*Hemileia vastatrix*). Rev Bras Bot 15(2):145–150

Mayne WW (1932) Physiological specializalion of *Hemileia vastatrix*. B. & Br. Nature 129:150

McCook S, Peterson PD (2020) The geopolitics of plant pathology: Frederick Wellman, coffee leaf rust, and cold war networks of science. Annu Rev Phytopathol 58:181–199. https://doi.org/10.1146/annurev-phyto-082718-100109

McDonald L (1926) A preliminary account of a disease of green coffee berries in Kenya Colony. Trans Br Mycol Soc 11:145–154

Meng X, Zhang S (2013) MAPK cascades in plant disease resistance signaling. Annu Rev Phytopathol 51:245–266

Nnadi NE, Carter DA (2021) Climate change and the emergence of fungal pathogens. PLoS Pathog 17(4):e1009503. https://doi.org/10.1371/journal.ppat.1009503

Noronha-Wagner M, Bettencourt AJ (1967) Genetic study of the resistance of *Coffea* spp. to leaf rust. Can J Bot 45:2021–2031

Pei Y, Li X, Zhu Y, Ge X, Sun Y, Liu N, Jia Y, Li F, Hou Y (2019) GhABP19, a novel germin-like protein from *Gossypium hirsutum*, plays an important role in the regulation of resistance to *verticillium* and *fusarium* wilt pathogens. Front Plant Sci 10:583

Penel C (2000) The peroxidase system in higher plants. In: Greppin H et al (eds) Integrated plant systems. University of Geneva, pp 359–367

Periyannan S, Milne RJ, Figueroa M, Lagudah ES, Dodds PN (2017) An overview of genetic rust resistance: from broad to specific mechanisms. PLoS Pathog 13:e1006380

Petitot AS, Lecouls AC, Fernandez D (2008) Sub-genomic origin and regulation patterns of a duplicated WRKY gene in the allotetraploid species *Coffea arabica*. Tree Genet Genomes 4(3):379–390. https://doi.org/10.1007/s11295-007-0117-x

Petitot AS, Barsalobres-Cavallari C, Ramiro D, Albuquerque Freire E, Etienne H, Fernandez D (2013a) Promoter analysis of the WRKY transcription factors CaWRKY1a and CaWRKY1b homoeologous genes in coffee (*Coffea arabica*). Plant Cell Rep 32(8):1263–1276. https://doi.org/10.1007/s00299-013-1440-3

Petitot AS, Severino FE, Maia IG, Fernandez D (2013b) Host response profiling to fungal infection: molecular cloning, characterization and expression analysis of NPR1 gene from coffee (*Coffea arabica*). In: Microbial pathogens and strategies for combating them: science, technology and education, pp 411–418

Phiri AN, Baker P (2009) Coffee wilt in Africa. Final technical report of the regional coffee wilt programme (2000–2007). CABI

Pinard F, Makune SE, Campagne P, Mwangi J (2016) Spatial distribution of coffee wilt disease under roguing and replanting conditions: a case study from Kaweri Estate in Uganda. Phytopathology 106:1291–1299

Porto BN, Caixeta ET, Mathioni SM, Vidigal PMP, Zambolim L, Zambolim EM, Donofrio N, Polson SW, Maia TA, Chen C, Adetunji M, Kingham B, Dalio RJD, de Resende MLV (2019) Genome sequencing and transcript analysis of *Hemileia vastatrix* reveal expression dynamics of candidate effectors dependent on host compatibility. PLoS ONE 14:e0215598. https://doi.org/10.1371/journal.pone.0215598

Possa KF, Silva JAG, Resende MLV, Tenente R, Pinheiro C, Chaves I, Planchon S, Monteiro ACA, Renaut J, Carvalho MAF et al (2020) Primary metabolism is distinctly modulated by plant resistance inducers in *Coffea arabica* leaves infected by *Hemileia vastatrix*. Front Plant Sci 11:309. https://doi.org/10.3389/fpls.2020.00309

Quispe-Apaza C, Mansilla-Samaniego R, Espejo-Joya R, Bernacchia G, Yabar-Larios M, López-Bonilla C (2021) Spatial and temporal genetic diversity and population structure of *Hemileia vastatrix* from Peruvian coffee plantations. Plant Pathol J 37(3):280–290. https://doi.org/10.5423/PPJ.OA.10.2020.0192

Ramalho TO, Figueira AR, Sotero AJ, Wang R, Geraldino Duarte PS, Farman M, Goodin MM (2014) Characterization of coffee ringspot virus-lavras: a model for an emerging threat to coffee production and quality. Virology 464–465:385–396

Ramalho TO, Figueira AR, Wang R et al (2016) Detection and survey of coffee ringspot virus in Brazil. Arch Virol 161:335–343. https://doi.org/10.1007/s00705-015-2663-0

Ramiro D, Escoute J, Petitot AS, Nicole M, Maluf MP, Fernandez D (2009) Biphasic haustorial differentiation of coffee rust (*Hemileia vastatrix* race II) associated with defence responses in resistant and susceptible coffee cultivars. Plant Pathol 58(5):944–955. https://doi.org/10.1111/j.1365-3059.2009.02122.x

Ramiro D, Jalloul A, Petitot AS, Grossi de Sá MF, Maluf MP, Fernandez D (2010) Identification of coffee WRKY transcription factor genes and expression profiling in resistance responses to pathogens. Tree Genet Genomes 6(5):767–781. https://doi.org/10.1007/s11295-010-0290-1

Ramos AH, Shavdia LD (1976) A die-back of coffee in Kenya. Plant Dis Rep 60:831–835

Rhiney K, Guido Z, Knudson C, Avelino J, Bacon CM, Leclerc G, Aime MC, Bebber DP (2021) Epidemics and the future of coffee production. Proc Natl Acad Sci USA 118. https://doi.org/10.1073/pnas.2023212118

Ritschel A (2005) Monograph of the genus *Hemileia* (Uredinales). In: Bresinsky A, Butin H, Tudzinski P (eds) Bibliotheca mycologica, vol 200, pp 3–132

Robert-Seilaniantz A, Grant M, Jones JD (2011) Hormone crosstalk in plant disease and defense: more than just jasmonate-salicylate antagonism. Annu Rev Phytopathol 49:317–343

Rodrigues CJ Jr (1990) Coffee rusts: history, taxonomy, morphology, distribution and host resistance. Fitopatol Bras 15:5–9

Rodrigues CJ Jr, Bettencourt AJ, Rijo L (1975) Races of the pathogen and resistance to coffee rust. Annu Rev Phytopathol 13:49–70

Rodrigues J, Rodriguez C, Moreira L, Villalobos W, Rivera C, Childers C (2002) Occurrence of coffee ringspot virus a *Brevipalpus* miteborne virus in coffee in Costa Rica. Plant Dis 86:564–564. https://doi.org/10.1094/PDIS.2002.86.5.564B

Rodrigues ASB, Silva DN, Várzea V, Paulo OS, Batista D (2022) Worldwide population structure of the coffee rust fungus *Hemileia vastatrix* is strongly shaped by local adaptation and breeding history. Phytopathology 112(9):1998–2011. https://doi.org/10.1094/PHYTO-09-21-0376-R

Rodríguez CM, Obando W, Villalobos RC (2001) First report of *Xylella fastidiosa* infecting coffee in Costa Rica. Plant Dis 85:1027

Rojas ML, Montes de Gómez V, Ocampo CA (1993) Stimulation of lipoxygenase activity in cotyledonary leaves of coffee reacting hypersensitively to the coffee leaf rust. Physiol Mol Plant Pathol 43:209–219

Rozo Y, Escobar C, Gaitan A, Cristancho M (2012) Aggressiveness and genetic diversity of *Hemileia vastatrix* during an epidemic in Colombia. J Phytopathol 160:732–740. https://doi.org/10.1111/jph.12024

Sá M, Ferreira JP, Queiroz VT, Vilas-Boas L, Silva MC, Almeida MH, Guerra-Guimarães L, Bronze MR (2014) A liquid chromatography/electrospray ionisation tandem mass spectrometry method for the simultaneous quantification of salicylic, jasmonic and abscisic acids in *Coffea arabica* leaves. J Sci Food Agric 94:529–536. https://doi.org/10.1002/jsfa.6288

Saleem M, Fariduddin Q, Castroverde CDM (2021) Salicylic acid: a key regulator of redox signalling and plant immunity. Plant Physiol Biochem 168:381–397. https://doi.org/10.1016/j.plaphy.2021.10.011

Santana MF, Zambolim EM, Caixeta ET, Zambolim L (2018) Population genetic structure of the coffee pathogen *Hemileia vastatrix* in Minas Gerais, Brazil. Trop Plant Pathol 43:473–476. https://doi.org/10.1007/s40858-018-0246-9

Silva MC, Nicole M, Rijo L, Geiger JP, Rodrigues CJ (1999) Cytochemistry of plant–rust fungus interface during the compatible interaction *Coffea arabica* (cv. Caturra)–*Hemileia vastatrix* (race III). Int J Plant Sci 160:79–91

Silva MC, Loureiro A, Guerra-Guimarães L, Nicole M, Valente P, Rodrigues Jr CJ (2001) Active oxygen metabolism in the hypersensitive response of coffee–rust interaction. In: Proceedings of the 11th congress of the Mediterranean phytopathological union & 3rd congress of the Sociedade Portuguesa de Fitopatologia, Évora, Portugal, 17–20 Sept 2001, pp 346–348

Silva MC, Nicole M, Guerra-Guimarães L, Rodrigues CJ (2002) Hypersensitive cell death and post-haustorial defence responses arrest the orange rust (*Hemileia vastatrix*) growth in resistant coffee leaves. Physiol Mol Plant Pathol 60:169–183. https://doi.org/10.1006/pmpp.2002.0389

Silva MC, Várzea V, Guerra-Guimarães L, Azinheira HG, Fernandez D, Petitot A-S, Bertrand B, Lashermes P, Nicole M (2006) Coffee resistance to the main diseases: leaf rust and coffee berry disease. Braz J Plant Physiol 18:119–147. https://doi.org/10.1590/S1677-04202006000100010

Silva MC, Guerra-Guimarães L, Loureiro A, Nicole MR (2008) Involvement of peroxidases in the coffee resistance to orange rust (*Hemileia vastatrix*). Physiol Mol Plant Pathol 72:29–38. https://doi.org/10.1016/j.pmpp.2008.04.004

Silva DN, Várzea V, Paulo OS, Batista D (2018) Population genomic footprints of host adaptation, introgression and recombination in coffee leaf rust. Mol Plant Pathol 19:1742–1753. https://doi.org/10.1111/mpp.12657

Silva MC, Guerra-Guimarães L, Diniz I, Loureiro A, Azinheira H, Pereira AP, Tavares S, Batista D, Várzea V (2022) An overview of the mechanisms involved in coffee–*Hemileia vastatrix* interactions: plant and pathogen perspectives. Agronomy 12(2):326. https://doi.org/10.3390/agronomy12020326

Sultana T, Deeba F, Naz F, Rose RJ, Saqlan Naqvi SM (2016) Expression of a rice GLP in *Medicago truncatula* exerting pleiotropic effects on resistance against *Fusarium oxysporum* through enhancing FeSOD-like activity. Acta Physiol Plant 38:255

Talhinhas P, Batista D, Diniz I, Vieira A, Silva DN, Loureiro A, Tavares, S, Pereira AP, Azinheira HG, Guerra-Guimarães L, Várzea V, Silva MC (2017) The coffee leaf rust pathogen *Hemileia vastatrix*: one and a half centuries around the tropics. Mol Plant Pathol 18:1039–1051. https://doi.org/10.1111/mpp.12512

Tobias PA, Edwards RJ, Surana P, Mangelson H, Inácio V, Silva MC, Várzea V, Park R, Batista D (2022) A chromosome-level genome resource for studying virulence mechanisms and evolution of the coffee rust pathogen *Hemileia vastatrix*. bioRxiv. https://doi.org/10.1101/2022.07.29.502101

Torres MA, Jones JD, Dangl JL (2006) Reactive oxygen species signaling in response to pathogens. Plant Physiol 141(2):373–378. https://doi.org/10.1104/pp.106.079467

Van der Vossen HAM, Walyaro DJ (2009) Additional evidence for oligogenic inheritance of durable host resistance to coffee berry disease (*Colletotrichum kahawae*) in arabica coffee (*Coffea arabica* L.). Euphytica 165:105–111. https://doi.org/10.1007/s10681-008-9769-3

Vermerris W, Nicholson R (2006) Phenolic compound biochemistry. ISBN: 9781402051630

Voegele RT, Mendgen KW (2011) Nutrient uptake in rust fungi: how sweet is parasitic life? Euphytica 179(1):41–55. https://doi.org/10.1007/s10681-011-0358-5

Wang A, Avelino J (1999) El ojo de gallo (*Mycena citricolor*). In: Bertrand B, Rapidel B (eds) Desafíos de la caficultura en Centroamérica. IICA, CRI, San José, pp 243–260

Wellman F (1952) Peligro de introducción de la *Hemileia* del café a las Américas. Turrialba 2:47–50

Zhang H, Gao Z, Zheng X, Zhang Z (2012) The role of G-proteins in plant immunity. Plant Signal Behav 7:1284–1288

Zhu QH, Shan WX, Ayliffe MA, Wang MB (2016) Epigenetic mechanisms: an emerging player in plant-microbe interactions. Mol Plant Microbe Interact 29(3):187–196. https://doi.org/10.1094/MPMI-08-15-0194-FI

Zoccoli DM, Takatsu A, Uesugi CH (2011) Ocorrência de mancha aureolada em cafeeiros na região do Triângulo Mineiro e Alto Paranaíba. Bragantia 70:843–849

Induced Mutagenesis of in Vitro Tissues and Cells of *Coffea arabica* L.

In Vitro Plantlet Establishment of *Coffea arabica* from Cut Seed Explants

Florian Goessnitzer, Joanna Jankowicz-Cieslak, and Ivan L. W. Ingelbrecht

Abstract Arabica coffee is one of the most important products in the world market. As a perennial crop, conventional breeding of Arabica coffee is challenged by its long reproductive cycle and narrow genetic base. In vitro tissue culture in combination with mutation induction techniques provides an attractive alternative approach for the genetic improvement of coffee. In this chapter we describe a simple and robust method to rapidly establish in vitro Arabica coffee plantlets from cut seed explants. The method streamlines the germination process under in vitro environmentally controlled conditions and overcomes microbial contamination, often associated with coffee seed lots harvested from the field or greenhouse. Using this protocol, disease-free in vitro coffee plantlets can be generated within 5–6 weeks, useful for downstream tissue culture manipulations such as the production of friable embryogenic callus and cell suspension cultures or induced chemical or physical mutagenesis.

1 Introduction

World coffee production relies mostly on two species: *Coffea arabica* and *Coffea canephora*, of which Arabica is the most widely cultivated, primarily because of its superior quality. Somatic embryogenesis in coffee has been used for mass propagation, genetic engineering and, more recently, also for induced mutagenesis studies (Los Santos-Briones and Hernández-Sotomayor 2006; Menéndez-Yufá et al. 2010; Bolívar-González et al. 2018). Since the 1990s, somatic embryogenesis (SE) techniques have enabled clonal propagation of both *Coffea arabica* and *Coffea canephora* (Etienne et al. 2018). In coffee, indirect SE through an intermediate callus phase using semi-solid and liquid media is well established while direct somatic embryogenesis

F. Goessnitzer (✉) · J. Jankowicz-Cieslak · I. L. W. Ingelbrecht (✉)
Plant Breeding and Genetics Laboratory, Joint FAO/IAEA Centre of Nuclear Techniques in Food and Agriculture, IAEA Laboratories Seibersdorf, International Atomic Energy Agency, Vienna International Centre, Vienna, Austria
e-mail: f.goessnitzer@iaea.org

I. L. W. Ingelbrecht
e-mail: i.ingelbrecht@iaea.org

© The Author(s) 2023
I. L. W. Ingelbrecht et al. (eds.), *Mutation Breeding in Coffee with Special Reference to Leaf Rust*, https://doi.org/10.1007/978-3-662-67273-0_3

methods have also been described but to a lesser extent (Quiroz-Figueroa et al. 2006; Murvanidze et al. 2021). In combination with techniques for induced mutagenesis, in vitro cell and tissue culture methods can provide an alternative strategy for enhancing genetic diversity and improvement of Arabica coffee.

Coffee seeds have limited viability. When stored at ambient temperature viability decreases rapidly; after two months for *Coffea canephora* and after six months in case of *Coffea arabica* (Wintgens 2012). In addition, the germination of coffee seeds is a slow process, taking 30–60 days under the most favourable conditions. In practice, favourable conditions may not always be available, hence the need for standardized conditions for coffee seed germination. This is particularly relevant for induced mutagenesis studies and related radiosensitivity testing where lethal doses or growth reduction values need to be determined. Coffee cell and tissue culture applications require that starting explants, frequently leaves, are free from microbial contamination. Depending on the environment and conditions of storage, however, coffee seed can become contaminated with fungi or other microbes. Due to the shape and morphology of the coffee seed, such microbial contamination can be challenging to remove through classical sterilization procedures.

Here we present a simple and robust in vitro protocol that streamlines the germination of coffee seed using cut seed explants. This method is effective for overcoming potential fungal/microbial contamination often associated with coffee seed batches. Germination proceeds under stable environmental conditions. Under in vitro conditions the coffee cut seed explants germinated, i.e., the radicle breaking through the endosperm, within 14 days. Overall, the procedure generated in vitro plantlets within 5–6 weeks, useful for downstream in vitro experiments such as somatic embryogenesis or induced mutagenesis studies.

2 Materials

2.1 Plant Material

1. Seed from three Arabica coffee varieties; Venecia, Caturra, and Catuai.

2.2 Chemicals

1. Murashige and Skoog Basal Salts with minimal organics (Sigma Alrich, M6899).
2. Sucrose (household grade).
3. 1 mM Stock solution S1-Naphthaleneacetic acid (NAA).
4. 1 mM Stock solution 6-Benzylaminopurine (BAP).
5. L-Cysteine.
6. Gelling agent (Agar or Gelrite).
7. Distilled Water.
8. NaOCl 14% (Stock concentration).
9. Tween 20.

2.3 Tools and Labware

1. Forceps.
2. Surgical blades.
3. Sieves.
4. Beakers for solvent.
5. Beaker for waste.
6. Magnet stir bar.
7. Aluminium foil.
8. Sterile paper.
9. Sterile petri dish.
10. In vitro 50 ml culture test tubes and vessels.

2.4 Lab Equipment

1. Autoclave.
2. Stirring/hot plate.
3. Flow bench.
4. Analytical balance.
5. Balance.
6. pH meter.
7. In vitro growth room.

3 Methods

3.1 Media Preparation

1. Prepare growth regulator and chemical stock solutions according to common procedures.
2. Prepare the M5 medium as described in Table 1.
3. Take the desired amount of solutions and chemicals and mix well before autoclaving.
4. Adjust the pH to 5.7 as described.
5. Dispense the M5 medium in sterile culture test tubes after autoclaving.
6. Always prepare fresh media shortly before use.

Component	1 L
MS medium (Sigma M6899)	4.4 g
Sucrose	30 g
BAP (1 mM stock)	0.5 mL
NAA (1 mM stock)	0.1 mL
Cysteine	25 mg
Gelrite	3 g
pH	5.7

Table 1 Composition of M5 medium (modified from Quiroz-Figueroa et al. 2006)

3.2 Seed Preparation

1. Seeds shall not be older than 2 months and shall be processed as soon as possible after ripening.
2. Discard seeds with visible microbial contamination e.g. growth of fungi, or with visible damage (Fig. 1a).
3. Remove the outer parchment (seed coat) (Fig. 1b) and integument (silver skin) by hand (Fig. 1c, d).
4. Collect cleaned seeds (Fig. 1e, f).

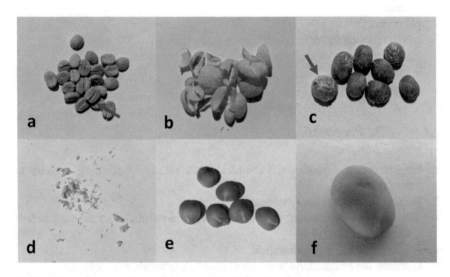

Fig. 1 *Coffea arabica* seed preparation for the isolation of cut seed explants

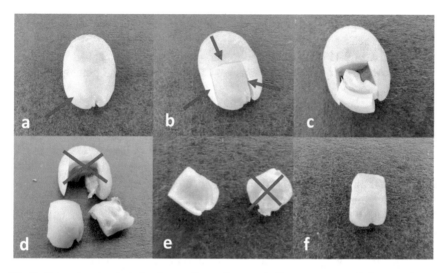

Fig. 2 Cut seed explant isolation from coffee seed; arrows indicate cuts; crosses indicate seed parts to be discarded

3.3   Explant Isolation

1. Place the seed on a clean surface suitable for cutting.
2. The embryo (3–4 mm) is located near the endosperm cap (Fig. 2a, and Note 1).
3. Hypocotyl and/or cotyledons are visible through the endosperm.
4. Carefully make 3 cuts with a surgical blade around the zygotic embryo (Fig. 2b), avoid damaging the embryo (see Note 2).
5. Isolate the cut explant measuring approximately 12 × 5 mm (Fig. 2c).
6. Remove the integument (silver skin) at the back of the explant, discard the remainder (endosperm) of the seed (Fig. 2d, e).
7. Collect the explant containing the zygotic embryo for surface sterilization (Fig. 2f).

3.4   Surface Sterilization of the Cut Seed

1. Place a 1 L beaker, a stirring/hot plate, 4 L autoclaved water and a waste beaker in the laminar air flow.
2. Calculate the concentration of NaOCl to reach a final concentration of 2.1% active chlorine.
3. Prepare 1 L of a 2.1% active-chlorine solution and add 0.2 mL Tween 20.
4. Place 10–15 explants in the sieve (Fig. 3).
5. Rinse the explants in the sieve under running tap-water for 5 min.

6. Place a magnetic stir bar and a maximun of two sieves containing the explants in a 1 L beaker.
7. Pour 1 L of 2.1% active chlorine solution into the 1 L beaker.
8. Remove any air bubbles in the sieves by gently lifting and putting down the beaker.
9. Cover the beaker with aluminium foil.
10. Gently swirl the beaker to wet the inner walls.
11. Spray the outer walls and top of the beaker with 70% EtOH.
12. Place the beaker on a stirring/hot plate in a laminar air flow.
13. Turn on the stirring/hot plate at approximately 200 rpm for 15 min.
14. Turn off the stirring/hot plate and partly open the cover.
15. Gently pour the solution into the waste beaker.
16. Refill the beaker with 1 L autoclaved water.
17. Gently stir and remove the water immediately.
18. Refill the beaker with autoclaved water.
19. Close the cover and place the beaker on the stirring/hot plate.
20. Turn on the stirring/hot plate at 200 rpm for 20 min.
21. Repeat the rinsing steps 18–20 twice.
22. Remove the water and leave the beaker inside the laminar air flow.
23. Take out the stirring/hot plate, waste beaker and water flasks.
24. Clean the laminar air flow surface with 70% EtOH.
25. Prepare the laminar for in vitro work (see below).

Fig. 3 Cut seed explants in the sieve

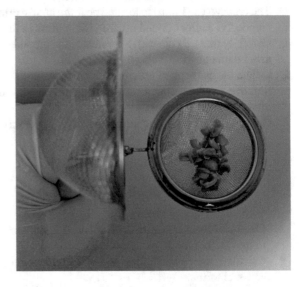

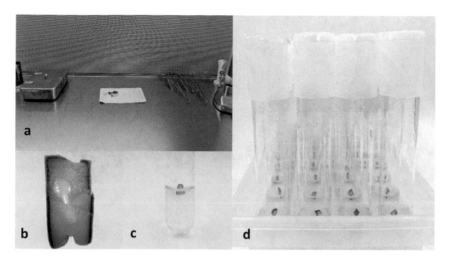

Fig. 4 Transfer of the cut seed explant to culture tubes

3.5 Transfer of Explants and Culture Conditions

1. Gently transfer the explants from the sieve onto a sterile working paper or petri dish in a laminar air flow (Fig. 4a).
2. Locate the embryo in the explant (Fig. 4b).
3. Insert the explant in the medium with the embryo facing upward, approximately halfway, one explant per culture tube (Fig. 4c).
4. Place the tubes on a suitable tray and move to the growth chamber (Fig. 4d).
5. Set the light conditions in the growth chamber to 16/8 light/dark photoperiod under fluorescent light, provided by SYLVANIA GRO-LUX 36 W ($60 \mu M \ m^{-2} \ s^{-1}$) at $26 \pm 2 \ °C$.

3.6 Germination and Plantlet Development

1. Monitor cultures daily for contamination, discard infected tubes (Fig. 5a).
2. After 1–2 weeks the radicle emerges (Fig. 5b).
3. After 2–3 weeks roots are developing, and the explant is lifted above the medium (Fig. 5c).
4. After 3–4 weeks the cotyledons break through the remaining endosperm (Fig. 5d).
5. After 5–6 weeks, when foliage leaves are formed, the germination process is complete.
6. Cut explant at hypocotyl, remove remaining endosperm and transfer plantlet to multiplication media (Fig. 5d, e).
7. Transfer plantlets into the growth chamber with 16/8 light/dark photoperiod under fluorescent light, provided by SYLVANIA GRO-LUX 36 W ($60 \mu M \ m^{-2} \ s^{-1}$) at $26 \pm 2 \ °C$ (Fig. 5e).

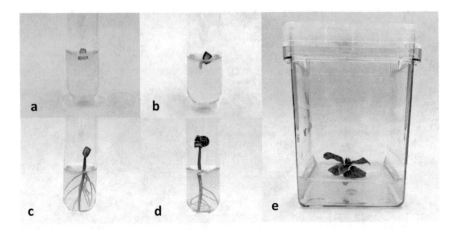

Fig. 5 Germination and plantlet development

4 Notes

1. The orientation of the embryo may differ, polyembryony might also occur.
2. In case of polyembryony extend the size of the cut area.

Acknowledgements The authors would like to thank Plant Breeding and Genetics Laboratory colleagues and participants of the Coordinated Research Project (CRP) D22005 'Efficient Screening Techniques to Identify Mutants with Disease Resistance for Coffee and Banana' for stimulating discussions. We would like to thank Dr Noel Arrieta Espinoza, Instituto del Café de Costa Rica (ICafe), Costa Rica for providing seed of Arabica coffee used in these experiments. This work was supported by the FAO and IAEA through the CRP D22005.

References

Bolívar-González A, Valdez-Melara M, Gatíca-Arias A (2018) Responses of Arabica coffee (*Coffea arabica* L. var. Catuaí) cell suspensions to chemically induced mutagenesis and salinity stress under in vitro culture conditions. In Vitro Cell Dev Biol Plant 54(6). https://doi.org/10.1007/s11627-018-9918-x

Etienne H, Breton D, Breitler J-C, Bertrand B, Déchamp E, Awada R, Marraccini P, Léran S, Alpizar E, Campa C, Courtel P, Georget F, Ducos J-P (2018) Coffee somatic embryogenesis: how did research, experience gained and innovations promote the commercial propagation of elite clones from the two cultivated species? Front Plant Sci 9:1630

Los Santos-Briones D, Hernández-Sotomayor SM (2006) Coffee biotechnology. Braz J Plant Physiol 18(1):217–227

Menéndez-Yufá A, Barry-Etienne D, Bertrand B, Georget F, Etienne H (2010) A comparative analysis of the development and quality of nursery plants derived from somatic embryogenesis and from seedlings for large-scale propagation of coffee (*Coffea arabica* L.). Plant Cell Tiss Organ Cult 102(3):297–307

Murvanidze N, Nisler J, Leroux O, Werbrouck SPO (2021) Cytokinin oxidase/dehydrogenase inhibitors stimulate 2iP to induce direct somatic embryogenesis in *Coffea arabica*. Plant Growth Regul 94:195–200. https://doi.org/10.1007/s10725-021-00708-6

Quiroz-Figueroa F, Monforte-González M, Galaz-Ávalos RM, Loyola-Vargas VM (2006) Direct somatic embryogenesis in *Coffea canephora*. In: Loyola-Vargas VM, Vázquez-Flota F (eds) Plant Cell Culture Protocols. Methods in Molecular Biology™, vol 318. Humana Press. https://doi.org/10.1385/1-59259-959-1:111

Wintgens JN (2012) Coffee: growing, processing, sustainable production. A guidebook for growers, processors, traders, and researchers. Ed. Wiley-VCH Verlag GmbH & Co.

Somatic Embryogenesis and Temporary Immersion for Mass Propagation of Chimera-Free Mutant Arabica Coffee Plantlets

Samira Tajedini, Florian Goessnitzer, and Ivan L. W. Ingelbrecht

Abstract Coffee is one of the most valuable cash crops providing employment for millions of people worldwide. Arabica coffee is widely grown in Latin America where it is under threat of leaf rust. Conventional breeding of Arabica coffee is challenged by its narrow genetic base and long reproductive cycle, and it can take up to 30 years for variety development and release. In vitro somatic embryogenesis is a propagation technique whereby a single plant somatic cell can give rise to a somatic embryo under appropriate culture conditions. For tree crops such as Arabica coffee, single-cell mutagenesis using embryogenic cell cultures provides a powerful approach to produce chimera-free mutant lines directly from cells. Here we describe protocols to induce friable embryogenic callus, establish embryogenic cell suspensions, and convert somatic embryos into plantlets using a RITA® bioreactor for *Coffea arabica* var. Venecia. In addition, methods for gamma-ray mutagenesis of regenerable cell suspensions are described.

1 Introduction

Coffee is a global commodity providing employment for millions of people worldwide (FAOSTAT 2021). The *Coffea* genus belongs to the family *Rubiaceae* and the two main cultivated species are *Coffea arabica* L. ($2n = 4x = 44$) and *Coffea canephora* Pierre ex A. Froehner ($2n = 2x = 22$). Arabica coffee is a self-pollinating species, widely cultivated in South America, Africa and Asia. During the past decade, coffee leaf rust, a fungal disease caused by *Hemileia vastatrix*, has devastated *C. arabica* plantations across Latin America (Avelino et al. 2015).

S. Tajedini (✉) · F. Goessnitzer · I. L. W. Ingelbrecht (✉)
Plant Breeding and Genetics Laboratory, Joint FAO/IAEA Centre of Nuclear Techniques in Food and Agriculture, IAEA Laboratories Seibersdorf, International Atomic Energy Agency, Vienna International Centre, Vienna, Austria
e-mail: stajedini@gmail.com

I. L. W. Ingelbrecht
e-mail: i.ingelbrecht@iaea.org

© The Author(s) 2023
I. L. W. Ingelbrecht et al. (eds.), *Mutation Breeding in Coffee with Special Reference to Leaf Rust*, https://doi.org/10.1007/978-3-662-67273-0_4

Conventional breeding of Arabica coffee is challenging due to its long reproductive cycle and narrow genetic base (Wintgens 2012; Scalabrin et al. 2020). Mutation-assisted breeding offers an attractive alternative to induce genetic diversity useful for coffee breeding and genetic studies. Since the 1990s in vitro tissue culture technologies have been developed for coffee, including methods to regenerate plants through somatic embryogenesis (Campos et al. 2017; Etienne et al. 2018). Both direct and indirect methods for somatic embryogenesis in coffee have been described (Quiroz-Figueroa et al. 2006; Murvanidze et al. 2021). However, to our knowledge so far, Arabica coffee single-cell micropropagation techniques have not been integrated with mutation induction techniques using gamma-ray irradiation.

Single cells or cell clusters derived from embryogenic callus or somatic cell suspensions are attractive targets for induced mutagenesis given that they are expected to (mostly) produce chimera-free, homo-histont plants, as opposed to mutagenesis of multicellular tissues such as seed which results in chimeras. Dissolving chimeras through successive cycles of selfing is possible in *C. arabica* as it is a self-compatible species, however this is a lengthy process given its long reproductive cycle. Here we present protocols to produce friable embryogenic callus, establish embryogenic cell suspensions and convert somatic embryos into plantlets using a RITA® bioreactor in Arabica coffee var. Venecia. Induced mutagenesis of embryogenic cell suspensions using gamma-ray irradiation is also described.

2   Materials

2.1   Plant Material

1. Mature coffee plants of *C. arabica* var. Venecia (see Note 1).

2.2   Supplies, Reagents and Basic Equipment

1. Aluminum foil.
2. Beakers (100, 500, and 1000 ml).
3. Bottles (100, 500, and 1,000 ml).
4. Casein hydrolysate.
5. Culture tubes (30 ml).
6. Culture vessels for liquid media.
7. Distilled Water.
8. 70% Ethanol (v/v).
9. 40—mesh filter (Sigma-Aldrich®).
10. Forceps.
11. Gelling agent (phytagel).

12. Glycine.
13. Magnet for stirring.
14. Microfuge tubes (e.g. Eppendorf).
15. Myo-inositol.
16. MS media (Murashige and Skoog media).
17. Petri dishes (60 × 15 mm; 100 × 20 mm).
18. Parafilm.
19. Pipettor.
20. Scissors.
21. Sieves.
22. Sterile distilled water.
23. Surgical blades.
24. Stirrer.
25. Thiamine.

2.3 Equipment

1. Analytical balance.
2. Autoclave.
3. Centrifuge.
4. Gammacell 220 irradiation unit (see Note 2).
5. In vitro growth room.
6. Laminar flow bench.
7. Orbital shaker.
8. pH meter.
9. RITA® bioreactor.
10. Stereomicroscope.

2.4 Stock Solutions and Tissue Culture Media

1. 1-Naphthaleneacetic acid (NAA) 1 mM.
2. 2,4-Dichlorophenoxyacetic acid (2,4 D) 1 mM.
3. 6-(γ,γ-Dimethylallylamino)Purine (2iP) 1 mM.
4. 6-Benzylaminopurine (BAP) 1 mM.
5. Indole-3-butyric acid (IBA) 1 mM.
6. Sodium hydroxide 1 N.
7. Sodium hypochlorite solution 2% (v/v).
8. Semi-solid callus induction medium (C) (see Table 1).
9. Semi-solid embryogenic callus induction medium (E) (see Table 1).
10. Liquid proliferation medium, flasks (CP) (see Table 1).
11. Liquid regeneration media, RITA® (R) (see Table 1).
12. Liquid development media, RITA® (EG) (see Table 1).

Table 1 Media composition (mg/l) for somatic embryogenesis and plantlet regeneration of *Coffea arabica* var. Venecia (van Boxtel and Berthouly, 1996)

	C	E	CP	R	EG
Macrominerals	MS/2	MS/2	MS/2	MS/2	MS/2
Microminerals	MS/2	MS/2	MS/2	MS/2	MS/2
$FeSO_4*7H_2O$	13.9	13.9	13.9	13.9	13.9
Na_2EDTA	18.65	18.65	18.65	18.65	18.65
Thiamine HCl	10	20	5	10	8
Pyridoxine-HCl	1	–	0.5	1	3.2
Glycine	1	20	–	2	–
Nicotinic acid	1	–	0.5	1	–
Cystein	–	40	10	–	–
Myo-inositol	100	200	50	200	100
Adenine sulfate	–	60	–	40	–
Casein hydrolysate	100	200	100	400	–
Malt extract	400	800	200	400	–
2,4-D	0.5	1.0	1	–	–
IBA	1	–	–	–	–
IAA	–	–	–	–	0.45
Kinetin	–	–	1	–	–
2iP	2	–	–	–	–
6-BAP	–	4	–	2	0.25
Sucrose	30,000	30,000	15,000	40,000	20,000
Medium consistency	Semi-solid	Semi-solid	Liquid	Liquid	Liquid
phytagel	2,000	2,000	–	–	–
pH	5.6	5.6	5.6	5.6	5.6

3 Methods

The procedures described below utilize leaf discs as starting material to produce friable embryogenic callus on semi-solid medium. The friable embryogenic callus is then transferred to a liquid medium in Erlenmeyer flasks to establish a homogenous embryogenic cell (cluster) suspension culture. The cell suspension culture serves to multiply the embryogenic cells and cell clusters and is used for gamma-ray irradiation. Next, globular-stage somatic embryo cell cultures are transferred to a 1-L RITA® bioreactor for further development of the somatic embryos and conversion to rooted plantlets. The different steps of the procedure are illustrated in Table 2 and Figs. 1, 2, 3.

Table 2 Steps in the in vitro regeneration and mutagenesis procedure with indicative timeline

Time	Stage
Day 1	Leaf explant on callus induction medium (C)
22 d	Leaf explant with primary calli on callus induction medium (C)
2 ~ 5 m	Friable embryogenic callus (FEC) induction medium (E)
~ 3 m	Proliferation of somatic embryos (SE) in liquid proliferation medium, flasks (CP)
~ 1 m	Initiate cell suspensions using sieved and mutagenized FEC in regeneration medium (R)
2 ~ 3 m	Develop SE in regeneration medium (R) using a RITA® bioreactor
1 ~ 3 m	Convert SE to plantlets in development media (EG) using a RITA® bioreactor
2 ~ 3 m	Develop plantlets to a first pair of leaves in development media using a RITA® bioreactor

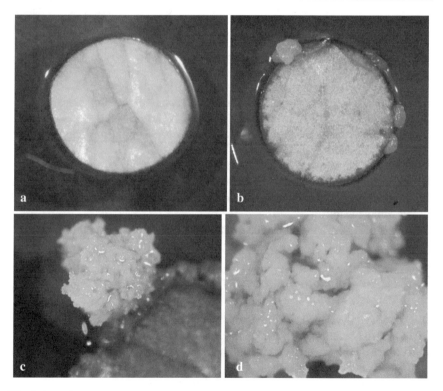

Fig. 1 Somatic embryogenesis process in *Coffea arabica* from ex vitro leaf disks. **a** Leaf disk in callus induction medium. **b** Leaf disk with primary calli. **c–d** Close-up of friable embryogenic calli

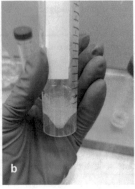

Fig. 2 Preparation of embryogenic cells/cell clusters for irradiation treatment. **a** 40-Mesh filter. **b** Embryogenic cell suspension after sieving. **c** Cell suspension in microcentrifuge tubes ready for gamma-ray irradiation

3.1 Media Preparation

1. Prepare the macro- and micro-nutrient stock solutions and iron stock solution according to Table 1 in advance and store at 4 °C (see Note 3).
2. Prepare the vitamin and growth regulator solutions according to Table 1 and keep in 15 ml plastic tubes and store at − 20 °C.
3. Autoclave the semi-solid and liquid media or filter sterilize (see Note 4).
4. For media preparations, start with less water of the required final volume in a flask or beaker and add the specified number of stocks, amino acids, sugar, and growth regulators.
5. Adjust pH to 5.6 and add distilled water to the required final volume.

3.2 Tissue Collection and Disinfection

1. Select well developed, healthy leaves from mature, greenhouse-grown plants.
2. In the lab or washroom, rinse the leaves with water and soap.
3. Perform all subsequent steps in a laminar flow cabinet; disinfect the cabinet with 70% ethanol prior to use.
4. In the laminar air flow cabinet place the leaves in a sterile beaker.
5. Disinfect the leaves with a sodium hypochlorite solution 2% (v/v) for 30 min.
6. Decant the solution, and rinse 3–4 times with sterile deionized water, each 5 min.
7. Transfer leaves to petri plates.

Fig. 3 *Coffea arabica* somatic embryogenesis process. **a** Friable embryogenic callus indicated with the arrow. **b** SE in proliferation medium. **c** Embryogenic cells for radiation treatment. **d** Germination of somatic embryos in a RITA® bioreactor. **e–g** Rooted plantlets

3.3 In Vitro Leaf Disc Preparation and Primary Callus Induction

1. Using a sterile cork-borer punch leaf discs measuring ca 5 mm diameter, avoiding main and secondary veins and leaf margins.
2. Place 6–7 leaf discs, upper surface down, in 60 × 15 mm petri plates containing callus induction medium (C); seal petri dish with parafilm or similar self-sealing plastic (Fig. 1a).
3. Place in culture room at 28 °C in complete darkness.
4. After 7–10 days callus starts to grow (Fig. 1b).
5. After 4–5 weeks leaf discs with primary calli can be moved to friable embryogenic callus induction medium.

3.4 Embryogenic Callus Induction

1. Subculture leaf disks from callus induction medium to friable embryogenic callus medium (E).
2. Seal petri dishes and place in dark at 28 °C.
3. Within 2–5 months, friable embryogenic callus (yellowish) develops on explants on E medium (Fig. 1c, d).

3.5 Cell Multiplication and Establishing the Embryogenic Cell Suspension

1. Isolate embryogenic calli and culture in liquid proliferation medium (CP) for multiplication purposes (Fig. 3b).
2. Culture about 10 mg calli per 1 ml CP medium in a 250 ml flask containing embryogenic callus medium (E).
3. Maintain flasks in an orbital shaker at 90–100 rpm at 28 °C in the dark.
4. Subculture every 3 weeks for 2–3 months.

3.6 Gamma-ray irradiation of the Embryogenic Cell Suspension

1. After 4 weeks of growth, pass the embryogenic suspension cultures through a 40—mesh filter (Sigma-Aldrich®) (Fig. 2a, b).
2. Once the material is obtained, divide it into sterile 1.5 ml microcentrifuge tubes.
3. Add 0.5 ml of regeneration media (R).
4. Seal the microcentrifuge tubes with Parafilm, place inside a petri dish and seal the petri dish (Fig. 2c).
5. Transport sealed petri dishes for irradiation in the dark at room temperature and irradiate using a Gammacell 220 irradiaton unit (see Note 2).

3.7 Somatic Embryo Differentiation and Germination

1. Maintain flask of irradiated cells in the dark on a rotary shaker (100 rpm) at 26 ± 2 °C.
2. Embryos can be transferred to the RITA® bioreactor after developing to globular stage (Fig. 3d).

3.8 Development of Somatic Embryos and Conversion to Plantlets

1. Place about 200 mg of the embryogenic aggregates in the RITA® bioreactor along with 200 ml of the regeneration medium (R).
2. Subculture embryos once every 2 months for embryo development and germination under light conditions (12 h photoperiod, 50 μmol photons m^{-2} s^{-1}) at 26 ± 2 °C.
3. Set the immersion frequency to 1 min every 12 h.
4. After 3–4-months cotyledonary shaped embryos will develop.
5. After 3 months, rooted plantlets will develop (Fig. 3e–g).

4 Notes

1. Well-developed leaves were harvested from mature, greenhouse-grown *Coffea arabica* var. Venecia plants, second node from the top.
2. The in vitro cultures were irradiated using a self-contained Gammacell 220 Cobalt-60 research irradiator at a dose rate of ~ 53.3 Gy/min. The following doses were applied: 0, 5, 10, 20, 40 and 80 Gy.
3. The MS micro and micro mineral solutions as well iron solutions were prepared separately as individual stock solutions and then combined to half- or full-strength MS (Murashige and Skoog 1962). The B5 vitamins were purchased as a commercial ready-for-use powder (Gamborg et al. 1968).
4. Check the liquid media for any contamination by keeping the freshly made liquid media at room temperature overnight and then store in the fridge.

Acknowledgements The authors thank PBG Laboratory colleagues and participants of the Coordinated Research Project (CRP) D22005 'Efficient Screening Techniques to Identify Mutants with Disease Resistance for Coffee and Banana' for stimulating discussions. We would like to thank Dr Noel Arrieta Espinoza, ICafe, Costa Rica for providing seed of *Coffea arabica* and sharing insights on coffee breeding and challenges. We further thank the IAEA and the Belgian Government for financial support through the CRP D22005 and the Peaceful Use Initiative 'Enhancing Climate Change Adaptation and Disease Resilience in Banana-coffee Cropping Systems in East Africa', respectively.

References

Avelino J, Cristancho M, Georgiou S, Imbach P, Aguilar L, Bornemann G, Läderach P, Anzueto F, Hruska AJ, Morales C (2015) The coffee rust crises in Colombia and Central America (2008–2013): impacts, plausible causes, and proposed solutions. Food Secur 7:303–321

Campos NA, Panis B, Carpentier SC (2017) Somatic embryogenesis in coffee: the evolution of biotechnology and the integration of omics technologies offer great opportunities. Front Plant Sci 8:1460

Etienne H, Breton D, Breitler J-C, Bertrand B, Déchamp E, Awada R, Marraccini P, Léran S, Alpizar E, Campa C, Courtel P, Georget F, Ducos J-P (2018) Coffee somatic embryogenesis: how did research, experience gained and innovations promote the commercial propagation of elite clones from the two cultivated species? Front Plant Sci 9:1630

FAOSTAT (2021) FAOSTAT data. Available at: Coffee | FAO | Food and Agriculture Organization of the United Nations

Gamborg OL, Miller RA, Ojima K (1968) Nutrient requirements of suspension cultures of soybean root cells. Exp Cell Res 50:151–158. https://doi.org/10.1016/0014-4827(68)90403-5

Murashige T, Skoog F (1962) A revised medium for rapid growth and bio assays with tobacco tissue cultures. Physiol Plant 15:473–497. https://doi.org/10.1111/j.1399-3054.1962.tb08052.x

Murvanidze N, Nisler J, Lerou O, Werbrouck SPO (2021) Cytokinin oxidase/dehydrogenase inhibitors stimulate 2iP to induce direct somatic embryogenesis in *Coffea arabica*. Plant Growth Regul 94:195–200. https://doi.org/10.1007/s10725-021-00708-6

Quiroz-Figueroa F, Monforte-González M, Galaz-Ávalos RM, Loyola-Vargas VM (2006) Direct somatic embryogenesis in *Coffea canephora*. In: Loyola-Vargas VM, Vázquez-Flota F (eds) Plant cell culture protocols. Methods in molecular biology™, vol 318. Humana Press. https://doi.org/10.1385/1-59259-959-1:111

Scalabrin S, Toniutti L, Di Gaspero G, Scaglione D, Magris G, Vidotto M, Pinosio S, Cattonaro F, Magni F, Jurman I, Cerutti M (2020) A single polyploidization event at the origin of the tetraploid genome of *Coffea arabica* is responsible for the extremely low genetic variation in wild and cultivated germplasm. Sci Rep 10(1):1–13. https://doi.org/10.1038/s41598-020-61216-7

van Boxtel J, Berthouly M (1996) High frequency somatic embryogenesis from coffee leaves. Plant Cell Tiss Organ Cult 44:7–17

Wintgens JN (2012) Coffee: growing, processing, sustainable production. A guidebook for growers, processors, traders, and researchers, 2nd edn. Wiley-VCH Verlag GmbH & Co. KGaA, Weinheim, Germany

Protocol on Mutation Induction in Coffee Using In Vitro Tissue Cultures

Margit Laimer, Rashmi Boro, Veronika Hanzer, Emmanuel Ogwok, and Eduviges G. Borroto Fernandez

Abstract Pathogens are the major limiting factors in coffee production. Approximately 26% of the global annual coffee production is lost to diseases, threatening the income of approx. 125 million people worldwide. Therefore, reducing coffee yield losses by improving coffee resistance to diseases and insect attacks through breeding can make a major contribution to agricultural sustainability. Mutation breeding in vegetatively propagated and perennial crops is hampered in large part due to bottlenecks in the induction of variation (lack of recombination) and challenges in screening. Tissue culture approaches using alternative types of material were developed. This offers a clear advantage of providing the required sample size for mutation induction and subsequent screening within a reasonable time frame. The protocols developed compare different tissue culture systems for mutation induction involving unicellular and multicellular explants requiring different numbers of subsequent subcultures to reduce the impact of chimerism: (a) axillary shoot culture for the provision of donor material for mutation induction and regeneration; (b) leaf disc cultures for the induction of calli; (c) direct and indirect somatic embryogenesis for the production of somatic embryos; (d) the irradiation of somatic embryos at the globular and cotyledonary stage. Mutagenesis was induced by irradiation with a Cobalt-60 Gamma-source at the FAO/IAEA Laboratories in Seibersdorf, Austria. A comparison of the time required for the regeneration of high numbers (hundreds) of plantlets from irradiated in vitro shoots versus irradiated embryogenic calli is clearly in favor of embryogenic calli, since the plantlets regenerate from individual cells and can be used for genotypic and phenotypic analyses directly. This chapter describes the general methods for mutation induction using gamma irradiation and the procedures that can be used to generate large numbers of induced mutants in different tissues of coffee under in vitro conditions.

M. Laimer (✉) · R. Boro · V. Hanzer · E. G. Borroto Fernandez
Plant Biotechnology Unit, Department of Biotechnology, University of Natural Resources and Life Sciences (BOKU), Muthgasse 18, 1190 Vienna, Austria
e-mail: margit.laimer@boku.ac.at

E. Ogwok
Department of Science and Vocational Education, Faculty of Education, Lira University, 1035, Lira, Uganda

© The Author(s) 2023
I. L. W. Ingelbrecht et al. (eds.), *Mutation Breeding in Coffee with Special Reference to Leaf Rust*, https://doi.org/10.1007/978-3-662-67273-0_5

1 Introduction

After petroleum, coffee beans are the second most economically important product traded worldwide. Approximately 11 million ha of coffee trees are cultivated in tropical regions (Déchamp et al. 2015), providing income to 125 million people. Despite the importance of coffee, several factors, which could be amplified by changing climatic conditions, hamper its production and influence the extent of yield losses, including agronomic management, growing environment, cultivar selection affected by diseases and pathogens. Pathogens are the major limiting factors in coffee production. Approximately 26% of the global annual coffee production is lost to diseases.

Cultivated coffee originated from wild populations in Africa, Madagascar, the Comoro and the Mascarene islands, the Indian subcontinent, Southern tropical Asia, South-East Asia and Australasia (Razafinarivo et al. 2013; Davis et al. 2006, 2007).

Among different coffee species, the two economically most important are the diploid *C. canephora* Pierre ex Froehn (2n = 2x = 22), native to Central and Western sub-Saharan Africa, and the allotetraploid *C. arabica* L. (2n = 4x = 44) from the Southwestern Ethiopian highlands, Mount Marsabit of Kenya and the Boma Plateau of Sudan (Anthony et al. 2002). *C. arabica* resulted from ancestral hybridization—dated approximately 50,000 years ago—between two diploid ecotypes, namely *C. eugenioides* and *C. canephora* (Anthony et al. 2010; Lashermes et al. 1997; 1999; Ribas et al. 2011). *C. arabica* is self-pollinated, while *C. canephora* is cross-pollinated. A recent study by Sant'Ana et al. (2018) showed a high allelic variation in wild accessions from Ethiopia, however, the mode of pollination and the history of coffee cultivation resulted in a reduction of genetic diversity in *C. arabica*. According to different authors, coffee was introduced from Ethiopia to Yemen between 1,500 and 300 years ago. From this point, the first reduction of diversity happened within Arabica cultivars. Genetic data analyses showed that two genetic bases spread from Mocha (Yemen) in the early eighteenth century (Chevalier and Dagron 1928; Haarer 1956). *C. arabica* var. *arabica* (also called var. *typica* Cramer) originated from a single plant that was introduced to Java (Indonesia) and later cultivated in the Botanical Garden of Amsterdam. *C. arabica* var. Bourbon (B. Rodr.) Choussy (Carvalho et al. 1969; Krug et al. 1939) was introduced to the Bourbon Island (Réunion). These were the starting points of coffee cultivars spreading rapidly to the American continent and Indonesia by the use of seeds produced by the auto-fertilization of coffee trees, which caused a further reduction in genetic diversity.

There are several major constraints in coffee breeding. As already mentioned, the vast majority of the world coffee production is based on two species, *Coffea arabica* (2n = 4x = 44 chromosomes) and *C. canephora* (2n = 2x = 22 chromosomes). This results in low genetic diversity among coffee cultivars and represents a massive limitation in case of control and management of pest and disease under climatic changes. The absence of pest resistance in the most preferred *Coffea arabica* cultivars can be overcome by cross-breeding, but due to the long juvenile period of tree crops this is a time-consuming process (Silva et al. 1999; 2006; Várzea et al. 2000).

Plant biotechnological interventions in coffee improvement are used to develop uniform planting material through cell and tissue culture (Krishnan 2011) since the pioneering attempts in mutation induction by Carvalho in the 1950s (Carvalho 1988).

In recent years mutation breeding programs have been initiated within the FAO/IAEA funded Coordinated Research Project (CRP) D22005 on mutation induction for coffee improvement (Dada et al. 2014, 2018; Bolívar-González et al. 2018; Bado et al. 2018a, b). In contrast to conventional breeding, taking at least 20 years to release a new cultivar, biotechnological methods offer valuable tools for coffee improvement and for speeding up the selection process of superior plants (Bado et al. 2017; Campos et al. 2017). Micropropagation by organogenesis is used for plant multiplication mainly from shoot tips and axillary buds allowing the production of large-scale populations for mutation induction and subsequent mutant line propagation and is mainly suitable for vegetatively propagated crops with a long juvenile period. This allows to reduce the time required and to accelerate mutation breeding when using single cell explants. A considerable acceleration of mutation breeding can be achieved by using single cell explants like double haploids or somatic embryos, which mark the shortest route to produce homozygous lines from heterozygous plant material.

Somatic embryos can be produced on a large-scale in suspension cultures and in bioreactors. As a matter of fact, somatic embryogenesis is an excellent system for mutation induction, since somatic embryos originate from single cells (da Câmara Machado et al. 1995).

Among the physical mutagens, gamma rays are the most commonly used for mutation breeding (Mba et al. 2012), resulting in small to large deletions, point mutations, single and double strand brakes and even chromosome deletions. When applying physical mutagens to different types of plant material, care should be taken with soft material such as in vitro shoot cultures as well as callus and embryogenic callus cultures, which require lower doses in comparison to seeds. In fact, the water content, storage time, applied mutagen dose and temperature represent important factors influencing mutagens in all types of plant material (Mba et al. 2010).

Depending on the explant type subjected to mutation induction different approaches are required for chimera dissolution. Plants originating either from unicellular or multicellular explants require different time frames for chimera dissolution ranging from 0 for plantlets stemming from somatic embryos to several generations up to M_1V_4 for plantlets originating from multicellular explants (Novak and Brunner 1992). Entire mutant populations are screened by either phenotypic evaluation to select the phenotype of interest or by genotypic evaluation to detect novel alleles in genes of interest (Fig. 1).

Pathogens are the major limiting factor in coffee production. New approaches are available to breed varieties that are resistant to a broad-spectrum of pathogens, genetically stable and high-yielding. Recently developed tools in genomic technologies allow to better understand coffee-pathogen interaction and help to identify the genes and mechanisms involved in pathogen resistance or susceptibility. Understanding the influence of individual factor and their interaction will help to select realistically interesting accessions and to accelerate breeding strategies.

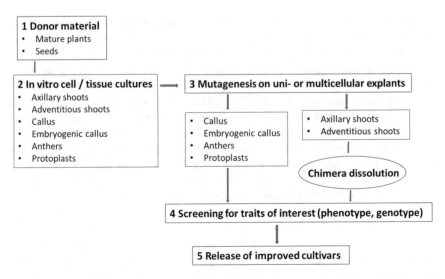

Fig. 1 Mutation breeding scheme for mutagenesis in *Coffea* sp. **1** Mature donor plants provide vegetative buds, flower buds, leaves, while seeds may be established directly in vitro. **2** In vitro cell and tissue cultures may involve somatic and gametic cells. **3** Coffee explants for mutation induction may be of uni- or multicellular origin. Note that multicellular systems require an additional step for chimera dissolution, while single cell systems do not require this step. **4** Screening of mutants can be conducted either before acclimatization through early screening of irradiated cells and plantlets or after acclimatization of plantlets to greenhouse or under open field conditions. **5** Selected improved cultivars can be released as direct mutants or be further used in breeding programs by hybridization

2 Materials

2.1 Establishment of a Collection of Donor Material In Vivo

A collection of starting material of the two species of coffee—both as seeds and potted plants—should be initiated in the greenhouse to allow the maintenance of mother plant under controlled conditions (Fig. 2). In order to rule out major genotypic differences, in our experiments this included fourteen different cultivars of the leaf rust susceptible *C. arabica* (self-pollinating) as well as two genotypes of the leaf rust resistant *C. canephora* (self-incompatible) with different climate requirements and tolerance/susceptibility to different pathogen races. The collection of *C. arabica* cultivars (https://sca.coffee/research/coffee-plants-of-the-world) comprised:

- **Bourbon**: A common cultivar *C. arabica* that developed naturally on Île Bourbon (an island in the Indian Ocean, east of Madagascar, now known as Réunion) from coffee brought to the island from Yemen by the French. Depending on the specific sub-group, this coffee can be red (Vermelho) or yellow (Amarelo). These plants generally have broader leaves and rounder fruit and seeds than Typica varieties. Stems are stronger and stand more upright than Typica. They are susceptible to all major diseases and pests.

- **Catimor**: A group of pure-line cultivars originating from crosses between Hibrido de Timor and Caturra. It has been distributed since the 1980s. It is known to be highly productive and shows resistance to coffee leaf rust and to coffee berry disease (CBD).
- **Caturra**: A pure-line dwarf spontaneous mutant of red Bourbon that has short internodes. It was found in 1937 in Brazil, and is highly productive. Its leaf and fruit characteristics are similar to Bourbon varieties and can produce red or yellow cherries. Like Bourbon, it is known to be susceptible to all main diseases and pests.
- **HDT** (Clones 1–4) **Hibrido de Timor** (Timor Hybrid): A spontaneous cross of *C. canephora* and *C. arabica* var. Typica that occurred naturally on the island of Timor in Southeast Asia. These "Arabusta"-type hybrids likely originated from a single Robusta parent plant. It became popular in Timor in the 1950s due to its natural resistance to leaf rust. These hybrids were collected in Timor in 1978 and planted on the islands of Sumatra and Flores shortly thereafter, and since then some changes and mutations have occurred. Different versions of this hybrid have been utilized in breeding programs to introduce the rust resistance into new varieties, such as Catimor and Sarchimor.
- **Java** (Clones 1, 2, 8, 9, 10, 12): A Typica selection suspected to be the progeny of coffee introduced from Yemen to the island of Java. From Java, this plant was first brought to neighboring islands (Timor) and later to East Africa (Cameroon), where it was released for cultivation in 1980. It has since been introduced in Central America by the Centre de Cooperation Internationale en Recherche Agronomique pour le Développement (CIRAD). It is known to be vigorous with moderate yield and shows good resistance to coffee berry disease in Cameroon. Java has elongated fruit and seeds and bronze-colored young leaves.
- **Kent**: A tall Typica selection that likely arose on or from coffee bred on the Kent Estate in India. It has been widely planted in India since the 1930s and a selection from this cultivar, known as K7, is more common in Kenya. It is known as the first coffee selected for rust resistance.
- **Pacamara**: A cross between Maragogype and Pacas developed in El Salvador. Similarly to Pacas, it is known to be susceptible to all main diseases and pests. Pacamara was released in 1958 but is genetically unstable, with 10–12% of plants reverting to Pacas.
- **Sarchimor**: A group of pure-line cultivars originating from a cross between Villa Sarchi and one Hibrido de Timor. Some Sarchimor lines show good resistance to coffee leaf rust; some are also resistant to coffee berry disease.
- **Typica** (Clones 1–3): This is a tall cultivar of *Coffea arabica*, originating from the coffee brought to Java from Yemen (possibly via India). The plants, most similar to what we today call Java, were spread from the island of Java in the early 1700s. It has bronze-tipped young leaves, and the fruit and seeds are large. Typica plants are known to have relatively low productivity and are susceptible to all main pests and diseases.
- **Villa Sarchi**: A dwarf mutation of Bourbon found in Costa Rica and released in 1957. It is known to be susceptible to most pests and diseases.

Fig. 2 Maintenance of
donor plants of *Coffea* sp.
under greenhouse conditions

For the maintenance of donor plants of *Coffea* sp. under greenhouse conditions
the following items should be available:

1. High quality, disease-free seeds and plants, uniform in size (see Notes 1 and 2).
2. Sterilized soil mixture.
3. Glasshouse facility.
4. Regular fertilization and irrigation.

2.2 Establishment of In Vitro Shoot Cultures as Donor Material

Axenic cultures were established and prepared as source material for the specific
regeneration and mutation programs. Different in vivo donor material was used with
the intention to establish micropropagation for media optimization and induction
of callus, somatic embryogenesis and suspension cultures: seeds, seedlings, leaves,
stems, roots, flowers. Serving both as material for mutagenesis treatment, as well
as for the recovery of individual mutagenized cells, these culture systems should be
maintained throughout the entire period of the experiment.

The first 4 nodes of orthotropic shoots growing in the greenhouse under controlled
conditions were excised as explants. Surface sterilization was achieved with a 15%
sodium hypochlorite solution containing 1% Tween 80 for 20 min followed by 4
washes with distilled water. To carry out these steps, the following facilities and
items are required:

1. High quality greenhouse grown plants of defined genotypes (see Note 1).
2. Laminar air flow cabinet for in vitro work (in vitro culture facility).
3. 70% ethanol.
4. 15% Danchlor solution.
5. Sterile Aqua dest.
6. Magenta boxes.
7. Culture media (see Note 3).

Fig. 3 Axenic in vitro cultures of *Coffea* sp. serving as donor material for mutation induction

Axillary shoot cultures of different cultivars were established from 2 genotypes of *C. canephora*, Niaoulli (14 clones) and Quillou (6 clones), as well as from 15 genotypes of *C. arabica*: HDT (4 clones), Caturra, Catimor, Kent, Sarchimor, Typica (3 clones), Villa Sarchi, Java (6 clones) and Bourbon (Fig. 3).

2.3 Establishment of Tissue Culture Material for Mutation Induction

Somatic embryogenesis is an excellent system for plant propagation and mutation induction, since somatic embryos originate from single cells and therefore reduce chimerism. Somatic embryos can be produced on a large-scale in suspensions in Erlenmeyer flasks and in bioreactors. Somatic embryos of *Coffea* can be obtained either by direct or by indirect somatic embryogenesis, the difference being the intermediate callus induction.

For the induction of coffee callus cultures and somatic embryogenesis additional multicellular explants, like leaves, stems and roots of in vitro grown seedlings (Fig. 4) and from in vitro shoots from selected cultivars are used.

The conversion of embryos to plantlets from a number of selected cultivars after transfer of emerging embryos from embryogenic calli to regeneration medium yielded a high number of mutant plantlets (Fig. 6).

To establish an efficient regeneration from embryogenic calli into plantlets the following items are needed:

1. High quality embryogenic callus cultures of defined genotypes.
2. Laminar air flow cabinet for in vitro work (in vitro culture facility).
3. Culture media for indirect and direct embryogenesis (see Note 4).
4. 70% (v/v) ethanol.
5. Parafilm.

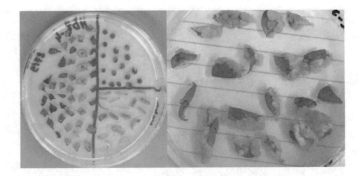

Fig. 4 Induction of direct and indirect embryogenesis from different in vitro explants of *Coffea* sp.

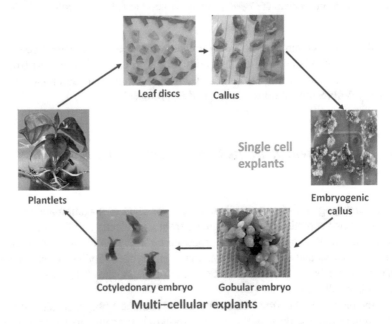

Fig. 5 Induction of direct or indirect embryogenesis from different in vitro explants

6. Petri dishes (9 cm diameter).
7. Magenta boxes.
8. Regeneration media for the recovery of plantlets (see Note 5).

Fig. 6 Regeneration of plantlets/somatic embryos from irradiated embryogenic calli

2.4 Mutagenesis by Physical Agents

Gamma ray mutagenesis may be performed using different facilities, such as gamma cell irradiator, gamma phytotron, gamma house, gamma field. The gamma cell irradiator with Cobalt-60 (or Cesium-137) as radioactive source is the most commonly available equipment worldwide (IAEA 1975, 1977).

However, the radioactive source remains the major consideration and constraint in plant mutagenesis (Bado et al. 2015).

1. Gamma radiation source.
2. Magenta boxes or petri dishes (9 cm diameter).
3. Parafilm.
4. Culture media (see Notes 4 and 5).

3 Methods

3.1 Mutagenesis by Physical Agents

Mutations were induced by gamma-irradiation of different explants of selected genotypes of *C. arabica* and *C. canephora* at different intervals with several repetitions. A Cobalt-60 Gamma irradiator was used and the irradiation was performed at the FAO/

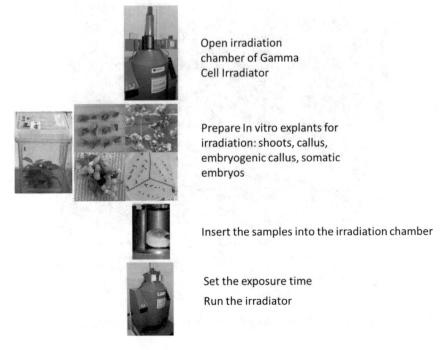

Open irradiation
chamber of Gamma
Cell Irradiator

Prepare In vitro explants for
irradiation: shoots, callus,
embryogenic callus, somatic
embryos

Insert the samples into the irradiation chamber

Set the exposure time
Run the irradiator

Fig. 7 Working steps for irradiation using a Cobalt-60 gamma cell

IAEA Laboratories in Seibersdorf, Austria. The workflow shown in Fig. 7 can be carried out with different plant cell and tissue cultures either in Magenta boxes or petri dishes.

3.2 Selection and Treatment of Explants

The first type of explants to be subjected to irradiation are axenic shoot cultures in order to determine the radiosensitivity of different *C. canephora* and *C. arabica* cultivars (Fig. 3, see Notes 6 and 7).

1. Prepare Magenta boxes with freshly micropropagated plantlets.
2. Place 20–25 shoots into a petri dish and a humid Whatman filter paper.
3. Seal the petri dishes with Parafilm to avoid contamination outside the tissue culture laboratory.
4. Label each petri dish with the sample ID and required dose (0, 10, 20, 40 and 60 Gy).
5. Prepare the regeneration media for the subsequent subculture.
6. Transport the next day to gamma irradiator facility for mutagenesis (see Notes 8 and 9).

7. Put the chamber at the irradiation stage by downing the elevator (see Note 10).
8. Start the time discount of exposure time for watch monitoring after the click of the elevator. In automatic conditions the irradiator timer will monitor to the full time (see Note 11).
9. Raise immediately the chamber to the loading stage when exposure time is completed. In automatic conditions the chamber raises when full time is reached.
10. Open the lead shielding collar and then sample chamber door.
11. Remove the irradiated plant material (see Note 12).
12. If necessary, repeat the treatment at defined time intervals to reach the required dose, which is function of exposure time base on the source dose rate.
13. Subculture irradiated plant material to fresh culture medium (see Note 13).

After having determined the dose range for entire shoots, callus cultures, embryogenic callus cultures, somatic embryos at the globular, torpedo or cotyledonary stage can be irradiated in a similar way.

1. Prepare petri dishes containing semi-solid of Zamarripa M3 medium with freshly subcultured callus or embryogenic callus cultures.
2. Alternatively plate 100 mg of somatic embryos at the globular, torpedo or cotyledonary stage per petri dish (with clearly assigned sample ID) and per replication.
3. Prepare three petri dishes per dose for a range of gamma irradiation doses of 0, 10, 15, 20, 25 and 30, 40, 60 and 80 Gy.
4. Label each petri dish with the sample ID and required dose (Fig. 8).
5. Seal the petri dishes with Parafilm to avoid contamination outside the tissue culture laboratory.
6. Prepare the regeneration media for the subsequent subculture.
7. Transport the next day to gamma irradiator facility for mutagenesis.
8. Expose petri dishes to irradiation by applying the required dose for mutation induction.
9. Subculture irradiated plant material to new petri dishes containing semi-solid of Zamarripa M4 medium.

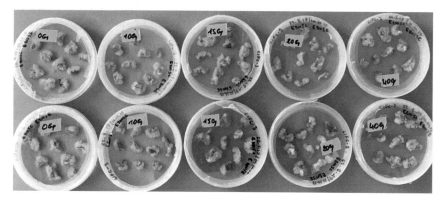

Fig. 8 Callus cultures of *Coffea* sp. prepared for mutation induction

3.3   Regeneration of Mutant Plant Lines

Plant cell and tissue cultures from these irradiation experiments were cultivated further and resulted in shoot formation and plantlet regeneration. These tissues have to undergo rigorous scrutiny for visual detection of altered phenotypes and are evaluated for a range of parameters (Table 1). Additional parameters to be evaluated for regenerated plantlets are active shoot growth, axillary bud formation, secondary root formation.

1. Transfer irradiated samples to tissue culture laboratory.
2. Surface sterilize every petri dish with 70% ethanol before removing the Parafilm (see Note 14).
3. Transfer the irradiated material into appropriate culture media.
4. Cultivate irradiated shoots in the incubation room with 28 °C and 12 h light.
5. Collect data on the further development and survival rates after transplanting on a regular basis (Table 2).
6. Subculture the growing in vitro shoot cultures for chimera dissolution by producing M_2 or higher mutant populations.
7. Screen mutant populations by either phenotypic or genotypic evaluation (Figs. 9 and 10).
8. Transfer plants to rooting and acclimatization phase and subsequently to the glasshouse for further mutant evaluation (Fig. 1).

Table 1 Overview of recommended Gamma doses used for different explants of *Coffea* sp. and evaluation parameters applied within this study

Explants	Sample size/ replicates	Replicates	Gamma dose (Gy)	Repetitions	Evaluation parameters
Shoot	5	3	0, 10, 15, 20, 40 and 60	2	Emerged nodes, shoot length, number of roots, longest root length, leaf area
Callus	10	2	0, 10, 15, 20 and 40	1	Survival and embryogenic callus induction rate
Embryogenic callus	5–10	3	0, 10, 15, 20, 40, 60 and 80	2	Survival and embryos induction rate
Globular embryo	10	3	0, 10, 15, 20, 40 and 60	2	Survival, torpedo induction and plantlet formation rate
Cotyledonary embryo	10	3	0, 10, 15, 20, 40 and 60	2	Survival and plantlet formation rate

Table 2 GR_{30} and GR_{50} determined according to effects observed from different Gamma doses used for *Coffea* sp. shoot cultures

Dose (Gy)	Growing buds	Leaf area	No. of roots	Root length (mm)	Shoot length (mm)
0	100	100	100	100	100
10	95.00	117.19	95.00	110.76	95.00
15	97.5	100.13	97.50	92.91	97.50
20	98.75	69.20	98.75	83.13	98.75
40	55.00	25.75	55.00	38.14	55.00
60	0	0	0	0	0
GR_{30}–GR_{50} (Gy)	34–42	20–26	34–42	26–34	34–42

Values are calculated in relation to control plants
Note GR growth reduction

Fig. 9 Phenotypic analyses of in vitro development of irradiated shoots with focus on root development at 0, 10, 15, 20, 40 and 60 Gy

Since roots are known to respond more sensitively to different stresses, their development was carefully evaluated. The optimal dose range for shoot cultures was identified between 20 and 42 Gy (Table 2).

From the original 75 irradiated shoots finally after a period of approx. 18 months more than 600 plants could be recovered (Table 3). Interestingly, no shoot survived the treatment with 60 Gy.

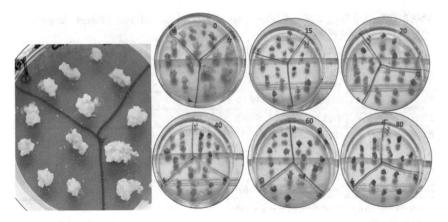

Fig. 10 Irradiated embryogenic callus of coffee

Table 3 Number of shoots of *Coffea canephora* cv. Quillou recovered after irradiation of shoots with 0, 15, 20 and 40 Gy

Dose (Gy)	Irradiated explants	Regenerated shoots after 12 months	Regenerated shoots after 18 months
0	25	33	74
10	45	62	175
15	35	54	151
20	30	52	127
40	40	35	101

Data are given from 3 replicates

Following the dose range determined for shoot cultures, single cell explants should be handled (Fig. 10).

1. Transfer irradiated samples to tissue culture laboratory.
2. Surface sterilize petri dishes with 70% ethanol before removing the Parafilm (see Note 14).
3. Transfer the irradiated material into appropriate culture media.
4. Take the cultures to the incubation room with 28 °C under light and dark conditions.
5. At regular intervals record survival rates of the mutagenized tissues, the number of observed embryos per petri dish, per dose and per genotype.
6. Subculture the growing embryogenic callus and transfer individual embryos to Zamarripa M5 for plant regeneration at regular intervals until development of the plantlets.
7. Record the number of plants regenerated per replication, per dose and per genotype.
8. Transfer plants to rooting and acclimatization phase and subsequently to the glasshouse for further mutant evaluation (Fig. 1).

After irradiation, initial growth was observed only in untreated calli for the first month. However, with a delay, calli from all treatments recovered and survived. All treated calli showed a change in colour as response to gamma irradiation compared to control which maintained the yellow colour. In the third month of incubation cotyledonary embryos were observed with the doses up to 20 Gy, whereas from 40 to 80 Gy no embryo development was observed. The irradiation of embryogenic callus of *Coffea canephora* irradiated on 24.01.2018 led to the recovery of hundreds of shoots, this time of single cell origin (Table 4). Again, it was noted that only very few shoots survived the treatment with 60 Gy.

Irradiation of different developmental stages of somatic embryos revealed, that globular stage and cotyledonary stage embryos besides not growing anymore after being irradiated with 40 and 60 Gy, did not develop directly into actively growing plantlets. However, the circuit through a repetitive embryogenesis allows to recover plantlets also through this process. In fact, irradiation of globular and cotyledonary embryos of *Coffea arabica* cv. Java after 9 months led to recallusing and from there again to embryogenic calli producing new embryos and finally after 12 months approximately 200 shoots.

Globular embryos were relatively more resistant to gamma irradiation than cotyledonary and torpedo shaped embryos (Fig. 11).

Table 4 Number of shoots of N20 (*Coffea canephora* cv. Quillou) recovered after irradiation of embryogenic calli with 0, 15, 20 and 40 Gy

Dose (Gy)	Irradiated explants	Regenerated shoots after 12 months	Regenerated shoots after 18 months
0	25	71	110
10	45	238	350
15	35	29	80
20	30	8	31
40	40	19	44
60	20	2	5

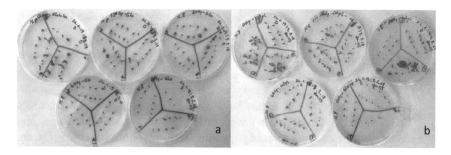

Fig. 11 Irradiated somatic embryos of coffee **a** globular stage, **b** cotyledonary stage

Table 5 Response of different in vitro explants to gamma irradiation

Explant	Mutation induction dose (LD_{30}-LD_{50}) (Gy)	Explant type
Shoot (GR30-GR50)	20–42	Multi-cellular
Globular embryos	6.77–19.82	
Cotyledonary embryos	16.37–30.23	
Callus	55–100	Single cell
Embryogenic callus	> 80	

As anticipated, the experiments allowed to confirm the higher radio-sensitivity of multi-cellular when compared to uni-cellular explants under in vitro conditions (Table 5). It was possible to:

- define an optimum mutation induction dosage range for several in vitro explants
- produce high numbers of different putative mutants generated of various in vitro explants
- determine the effectiveness of mutation induction by phenotypic analyses
- identify the most efficient in vitro explants for mutation induction in coffee.

According to the mutagenesis objectives starting from the second generation and higher after chimera dissolution, in vitro plants can be screened for the selection of candidate based on phenotypes or genotypes. Mutations can be detected with various direct and indirect methods. Direct methods such as sequencing, exome capture sequencing, restriction site associated DNA (RAD) sequencing and genotyping by sequencing (GBS) provide the necessary information for mutation detection and confirmation (Denoeud et al. 2014, Dereeper et al. 2015). Additionally, the generation of various EST sequences in *C. arabica* (Anthony et al. 2001; Mishra and Slater 2012; de Moro et al. 2009; Krishnan 2014; Vieira et al. 2006; Leroy et al. 2005; Lin et al. 2005; Noir et al. 2004) will allow to identify genes and their regulatory sequences responsible for mutated traits and estimate their value for further breeding programs.

4 Notes

1. It is advisable to grow the donor material in a greenhouse to reduce contamination with fungi and bacteria. Plants should be grown under ideal conditions to improve the establishment rate of tissue cultures (Debergh 1987).
2. Consider that the different genotypes and explant types, e.g. seeds, in vitro cuttings, or embryogenic callus have different requirements and capacities. This is especially important in the case of long lived organisms like trees, and has consequences at the level of population size, dissolution of chimerism and frequency of mutation. Therefore, it is advisable to use several genotypes as control material.

3. Consider that the different genotypes have different requirements. Different media should therefore be compared for efficient micropropagation: Medium 1 (Priyono et al. 2010), Medium 2 (Ebrahim et al. 2007) and Medium 3 (Abd El Gawad et al. 2012). Visual observations of different micropropagation media indicated, that Medium 1 induced small plantlets, small, light green leaves, many short roots. Medium 2 induced quite vigorous plantlets, axillary buds, but no roots. Medium 3 yielded the most vigorous plantlets, with large, dark green leaves, formation of a long root with secondary roots. Media were supplemented with 30 g/L sucrose and the pH adjusted to 5.7 prior to the addition of 7 g/L Agar (Sigma). Media were autoclaved at 120 °C and 1.1 kg/cm^2 for 20 min, and then 25 ml of medium was dispensed into each Magenta Box. The cultures were maintained at 26–27 °C in the dark. In vitro shoots are subcultured every 8–12 weeks by axillary cuttings.

4. Media according to Zamarripa et al. (1991), Etienne (2005) and Priyono et al. (2010) are indicated as suitable for indirect somatic embryogenesis, while for direct embryogenesis protocols were described by CATIE (1988), Hatanaka et al. (1991) and Lubabali et al. (2014).

5. Media M1 to M5 according to Zamarripa et al. (1991) are indicated as suitable for plantlet recovery from somatic embryos.

6. When the applied dose for the genotype is unknown, a radiation test should be performed to determine the optimal dose. To perform the radiosensitivity test on vegetative material like cuttings, select 30 cuttings per dose with a wide range from 0 to 100 Gy (Gy) for vegetatively propagated crops. However, the range of 0, 10, 20, 30, 40, 50 and 60 Gy of gamma rays may be sufficient to establish the optimal dose due to the high moisture content in comparison to seeds. The Gy unit used to quantify the absorbed dose of radiation (1 Gy = 1 J/kg).

7. When applying physical mutagens to different types of plant material, care should be taken with soft materials such as in vitro shoot cultures as well as callus and embryogenic callus cultures, which require lower doses in comparison to seeds. In fact, the water content, storage time, applied mutagen dose and temperature represents an important factor influencing mutagens in all types of plant material.

8. Radioactivity is mutagenic and carcinogenic. It should be operated by trained and authorized person and carried out in a defined lab. In fact, the safety precautions for exposing plant material to a gamma irradiation source have to be strictly observed.

9. Take care to observe all safety precautions before exposing tissues to irradiation.

10. A dose film can be included together with the samples to capture the absorbed dose.

11. Exposure time is equal to the required dose divided by the dose rate of the day.

12. The irradiated samples are safe to be held in hands because the sample chamber isolates the plant material from the source and there is no surface contamination.

13. Untreated samples (control) have to be prepared and kept in the same conditions as the treated samples.

14. Observe general rules for plant tissue culture practice.

Acknowledgements This work was supported by the Food and Agriculture Organization of the United Nations and the International Atomic Energy Agency through their Joint FAO/IAEA Program of Nuclear Techniques in Food and Agriculture as Coordination Research Project D22005. We also thank Dr. Bado S. for technical assistance and the Plant Breeding and Genetics Laboratory (PBGL) Seibersdorf, Austria for the irradiation services provided.

References

Abd El Gawad NMA, Mahdy HA, Boshra ES (2012) In vitro micropropagation protocol and acclimatization of coffee trees (*Coffea arabica* L.). J Plant Prod Mansoura Univ 3(1):109–116

Anthony F, Bertrand B, Quiros O, Wilches A, Lashermes P, Berthaud J, Charrier A (eds) (2001) Genetic diversity of wild coffee (*Coffea arabica* L.) using molecular markers. Euphytica 118:53–65

Anthony F, Combes C, Astorga C, Bertrand B, Graziosi G, Lashermes P (2002) The origin of cultivated *Coffea arabica* L. varieties revealed by AFLP and SSR markers. Theor Appl Genet 104:894–900

Anthony F, Diniz LEC, Combes M-C, Lashermes P (2010) Adaptive radiation in Coffea subgenus Coffea L. (Rubiaceae) in Africa and Madagascar. Plant Syst Evol 285:51–64

Bado S, Forster BP, Nielen S, Ghanim A, Lagoda PJL, Till BJ, Laimer M (2015) Plant mutation breeding: current progress and future assessment. In: Janick J (ed) Breed Rev 39:23–87

Bado S, Yamba NGG, Sesay JV, Laimer M, Forster BP (2017) Plant mutation breeding for the improvement of vegetatively propagated crops: successes and challenges. CAB Rev 12:1–21

Bado S, Maghuly F, Laimer M (2018a) Mutation induction in Coffea spp. to counteract the impact of a changing climate. In: 10th ÖGMBT annual meeting, "10 years of life, science and molecules", Vienna, Austria, 17–20 Sept 2018

Bado S, Maghuly F, Laimer M, Varzea V (2018b) Mutagenesis of in vitro explants of *Coffea arabica* to induce fungal resistance (No. IAEA-CN-263)

Bolívar-González A, Valdez-Melara M, Gatica-Arias A (2018) Responses of Arabica coffee (*Coffea arabica* L. var. Catuaí) cell suspensions to chemically induced mutagenesis and salinity stress under in vitro culture conditions. In Vitro Cell Dev Biol Plant 54:1–14

Campos NA, Panis B, Carpentier SC (2017) Somatic embryogenesis in coffee: the evolution of biotechnology and the integration of omics technologies offer great opportunities. Front Plant Sci 8:1460. https://doi.org/10.3389/fpls.2017.01460

Carvalho A, Monaco LC (1969) The breeding of arabica coffee. Outlines of perennial crop breeding in the tropics. Misc Pap Agric Univ, Wageningen, 4:198–216

Carvalho A (1988) Principles and practice of coffee plant breeding for productivity and quality factors: *Coffea arabica*. In: Clarke RJ, Macrae R (eds) Coffee, volume 4: agronomy. Elsevier Applied Science, London, pp 129–165

CATIE (Centro Agronómico Tropical de Investigación y Enseñanza) (1988) Curso teórico-práctico de tejidos tropicales. Unidad de Biotecnología, Programa de Mejoramiento de Cultivos Tropicales, Turrialba, Costa Rica, 80 pp

Chevalier A, Dagron M (1928) Recherches historiques sur les débuts de la culture du caféier en Amérique. In: Genetic diversity of wild coffee (*Coffea arabica* L.) using molecular markers. Communications et Actes de l'Académie des Sciences Coloniales, Paris

da Câmara Machado A, Puschmann M, Pühringer H, Kremen R, Katinger H, da Câmara Machado M (1995) Somatic embryogenesis of *Prunus subhirtella autumno rosa* and regeneration of transgenic plants after *Agrobacterium*-mediated transformation. Plant Cell Rep 14:335–340. https://doi.org/10.1007/BF00238592

Dada KE, Bado S, Anagbogu CF, Daniel MA, Forster BP (2014) Radio-sensitivity testing in coffee (*Coffea arabica*) as a prelude to coffee improvement through mutation breeding. In: The 25th international conference on coffee & science, ASIC 2014, Colombia, 8–13 Sept 2014, pp 177–178

Dada KE, Mustapha OT, Forster BP, Bado S (2018) Biological effect of gamma irradiation on vegetative propagation of *Coffea arabica* L. Afr J Plant Sci 12:122–128. https://doi.org/10.5897/AJPS2016.1504

Davis AP, Govaerts R, Bridson DM, Stoffelen P (2006) An annotated taxonomic conspectus of the genus *Coffea* (*Rubiaceae*). Bot J Linn Soc 142:465–512. https://doi.org/10.1111/j.1095-8339.2006.00584.x

Davis AP, Chester M, Maurin O, Fay M (2007) Searching for the relatives of Coffea (Rubiaceae, Ixoroideae): the circumscription and phylogeny of Coffeeae based on plastid sequence data and morphology. Am J Bot 94:313–329. https://doi.org/10.3732/ajb.94.3.313

Debergh P (1987) Recent trends in the application of tissue culture to ornamentals. In: Green CE, Somers DA, Hackett WP, Biesboer DD (eds) Plant tissue and cell culture. Alan R. Liss, New York, pp 383–393

Déchamp E, Breitler J-C, Leroy T, Etienne H (2015) Coffee (*Coffea arabica* L.). Methods Mol Biol 1224:275–291. https://doi.org/10.1007/978-1-4939-1658-0_22

de Moro G, Modonut M, Asquini E, Tornincasa P, Pallavicini A, Graziosi G (2009) Development and analysis of an EST databank of *Coffea arabica*. In: Proceedings of the 6th Solanaceae genome workshop, New Delhi, India, p 127

Denoeud F, Carretero-Paulet L, Dereeper A, Droc G, Guyot R, Pietrella M, Zheng C, Alberti A, Anthony F, Aprea G, Aury JM, Bento P, Bernard M, Bocs S, Campa C, Cenci A, Combes MC, Crouzillat D, Da Silva C, Daddiego L, De Bellis F, Dussert S, Garsmeur O, Gayraud T, Guignon V, Jahn K, Jamilloux V, Joët T, Labadie K, Lan T, Leclercq J, Lepelley M, Leroy T, Li LT, Librado P, Lopez L, Muñoz A, Noel B, Pallavicini A, Perrotta G, Poncet V, Pot D, Priyono, Rigoreau M, Rouard M, Rozas J, Tranchant-Dubreuil C, VanBuren R, Zhang Q, Andrade AC, Argout X, Bertrand B, de Kochko A, Graziosi G, Henry RJ, Jayarama, Ming R, Nagai C, Rounsley S, Sankoff D, Giuliano G, Albert VA, Wincker P, Lashermes P (2014) The coffee genome provides insight into the convergent evolution of caffeine biosynthesis. Science 345:1181–1184. https://doi.org/10.1126/science.1255274

Dereeper A, Bocs S, Rouard M, Guignon V, Ravel S, Tranchant-Dubreuil C, Poncet V, Garsmeur O, Lashermes P, Droc G (2015) The coffee genome hub: a resource for coffee genomes. Nucleic Acids Res 43:D1028–D1035. https://doi.org/10.1093/nar/gku1108

Ebrahim N, Shibli R, Makhadmeh I, Shatnawi M, Abu-Ein A (2007) In vitro propagation and in vivo acclimatization of three coffee cultivars (*Coffea arabica* L.) from Yemen. World Appl Sci J 2(2):142–150

Etienne H (2005) Somatic embryogenesis protocol: coffee (*Coffea arabica* L. and *C. canephora* P.). In: Jain SM, Gupta PK (eds) Protocols for somatic embryogenesis in woody plants. Springer, Dordrecht, The Netherlands, pp 167–179

Haarer AE (1956) Modern coffee production. Leonard Hill Limited, London, UK, pp 1–467

Hatanaka T, Arakawa O, Yasuda T, Uchida N, Yamaguchi T (1991) Effect of plant growth regulators on somatic embryogenesis in leaf cultures of *Coffea canephora*. Plant Cell Rep 10:179–182. https://doi.org/10.1007/BF00234290

IAEA (1975) Manual on mutation breeding. IAEA, Vienna

IAEA (1977) Manual on mutation breeding. IAEA, Vienna

Krishnan S (2011) Coffee biotechnology: implications for crop improvement and germplasm conservation. Acta Hortic 894:33–44. https://doi.org/10.17660/ActaHortic.2011.894.2

Krishnan S (2014) Marker-assisted selection in coffee, chap 9. In: Benkeblia N (ed) Omics technologies and crop improvement. CRC Press, Taylor & Francis Group, pp 209–218. https://doi.org/10.1201/b17573-10

Krug CA, Mendes JET, Carvalho A (1939) Taxonomia de *Coffea arabica* L. Bolétim Técnico no 62. Instituto Agronômico do Estado, Campinas, Brazil, pp 154–163

Lashermes P, Combes MC, Trouslot P, Charrier A (1997) Phylogenetic relationships of coffee-tree species (Coffea L.) as inferred from ITS sequences of nuclear ribosomal DNA. Theor Appl Genet 94:947–955. https://doi.org/10.1007/s001220050500

Lashermes P, Combes M, Robert J, Trouslot P, D'Hont A, Anthony F, Charrier A (1999) Molecular characterisation and origin of the *Coffea arabica* L. genome. Mol Gen Genet 261:259–266. https://doi.org/10.1007/s004380050965

Leroy T, Marraccini P, Dufour M, Montagnon C, Lashermes P, Sabau X, Ferreira LP, Jourdan I, Pot D, Andrade AC, Glaszmann JC, Vieira LG, Piffanelli P (2005) Construction and characterization of a *Coffea canephora* BAC library to study the organization of sucrose biosynthesis genes. Theor Appl Genet 111:1032–1041. https://doi.org/10.1007/s00122-005-0018-z

Lin C, Mueller LA, Carthy JM, Crouzillat D, Pétiard V, Tanksley SD (2005) Coffee and tomato share common gene repertoires as revealed by deep sequencing of seed and cherry transcripts. Theor Appl Genet 112:114–130. https://doi.org/10.1007/s00122-005-0112-2

Lubabali AH, Alakonya A, Gichuru EK, Kahia JW, Mayoli R (2014) In vitro propagation of the new disease resistant *Coffea arabica* variety Batia. Afr J Biotechnol 13(24):2424–2419. https://doi.org/10.5897/AJB2014.13735

Mba C, Afza R, Bado S, Jain SM (2010) Induced mutagenesis in plants using physical and chemical agents. In: Davey MR, Anthony P (eds) Plant cell culture: essential methods. Wiley, Chichester, UK, pp 111–130. https://doi.org/10.1002/9780470686522.ch7

Mba C, Afza R, Shu QY (2012) Mutagenic radiations: X-rays, ionizing particles and ultraviolet. In: Shu QY, Forster BF, Nakagawa H (eds) Plant mutation breeding and biotechnology. CABI, pp 83–106. www.cabi.org/cabebooks/ebook/20123349338

Mishra MK, Slater A (2012) Recent advances in the genetic transformation of coffee. Biotechnology Research International, p 17. https://doi.org/10.1155/2012/580857

Noir S, Patheyron S, Combes MC, Lashermes P, Chalhoub B (2004) Construction and characterisation of a BAC library for genome analysis of the allotetraploid coffee species (*Coffea arabica* L.). Theor Appl Genet 109:225–230. https://doi.org/10.1007/s00122-004-1604-1

Novak FJ, Brunner H (1992) Plant breeding: induced mutation technology for crop improvement. IAEA Bull 4:25–33

Priyono, Florin B, Rigoreau M, Ducos JP, Sumirat U, Mawardi S, Lambot C, Broun P, Pétiard V, Wahyudi T, Crouzillat D (2010) Somatic embryogenesis and vegetative cutting capacity are under distinct genetic control in *Coffea canephora* Pierre. Plant Cell Rep 29:343–357. https://doi.org/10.1007/s00299-010-0825-9

Razafinarivo NJ, Guyot R, Davis AP, Couturon E, Hamon S, Crouzillat D, Rigoreau M, Dubreuil-Tranchant C, Poncet V, De Kochko A, Rakotomalala J-J, Hamon P (2013) Genetic structure and diversity of coffee (Coffea) across Africa and the Indian Ocean islands revealed using microsatellites. Ann Bot 111:229–248. https://doi.org/10.1093/aob/mcs283

Ribas AF, Cenci A, Combes MC, Etienne H, Lashermes P (2011) Organization and molecular evolution of a disease-resistance gene cluster in coffee trees. BMC Genom 12:240. https://doi.org/10.1186/1471-2164-12-240

Sant'Ana GC, Pereira LFP, Pot D, Ivamoto ST, Domingues DS, Ferreira RV, Pagiatto NF, da Silva BSR, Nogueira LM, Kitzberger CSG, Scholz MBS, de Oliveira FF. Sera GH, Padilha L, Labouisse J-P, Guyot R, Charmetant P, Leroy T (2018) Genome-wide association study reveals candidate genes influencing lipids and diterpenes contents in *Coffea arabica* L. Sci Rep 8:465. https://doi.org/10.1038/s41598-017-18800-1

Silva MC, Várzea VMP, Rijo L, Rodrigues Jr CJ, Moreno G (1999) Cytological studies in Hibrido de Timor derivatives with resistance to *Colletotrichum kahawae*. In: Proceedings of the 18th international conference on coffee science (ASIC), Helsinki, Finland, Abstract A130

Silva MC, Várzea V, Guerra-Guimarães L, Azinheira HG, Fernandez D, Petitot AS, Bertrand B, Lashermes P, Nicole M (2006) Coffee resistance to the main diseases: leaf rust and coffee berry disease. Braz J Plant Physiol 18:119–147. https://doi.org/10.1590/S1677-04202006000100010

Várzea VMP, Rodrigues Jr CJ, Marques D, Silva MC (2000) Loss of resistance in interspecific tetraploid coffee varieties to some pathotypes of *Hemileia vastatrix*. In: International symposium on durable resistance: key to sustainable agriculture, Wageningen, The Netherlands, Abstract book, p 34

Vieira LGE, Andrade AC, Colombo C et al (2006) Brazilian coffee genome project: an EST-based genomic resource. Braz J Plant Physiol 18:95–108. https://doi.org/10.1590/S1677-042020060 00100008

Zamarripa A, Ducos JP, Tessereau H, Bollon H, Eskes A, Pétiard V (1991) Développement d'un procédé de multiplication en masse du caféier par embryogenèse somatique en milieu liquide. In: 14ème Colloque Scientifique International sur Le Café, San Francisco, 14–19 July 1991. ASIC, Paris, pp 392–402

Mutation Induction Using Gamma-Ray Irradiation and High Frequency Embryogenic Callus from Coffee (*Coffea arabica* L.)

Miguel Barquero-Miranda and Reina Céspedes

Abstract Mutation induction through chemical or physical mutagenesis has been widely used for crop improvement for more than 70 years. Coffee is one of the most important crops in Latin-America, and, as any other crop, it can be affected by pests and diseases. Coffee leaf rust (CLR), caused by the biotrophic fungus *Hemileia vastatrix*, is the most important disease affecting Arabica coffee leading to significant losses for growers. As a perennial crop, conventional breeding of Arabica coffee is time-consuming. Plant tissue culture in combination with mutation induction techniques can provide an alternative approach to increase genetic variability of Arabica coffee for breeding applications. The present chapter describes protocols to establish embryogenic callus suspensions from Arabica coffee cv Venecia and for gamma ray irradiation of callus suspension cultures to achieve genetic improvement in the crop.

1 Introduction

Coffea arabica L. (coffee) belongs to the Rubiaceae family which comprises about 500 genera and more than 6000 species, mostly tropical trees and shrubs (Jiménez and Carril 2014). The *Coffea* genus includes more than 100 species from which only *C. arabica* and *C. canephora* are grown commercially (Mishra and Slater 2012). Central America is the world's fifth largest Arabica coffee producer, where Costa Rica stands out in terms of production and quality (ICAFE 2016).

More than 80% of Arabica coffee produced in Latin America comes from varieties derived from a narrow genetic base, being highly susceptible to diseases and pests, caused by microorganisms such as fungi, bacteria, viruses and nematodes (Bertrand et al. 2011). In the region, the majority of the diseases are caused by phytoparasitic fungi and around 300 diseases affecting the crop have been detected worldwide (Canet Brenes et al. 2016). Coffee leaf rust (CLR), caused by the fungus *Hemileia*

M. Barquero-Miranda (✉) · R. Céspedes
Coffee Research Center, 37-1000, Heredia San Pedro, Barva, Costa Rica
e-mail: mbarquero@icafe.cr

© The Author(s) 2023
I. L. W. Ingelbrecht et al. (eds.), *Mutation Breeding in Coffee with Special Reference to Leaf Rust*, https://doi.org/10.1007/978-3-662-67273-0_6

vastatrix, is one of the main limiting factors of Arabica coffee production in all coffee growing countries (Mishra and Slater 2012).

Despite ongoing efforts for resistance breeding, chemical control is still the most widely used method to contain pests and diseases, including CLR (Neto and da Cunha 2016). Therefore, the development of alternative, environmentally friendly solutions for control of CLR is important. A long-term solution is through the development of resistant varieties, which is the focus of many breeding programmes in Arabica coffee (Mishra and Slater 2012). However, due to the perennial nature of coffee it can be difficult and time consuming to breed for disease resistance through conventional breeding methods (Barrueto Cid et al. 2004).

Plant breeders can use different tools to induce genetic variation in crops (Bermúdez-Caraballoso et al. 2016). Given the perennial nature of Arabica coffee, an effective way to induce variability, can be plant tissue culture in combination with mutation induction (Muthusamy et al. 2007). Combined, these techniques could increase genetic variability and reduce the time needed to develop new plant varieties (Bolívar-González et al. 2018).

Mutations can be induced by physical mutagens such as X-rays, gamma rays, neutrons and by chemical mutagens such as ethyl methanesulfonate (EMS). Physical mutagens appear more frequently used than chemical mutagens (Beyaz and Yildiz 2017). Since the 1960s gamma-ray mutagenesis has been the most commonly used method in plant mutation breeding (Li et al. 2019). Gamma rays are ionizing radiation (Beyaz and Yildiz 2017); they interact with atoms or molecules producing free radicals in cells that induce physiological, biochemical, cytological, genetic and morphogenetic changes in cells and tissues of plants (Chusreeaeom and Khamsuk 2019).

Somatic embryogenesis (SE) is a plant tissue culture technique where embryos are obtained from cells that are not the product of gametic fusion. Through SE thousands of seedlings identical to the mother plant can be produced (Bartos et al. 2018). Induced mutations are single cell events and thus the mutagenic treatment of seeds will result in chimeric M_1 plants, i.e., they may carry different mutations, each occupying a (small) part of the plant. Since somatic embryos regenerate from single cells, somatic embryogenesis is considered to be an effective method for eliminating chimeras (Roux et al. 2004).

The optimal irradiation dose(s) leading to genetic improvement of a specific crop or trait may vary depending on the genetic constitution of the plant species and cultivar. Until now only the work of Sari et al. (2019) has referred to the use of gamma rays for mutation induction of Robusta coffee embryogenic callus suspensions. It is necessary and essential to conduct radiosensitivity testing to determine the optimal dose(s) of gamma-ray irradiation of Arabica coffee embryogenic cell cultures before conducting bulk irradiation experiments (Bermúdez-Caraballoso et al. 2016; Spencer-Lopes et al. 2018). The present chapter describes a protocol on how to obtain the embryogenic callus suspensions of *Coffea arabica* and to determine the optimal irradiation dose.

2   Materials

2.1   Plant Material

1. In these experiments *Coffea arabica* cv Venecia.

2.2   Explants Collection and Disinfection

1. Young leaves from a donor plant.
2. Soap and distilled water.
3. Beakers (1,000 ml).
4. Sodium hypochlorite (NaOCl) 2% (v/v).
5. Antioxidant sterile solution (350 mg/L ascorbic acid, 200 mg/L citric acid y 20 g/L saccharose) + fungicide (3 g/L AMISTAR).
6. Sterile deionized water.
7. Laminar airflow cabinet.

2.3   Induction of Embryogenic Callus

1. Dissecting instruments (scalpels, blades, and forceps).
2. Glass jars (72.5 × 84.5 mm) with 15 ml of V1 media culture (see Table 1).
3. Glass jars (72.5 × 84.5 mm) with 20 ml of V2 media culture (see Table 1).
4. Laminar airflow cabinet.
5. Room with environmental control.

2.4   Embryogenic Callus Multiplication

1. Erlenmeyer flasks (25, 50, 125, 250, 500, 1,000 and 2,000 ml).
2. Aluminum foil.
3. Parafilm.
4. Graduated cylinder (100 ml).
5. Falcon tubes (50 ml).
6. Orbital shake.
7. Analytical balance.
8. Laminar airflow cabinet.
9. Room with environmental control.

Table 1 Composition of media culture (mg/l) for indirect somatic embryogenesis in Coffee (*Coffea arabica*) (see Note 2)

Source	V1	V2	V3	V4	V5	V6
Macrominerals	MS/2	MS/2	MS/2	MS/2	MS/2	MS
Microminerals	MS/2	MS/2	MS/2	MS/2	MS/2	MS
$FeSO_4 \cdot 7H_2O$	13.9	13.9	13.9	13.9	13.9	27.8
Na_2EDTA	18.65	18.65	18.65	18.65	18.65	37.3
Thiamine HCl	10	20	5	10	8	1
Pyridoxine-HCl	1	–	0.5	1	3.2	1
Glycine	1	20	–	2	–	–
Nicotinic acid	1	–	0.5	1	–	1
Cystein	–	40	10	–	–	–
Myo-inositol	100	200	50	200	100	100
Adenine	–	60	–	40	–	–
Casein hydrolysate	100	200	100	400	–	–
Malt extract	400	800	200	400	–	–
2,4-D	0.5	1.0	1	–	–	–
IBA	1	–	–	–	–	–
IAA	–	–	–	–	0.45	–
Kinetin	–	–	1	–	–	
2ip	2	–	–	–	–	–
6-BAP	–	4	–	2	0.25	0.3
Saccharose	30,000	30,000	15,000	40,000	20,000	30,000
Phytagel	3,600	3,600	None	None	3,600/or none	3,600

Source Van Boxtel and Berthouly (1996)

2.5 Embryogenic Callus Irradiation

1. Filter paper (Diameter: 9 cm).
2. Funnel.
3. Erlenmeyer (250 ml).
4. Microcentrifuge tubes (1.5 ml).
5. Micropipette (100–1,000 μl).
6. Micropipette tips (100–1,000 μl).
7. Media culture V4 (see Table 1).
8. Parafilm.
9. Petri dishes (100 × 15 mm).
10. Forceps.
11. Analytical balance.
12. Laminar airflow cabinet.
13. Irradiation source (see Note 1).

2.6 Regeneration of the Irradiated Embryogenic Callus

1. Micropipette tips (100–1,000 μl).
2. Micropipette (100–1,000 μl).
3. Erlenmeyer flasks (250 ml).
4. Media culture V4 (see Table 1).
5. Graduated cylinder (100 ml).
6. Laminar airflow cabinet.
7. Orbital shaker.
8. Room with environmental control.

2.7 Embryo Germination

1. Filter paper (Diameter: 9 cm).
2. Funnel.
3. Erlenmeyer (250 ml).
4. Forceps.
5. Glass jars (59.5 × 68.0 mm) with 15, 20, 25 ml of V5 media culture (see Table 1).
6. Plastic film.
7. Laminar airflow cabinet.
8. Room with environmental control.

2.8 Development of Somatic Embryos into Plantlets

1. Forceps.
2. Glass jars (59.5 × 68.0 mm) with 25 ml of V6 media culture (see Table 1).
3. Plastic film.
4. Laminar airflow cabinet.
5. Room with environmental control.

2.9 Media Culture: Preparation

1. pH meter.
2. Autoclave.
3. Glass jars.
4. Microwave.
5. Analytical balance.
6. Magnetic stirrers.

3   Methods

3.1   Explants Collection and Disinfection

1. Remove well developed and healthy leaves from the first or second node from the donor plant (see Note 3).
2. Rinse with tap water and soap.
3. Transfer into the laminar airflow cabinet and place leaves in a sterile beaker.
4. Add the fungicide-antioxidant solution and frequently swirl the solution by hand during 10 min.
5. Decant the solution and rinse three times with sterile deionized water.
6. Disinfect the explants with a sodium hypochlorite solution 2% (v/v) and swirl gently during 30 min.
7. Decant the solution and rinse three times with sterile deionized water and decant.

3.2   Induction of Embryogenic Callus

1. Cut square leaf segments of 0.5 cm^2 avoiding the main, secondary veins and leaf borders.
2. Place 7 segments in each flask, with the upper side in contact with V1 media culture.
3. Incubate in the dark at 28 °C for 22 days.
4. Remove the leaf segments from the V1 media culture and transfer on the V2 media.
5. Incubate in the dark at 28 °C until embryogenic callus is formed (friable callus, white color of dusty consistency) can be observed.

3.3   Embryogenic Callus Multiplication

1. After 4–5 months, in the laminar airflow cabinet, remove the somatic embryogenic callus and weigh using an analytical balance.
2. Once the weight is determined, choose an Erlenmeyer flask that can contain the necessary volume (10% of the actual capacity) maintaining a ratio of 10 mg embryogenic callus/ml of V3 culture medium. Seal with parafilm.
3. Place the flask containing the embryogenic callus in a liquid media on an orbital shaker and incubate in the dark at 90 rpm and 28 °C.
4. Subculture every 20 days.
5. Transfer the embryogenic callus into a 50 ml centrifuge tube, carefully decant the old media keeping a minimum amount to avoid loss of callus.
6. Duplicate the media culture volume and transfer to a new Erlenmeyer flask.

7. Seal with parafilm and incubate in the dark on an orbital shaker at 90 rpm and 28 °C.
8. Repeat steps 4–7 for a maximum of 5 subcultures (3 months incubation).

3.4 Irradiation of the Embryogenic Callus

1. Take an Erlenmeyer containing the embryogenic suspension culture from *Coffea arabica*, in multiplication stage, with a maximum of five subcultures (3 months) (see Fig. 1a).
2. Carefully open the Erlenmeyer in the laminar airflow cabinet and decant the material.
3. Filter the material in a filter paper-funnel-Erlenmeyer system to eliminate as much culture medium as possible (see Fig. 1a, b).
4. Once the material is obtained, divide it into sterile 1.5 ml microcentrifuge tubes with 20 mg each.
5. Add 0.5 ml of regeneration media culture (V4).
6. Seal the microcentrifuge tubes with Parafilm, place inside a petri dish and seal the petri dish for transportation (see Fig. 1d).
7. Transport the culture in dark condition and at room temperature to the irradiation facility (see Fig. 1e) and irradiate the samples using different doses (see Note 4).

3.5 Regeneration of the Irradiated Embryogenic Callus

1. Use a sterile micropipette to remove the irradiated material from the microcentrifuge tube.
2. Transfer to an Erlenmeyer (250 ml) maintaining the ratio 1 mg per ml of regeneration media (V4).
3. Place the Erlenmeyer flask on a rotary shaker and incubate in the dark at 100 rpm and 28 °C.
4. Incubate until the regeneration of the embryos is observed (1–3 months) (see Fig. 1f).

3.6 Embryo Germination

1. Remove the regenerated embryos from the Erlenmeyer flask as described in Sect. 3.4, step 3.
2. Place individual embryos on glass jars containing V5 media.
3. Incubate the material at 28 °C with a 12 h photoperiod for 30 days.
4. Subculture the viable material and discard the amorphous material. Keep track of the data to determine the LD_{50} (see Note 4).

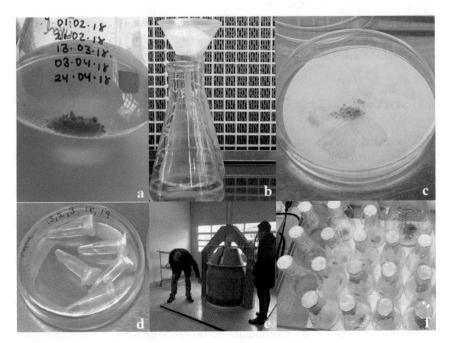

Fig. 1 *Coffea arabica* embryogenic callus irradiation; **a** embryogenic suspension; **b** filtration system; **c** filtrated embryogenic callus; **d** embryogenic callus divided into microcentrifuge tubes to irradiate; **e** callus irradiation on Ob-Servolgnis irradiator; **f** regeneration of the irradiated material

3.7 Regeneration of Plantlets

1. Once the embryos have been developed correctly (foliar and root development) subculture them in semi-solid V6 media culture.
2. Incubate the material at 28 °C with 12 h photoperiod, subculture every 30 days until the plantlets are developed (2–3 pairs of true leaves and 1–2 root cm) (Fig. 2).

4 Notes

1. The irradiation of the materials was conducted at the Gamma radiation laboratory at the facilities of the Instituto Tecnológico de Costa Rica (TEC), using a Ob-Servolgnis irradiator (Cobalt 60 radioactive source and an activity of 4.4×10^{14} Bq) (Becquerel).

Fig. 2 Somatic embryogenesis process in *Coffea arabica*

2. For the preparation of the media culture the following procedures are followed: The components for each media culture are specified in Table 1, each amount shown is in mg/L. Place all media components in a volumetric flask and stir until fully dissolved. Complete the volume with water until the mark. Decant to a beaker and adjust the pH to 5.6. If the media is semi-solid add phytagel (the amount of phytagel may vary between 3.6 and 5.0 g/L), microwave until boiled (8 min/L). Dispense in glass jars and sterilize at 121 °C 1.5 lb of pressure.
3. Healthy leaves were collected from a donor plant established in the field. Young leaves from the first or second internode were removed from branches of the coffee plant. The selected branches were positioned at the middle part of the plant.
4. The doses used in the experiment were: 5, 10, 15, 20, 25, 30, 35, 40, 60, 80 and 100 Gy. The LD_{50} is 40 Gy.

Acknowledgements Funding for this work was provided by the Costa Rican Coffee Institute-Coffee Center Research and the IAEA. This work is part of the IAEA Coordinated Research Project D22005 titled "Efficient Screening Techniques to Identify Mutants with Disease Resistance for Coffee and Banana", Contract Number 20475.

References

Barrueto Cid LP, Ramos Cruz AR, Rodrigues Castro LH (2004) Somatic embryogenesis from three coffee cultivars: "Rubi", "Catuaí Vermelho 81", and "IAPAR 59." HortScience 39(1):130–131

Bartos PMC, Gomes HT, do Amaral LIV, Teixeira JB, Scherwinski-Pereira JE (2018) Biochemical events during somatic embryogenesis in *Coffea arabica* L. 3 Biotech 8(4). http://doi.org/10.1007/s13205-018-1238-7

Bermúdez-Caraballoso I, Rodríguez M, Reyes M, Gómez-Kosky R, Chong-Pérez B, Rivero L (2016) Mutagénesis in vitro en suspensiones celulares embriogénicas de banano cv. Grande naine (Musa AAA). Biotecnología Vegetal 16(2):103–111. Retrieved from https://revista.ibp.co.cu/index.php/BV/article/view/515

Bertrand B, Alpizar E, Lara L, SantaCreo R, Hidalgo M, Quijano JM et al (2011) Performance of *Coffea arabica* F1 hybrids in agroforestry and full-sun cropping systems in comparison with American pure line cultivars. Euphytica 181(2):147–158. https://doi.org/10.1007/s10681-011-0372-7

Beyaz R, Yildiz M (2017) The use of gamma irradiation in plant mutation breeding. Plant Eng. https://doi.org/10.5772/intechopen.69974

Bolívar-González A, Valdez-Melara M, Gatica-Arias A (2018) Responses of Arabica coffee (*Coffea arabica* L. var. Catuaí) cell suspensions to chemically induced mutagenesis and salinity stress under in vitro culture conditions. In Vitro Cell Dev Biol Plant 54(6):576–589. http://doi.org/10.1007/s11627-018-9918-x

Canet Brenes G, Soto Víquez C, Ocampo Tomason P, Rivera Ramírez J, Navarro Hurtado A, Guatemala Morales G, Villanueva Rodríguez S (2016) La situación y tendencias de la producción de café en América Latina y el Caribe. In: IICA. Retrieved from http://www.iica.int/sites/default/files/publications/files/2017/BVE17048805e.pdf

Chusreeaeom K, Khamsuk O (2019) Effects of gamma irradiation on lipid peroxidation, survival and growth of turmeric in vitro culture. J Phys Conf Ser 1285(1). http://doi.org/10.1088/1742-6596/1285/1/012003

ICAFE (2016) Procedimiento para la reproducción de café por embriogénesis somática. Instituto del Café de Costa Rica

Jiménez ER, Carril EP (2014) Café I (G. Coffea) 7(2):113–132

Li F, Shimizu A, Nishio T, Tsutsumi N, Kato H (2019) Comparison and characterization of mutations induced by gamma-ray and carbon-ion irradiation in rice (*Oryza sativa* L.) using whole-genome resequencing. G3 (Bethesda, Md.) 9(11):3743–3751. http://doi.org/10.1534/g3.119.400555

Mishra MK, Slater A (2012) Recent advances in the genetic transformation of coffee. Biotechnol Res Int 2012:1–17. https://doi.org/10.1155/2012/580857

Muthusamy A, Vasanth K, Sivasankari D, Chandrasekar BR, Jayabalan N (2007) Effects of mutagens on somatic embryogenesis and plant regeneration in groundnut. Biol Plant 51(3):430–435. https://doi.org/10.1007/s10535-007-0092-y

Neto JG, da Cunha JPAR (2016) Spray deposition and chemical control of the coffee leaf-miner with different spray nozzles and auxiliary boom. Engenharia Agricola 36(4):656–663. https://doi.org/10.1590/1809-4430-Eng.Agric.v36n4p656-663/2016

Roux N, Toloza A, Dolezel J, Panis B, Jain S, Swennen R (2004) Usefulness of embryogenic cell suspension cultures for the induction and selection of mutants in Musa spp. Banana improvement: cellular, molecular biology, and induced mutations. In: Proceedings of a meeting held in Leuven, Belgium, 24–28 Sept 2001, pp 33–43. Retrieved from http://scholar.google.com/scholar?hl=en&btnG=Search&q=intitle:Usefulness+of+embryogenic+cell+suspension+cultures+for+the+induction+and+selection+of+mutants+in+Musa+spp#0

Sari M, Ibrahim D, Randriani E, Sari L, Nuraini A, Penelitian B et al (2019) Radiosensitivity of embryogenic callus of Robusta coffee against irradiation of gamma rays. J Industrial and Beverage Crops vol 6, pp 41–50

Spencer-Lopes MM, Forster BP, Jankuloski L (2018) Manual on mutation breeding. J Nucl Energy 26. http://doi.org/10.1016/0022-3107(72)90060-3

Van Boxtel J, Berthouly M (1996) High frequency somatic embryogenesis from coffee leaves. Plant Cell Tissue Organ Cult 44:7–17

Chemical Mutagenesis of Embryogenic Cell Suspensions of *Coffea arabica* L. var. Catuaí Using EMS and NaN$_3$

Andrés Gatica-Arias and Alejandro Bolívar-González

Abstract Chemical mutagens, such as ethyl methanesulfonate (EMS) and sodium azide (NaN$_3$), interact with DNA and can primarily induce single base modifications along the genome. Populations derived from chemical mutagenesis experiments are presumed to harbor high density of point mutations in the genome. Therefore, this technique, along with in vitro culture methods such as somatic embryogenesis (SE), can introduce genetic variation in otherwise genetically homogeneous populations. In vitro mutagenesis of embryogenic cell suspension cultures represents an efficient method to quickly develop mutant plantlets of unicellular origin. The development of mutant populations in this important crop represents a fundamental steppingstone in the development of novel varieties and the characterization of candidate genes involved in traits such as disease resistance, grain metabolite content and flowering induction. This chapter describes the protocol for establishment of embryogenic cell suspension cultures as well as methods of mutation induction using EMS and NaN$_3$ on embryogenic cell suspensions of *C. arabica*, variety Catuaí. Furthermore, this chapter includes a protocol for mutant plant regeneration in in vitro conditions.

1 Introduction

The combination of chemical mutagenesis with in vitro culture techniques offers advantages to improve the efficiency of mutagenic treatments. The easier management of large populations of plants and the independence of agronomic and environmental factors can be listed among these advantages (Xu et al. 2011). In vitro selection procedures may also be applied to accelerate some screening steps and mutant lines can be quickly micropropagated. Success of these protocols depends on the establishment of robust in vitro regeneration procedures. It is advisable to apply mutagenic treatments on culture methods that involve regeneration via individual cells. This way, chimeric events could be avoided in most cases, or at least, can

A. Gatica-Arias (✉) · A. Bolívar-González
Laboratorio Biotecnología de Plantas, Escuela de Biología, Universidad de Costa Rica, 2060, San Pedro, Costa Rica
e-mail: andres.gatica@ucr.ac.cr

© The Author(s) 2023
I. L. W. Ingelbrecht et al. (eds.), *Mutation Breeding in Coffee with Special Reference to Leaf Rust*, https://doi.org/10.1007/978-3-662-67273-0_7

be dissolved more rapidly. Somatic embryogenesis (SE) is one of the ideal systems which could be incorporated in a mutation breeding program. It can be described as a morphogenic process characterized by the formation of embryos from somatic cells without fecundation (Campos et al. 2017). Somatic embryos or embryogenic calli usually are formed from a limited number of cells on the plant tissue, thus rendering a mostly unicellular origin for the regenerated plantlets. Chimeric events can be reduced when embryogenic cultures are mutagenized and time limitations in breeding programs can be overcome. The predominant unicellular origin of embryogenic structures facilities the early and direct screening of M_1V_1 plantlets regenerated from M_1V_1 treated calli or tissues without the need to develop an M_2 generation (Serrat et al. 2014).

When using embryogenic cultures, variables such as survival of cells and regeneration capacity after mutagenic treatment must be assessed to optimize the mutagen dose(s). Dual tests that allow qualitative and quantitative viability analysis are advised. One of the most widely used assays for checking the viability of in vitro cultures is the 2,3,5-triphenyltetrazolium chloride (TTC) test used to differentiate between metabolically active and inactive tissues. In living tissues TTC is converted to a red colored precipitate 1,3,5-triphenylformazan (TPF) that can be easily detected and quantified. Somatic embryos regenerated from mutagenized tissues can show germination and growth delay, germination inhibition can also appear.

Coffee is a key driver in social development and cultural identity of many tropical and subtropical regions. Worldwide production of coffee relies on two species, *Coffea arabica* L. (60%) and *C. canephora* (40%). The better cup quality and higher market value are associated to *C. arabica* L., the only allotetraploid ($2n = 4x = 44$) species among *Coffea*. *C. arabica* is an autogamous plant mostly incompatible with the remainder of *Coffea* species (Anthony et al. 2002). These characteristics, along with severe bottlenecks that happened during coffee domestication led to reduced genetic variability in *C. arabica* populations; this reduction enhances the general susceptibility of many *C. arabica* L. genotypes to diseases (Hendre and Aggarwal 2007).

SE has been developed in coffee, both directly through proembryogenic cells and indirectly through embryogenic calli from leaf explants. Embryogenic cultures are induced on an auxin containing medium, thereafter, subculture on auxin free medium induces embryo regeneration (Gatica-Arias et al. 2008; Quiroz-Figueroa et al. 2002; van Boxtel and Berthouly 1996). Coffee SE has been widely studied and represents an useful tool in breeding (reviewed by Campos et al. 2017). In this chapter, we describe methods for the mutagenic treatment of in vitro coffee embryogenic cultures to induce genetic variability. The chapter covers the application of the chemical mutagens EMS and sodium azide on the *C. arabica* var. Catuaí embryogenic cell suspensions as well as the regeneration of mutant plantlets.

2 Materials

2.1 Plant Material

1. Mature coffee cherries (e.g. *Coffea arabica* L. var. Catuaí) (see Note 1).
2. Embryogenic cell suspension cultures (e.g. *Coffea arabica* L. var. Catuaí) (see Note 1).

2.2 Reagents

1. 10% (w/v) sodium thiosulfate (e.g. Sigma Cat Nr.: 217263).
2. 100 mM phosphate buffer (pH 3.0).
3. 1 N HCl (e.g. Phytotechnology Cat Nr.: H245).
4. 1 N KOH (e.g. Phytotechnology Cat Nr.: P682).
5. 1 N NaOH (e.g. Phytotechnology Cat Nr.: S835).
6. 2,3,5-triphenyltetrazolium (TTC) (e.g. Phytotechnology Cat Nr.: T8164).
7. 2,4-Dichlorophenoxyacetic acid (2,4-D) (e.g. Sigma Cat Nr.: D7299).
8. 3-Indoleacetic acid (IAA) (e.g. Sigma Cat Nr.: I2886).
9. 6-(γ,γ-dimethylallylamino)-purine (2-iP) (e.g. Sigma Cat Nr.: D7257).
10. 6-Benzylaminopurine (BAP) (e.g. Sigma Cat Nr.: D3408).
11. Absolute ethanol.
12. Adenine sulfate (e.g. Phytotechnology Cat Nr.: A545).
13. Biotin (e.g. Phytotechnology Cat Nr.: B140).
14. Calcium pantothenate (e.g. Phytotechnology Cat Nr.: C186).
15. Casein hydrolysate (e.g. Phytotechnology Cat Nr.: C184).
16. Citric acid (e.g. Phytotechnology Cat Nr.: C277).
17. Ethyl methanesulphonate (EMS) (e.g. Sigma Cat Nr.: M0880).
18. Gelling agent (e.g. Phytagel: Sigma Cat Nr.:P8169).
19. Glycine (e.g. Phytotechnology Cat Nr.: G503).
20. Phosphorus acid (H_3PO_3) (e.g. Sigma Cat Nr.: 176680).
21. Indole-3-butyric acid (IBA) (e.g. Sigma Cat Nr.: I5386).
22. KH_2PO_4 (e.g. Sigma Cat Nr.: P5655).
23. Kinetin (e.g. Sigma Cat Nr.: K1885).
24. L-cysteine (e.g. Phytotechnology Cat Nr.: C204).
25. Malt extract (e.g. Phytotechnology Cat Nr.: M474).
26. Murashige and Skoog (MS) basal salt mixture (e.g. Phytotechnology Cat Nr.: M524).
27. Myo-inositol (e.g. Phytotechnology Cat Nr.: I703).
28. Nicotinic acid (e.g. Phytotechnology Cat Nr.: N765).
29. Phosphorus acid (H_3PO_3) (e.g. Sigma Cat Nr.: 176680).
30. Pyridoxine HCl (e.g. Phytotechnology Cat Nr.: P866).
31. Sodium azide (NaN_3) (e.g. Sigma Cat Nr.: S2002).

32. Sodium hypochlorite.
33. Sterile distilled water.
34. D-Sucrose (e.g. Phytotechnology Cat Nr.: S829).
35. Thiamine HCl (e.g. Phytotechnology Cat Nr.: T390).
36. Tissue culture grade water.
37. Tween 20 (e.g. Phytotechnology Cat Nr.: P720).

2.3 Glassware and Minor Equipment

1. 0.22 μm millipore filter.
2. Baby food jars (e.g. Phytotechnology Cat Nr.: C1770).
3. Beakers (100 ml, 500 ml, and 1,000 ml).
4. Bottles (100 ml, and 500 ml).
5. Box for dry hazardous material disposal.
6. Closures for culture tubes (e.g. Phytotechnology Cat Nr.: C1805).
7. Culture tubes (25 mm $\times$ 150 mm).
8. Disposable pipettes (1 ml, 5 ml, 10 ml, and 25 ml).
9. Erlenmeyer flasks (250 ml).
10. Forceps.
11. Glass or disposable pipettes (1 ml, 5 ml, 10 ml, 25 ml).
12. Graduated cylinder (50 ml, 100 ml, 500 ml and 1,000 ml).
13. Hazardous liquid waste receptacle (collection vessels for NaN_3 and EMS waste solution).
14. Magnetic stir bar.
15. Personal protective equipment (disposable laboratory coat dedicated only to mutagenesis experiments, eye protection/goggles, shoe protection, nitrile gloves).
16. Petri dishes (100 mm $\times$ 20 mm).
17. Pipette rubber bulb or electronic pipette controller.
18. Reaction tubes (2 ml, 15 ml).
19. Scalpels.
20. Scalpel blades.
21. Spatula.
22. 1 ml syringe.
23. Volumetric flasks (50 ml, 100 ml, and 1,000 ml).
24. Weighing trays.

2.4 Equipment

1. Analytical balance.
2. Autoclave.

3. Centrifuge.
4. Chemical mutagen laboratory equipped with fume hood and flow bench (see Notes 2 and 3).
5. Hot plate shaker.
6. Magnetic stir bar.
7. Medium dispenser.
8. Orbital shaker.
9. pH meter.
10. Spectrophotometer.
11. Stereoscope.
12. Water bath.

2.5 Tissue Culture Media

1. Semi-solid development medium (DEV), pH 5.6.
2. Semi-solid germination medium (EG), pH 5.6.
3. Semi-solid regeneration medium (R), pH 5.6.
4. TEX liquid culture medium, pH 5.6.
5. TEX liquid culture medium (pH 3.0 and pH 5.6).

2.6 Software

1. Standard spreadsheet software (e.g. Microsoft Excel or Open Office Excel).

3 Methods

3.1 Preparation of Stock Solutions

1. 100 mM Phosphate buffer: add 680.5 mg of KH_2PO_4 to 250 ml volumetric flask and dissolve by adding 25 ml of tissue culture grade water. Once completely dissolved, stir the solution while adding tissue culture grade water and bring to 50 ml. Adjust pH to 3.0 using phosphorus acid (H_3PO_3).
2. 2,4-Dichlorophenoxyacetic acid (2,4-D, 1 mg/ml stock solution): add 50 mg of the powder to a 100 ml volumetric flask and dissolve by adding 2–5 ml of 1 N NaOH or 95% v/v ethanol. Once completely dissolved, stir the solution while adding tissue culture grade water and bring to 50 ml. Sterilize by filtering through a 0.2 μm filter and store aliquots (1 ml) at 4 °C.

3. 3-Indoleacetic acid (IAA; 1 mg/ml stock solution): add 50 mg of the powder to a 100 ml volumetric flask and dissolve by adding 2–5 ml of 1 N NaOH. Once completely dissolved, stir the solution while adding tissue culture grade water and bring to 50 ml. Sterilize by filtering through a 0.2 μm filter and store aliquots (1 ml) at − 20 °C.

4. 6-(γ,γ-dimethylallylamino)-purine (2-iP; 1 mg/ml stock solution): add 50 mg of the powder to a 100 ml volumetric flask and dissolve by adding 2–5 ml of 1 N NaOH. Once completely dissolved, stir the solution while adding tissue culture grade water and bring to 50 ml. Sterilize by filtering through a 0.2 μm filter and store aliquots (1 ml) at − 20 °C.

5. 6-Benzylaminopurine (BAP; 1 mg/ml stock solution): add 50 mg of the powder to a 100 ml volumetric flask and dissolve by adding 2–5 ml of 1 N NaOH. Once completely dissolved, stir the solution while adding tissue culture grade water and bring to 50 ml. Sterilize by filtering through a 0.2 μm filter and store aliquots (1 ml) at 4 °C.

6. Indole-3-butyric acid (IBA; 1 mg/ml stock solution): add 50 mg of the powder to a 100 ml volumetric flask and dissolve by adding 2–5 ml of 1 N NaOH or 95% (v/v) ethanol. Once completely dissolved, stir the solution while adding tissue culture grade water and bring to 50 ml. Sterilize by filtering through a 0.2 μm filter and store aliquots (1 ml) at 4 °C.

7. KH_2PO_4 (0.07 M): add 907.8 mg of the powder to 250 ml volumetric flask and dissolve by adding 50 ml of tissue culture grade water. Once completely dissolved, stir the solution while adding tissue culture grade water and bring to 100 ml.

8. Kinetin (KIN; 1 mg/ml stock solution): add 50 mg of the powder to a 100 ml volumetric flask and dissolve by adding 2–5 ml of 1 N NaOH. Once completely dissolved, stir the solution while adding tissue culture grade water and bring to 50 ml. Sterilize by filtering through a 0.2 μm filter and store aliquots (1 ml) at − 20 °C.

9. Na_2HPO_4 (0.08 M): add 1.1876 g of the powder to 250 ml volumetric flask and dissolve by adding 50 ml of tissue culture grade water. Once completely dissolved, stir the solution while adding tissue culture grade water and bring to 100 ml.

10. Sodium azide (NaN_3) (500 mM): add 1.625 g of the powder to 250 ml volumetric flask and dissolve by adding 50 ml phosphate buffer (100 mM). Once completely dissolved, stir the solution while adding phosphate buffer (100 mM) and bring to 50 ml. Sterilize by filtering through a 0.2 μm filter and store at 4 °C and protect it from light.

11. 2,3,5-triphenyltetrazolium (TTC) stock solution: add 4 ml of KH_2PO_4 (0.07 M) and 6 ml of Na_2HPO_4 (0.08 M). Adjust pH to 7. Add TTC to reach 1% (m/v) (100 mg for 10 ml of solution). Sterilize by filtering through a 0.2 μm filter and store aliquots in the dark at 4 °C.

3.2 Preparation of Tissue Culture Media

1. Prepare stock solutions of IAA (1 mg/ml), 2,4-D (1 mg/ml), 2-iP (1 mg/ml), BAP (1 mg/ml), Biotin (1 mg/ml), Calcium pantothenate (1 mg/ml), Glycine (1 mg/ml), IBA (1 mg/ml), KIN (1 mg/ml), Nicotinic acid (1 mg/ml), Pyridoxine HCl (1 mg/ml), Thiamine HCl (1 mg/ml), adenine sulfate (10 mg/ml), L-cysteine (1 mg/ml).
2. Place a beaker containing 400 ml tissue culture grade water on a hot plate shaker and mix:

   (a) For 1 L of callus induction medium: half-strength MS salts, 10 ml thiamine HCl, 1 ml pyridoxine HCl, 1 ml nicotinic acid, 1 ml glycine, 100 mg myo-inositol, 100 mg casein hydrolysate, 400 mg malt extract, 0.5 ml 2,4-D, 1 ml IBA, 2 ml 2-iP, 30 g sucrose, 2 g Phytagel™, pH 5.6.

   (b) For 1 L of embryo induction medium: half-strength MS salts, 20 ml thiamine HCl, 20 ml glycine, 40 ml L-cysteine, 200 mg myo-inositol, 6 ml adenine hemisulfate salt, 200 mg casein hydrolysate, 800 mg malt extract, 1 ml 2,4-D, 4 ml BAP, 30 g sucrose, 2 g Phytagel™, pH 5.6.

   (c) For 1 L of liquid proliferation medium (CP): half-strength MS salts, 5 ml thiamine HCl, 0.5 ml pyridoxine HCl, 0.5 ml nicotinic acid, 10 m L-cysteine, 50 mg myo-inositol, 100 mg casein hydrolysate, 200 mg malt extract, 2 ml 2,4-D, 1 ml KIN, 30 g sucrose, pH 5.6.

   (d) For 1 L of regeneration medium (R): half-strength MS salts, 10 ml thiamine HCl, 1 ml pyridoxine HCl, 1 ml nicotinic acid, 2 ml glycine, 200 ml myo-inositol, 4 ml adenine hemisulfate salt, 400 mg casein hydrolysate, 400 mg malt extract, 4 ml BAP, 40 g sucrose, 2.5 g Phytagel™, pH 5.6.

   (e) For 1 L of germination medium (EG): half-strength MS salts, 8 ml thiamine HCl, 3.2 ml pyridoxine HCl, 100 mg myo-inositol, 0.45 ml IAA, 0.25 ml BAP and 2.5 g Phytagel™, pH 5.6.

   (f) For 1 L of development medium (DEV): full-strength MS salts 1 ml thiamine HCl, 1 ml pyridoxine HCl, 1 ml nicotinic acid, 1 ml calcium pantothenate, 0.01 ml biotin, 100 mg myo-inositol, 0.3 ml BAP, 30 g sucrose, 2.5 g Phytagel™, pH 5.6.

   (g) For 1 L of TEX medium (Teixeira et al. 2004): half-strength MS salts, 10 ml thiamine HCl, 1 ml pyridoxine HCl, 1 ml glycine, 250 mg citric acid, 10 ml L-cysteine, 100 mg/L myo-inositol, 100 mg/L casein hydrolysate, 200 mg/L malt extract, 1 ml 2,4-D, 1 ml IBA, 2 ml 2-iP, 20 g sucrose, pH 5.6.

3. While stirring, add tissue culture grade water to a final volume of 1,000 ml.
4. Stir until the solution is homogenous and clear.
5. Calibrate the pH meter as per manufacturer instructions.
6. While stirring, adjust medium to pH 5.6 using 1 N NaOH or 1 N HCl.
7. For semi-solid medium, add the gelling agent and heat while stirring until the solution is homogenous and clear.

8. Dispense the culture medium in the respective culture vessel before or after autoclaving (depending on the application).
9. Sterilize all media in a validated autoclave at 1 kg/cm^2 for 21 min at 121 °C.
10. Allow the medium to cool prior to use.
11. Store the medium for up to a week in a cold room.

3.3 Germination of Coffee Zygotic Embryos Under In Vitro Conditions

1. Collect mature cherries from genetically homogenous mother plants maintained in the greenhouse or in the field (see Note 4).
2. Remove the pulp, the mucilage, and the parchment by hand.
3. Soak the seeds for 24 h in distilled water with two drops of Tween 20 with orbital rotation.
4. Disinfect the seeds with 3.5% (v/v) sodium hypochlorite for 1 h and finally rinse three times with sterile distilled water.
5. Remove the endosperm over the embryo using a scalpel and forceps and extract the embryo levering it out from the root pole with the same scalpel blade.
6. Culture the zygotic embryos in test tubes containing 20 ml of germination medium (GER) and place them in the dark at 27 ± 2 °C for 8 weeks.
7. Transfer the in vitro plantlets developed from these embryos to baby food jars with 20 ml of the above medium under a 16 h light photoperiod at 27 ± 2 °C.
8. Subculture the in vitro plantlets to fresh development medium (DEV) every 90 days.

3.4 Protocol for Plant Regeneration via Somatic Embryogenesis

This protocol involves a series of sequential stages: callus formation with embryogenic structures; establishment and multiplication of embryogenic suspension cultures; formation, maturation, and germination of somatic embryos; and conversion to plants; field evaluation (van Boxtel and Berthouly 1996) (see Fig. 1).

3.4.1 Embryogenic Callus Culture Initiation

1. Take the first and second completely developed leaf from the in vitro plantlets and cut leaf sections measuring 0.5 cm^2, without the midvein and the margins.
2. Culture the leaf pieces with the abaxial surface upwards on 20 ml of callus induction medium (van Boxtel and Berthouly 1996) contained in baby food jars for 4 weeks in the dark at 27 ± 2 °C.

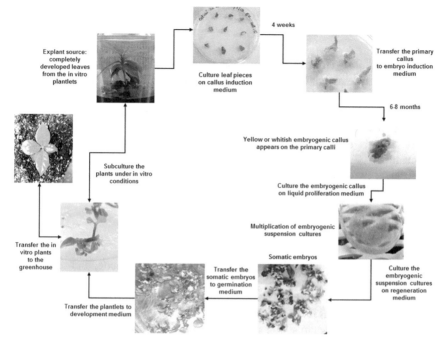

Fig. 1 Schematic representation of the steps for indirect somatic embryogenesis in coffee (*Coffea arabica* L. var. Catuaí)

3. Then, transfer the necrotic primary callus and explants to baby food jars containing 20 ml of the embryo induction medium (van Boxtel and Berthouly 1996).

4. Incubate the baby food jars during 6–8 months under a 16 h low-light photoperiod (10 µE/m²/s) at 27 ± 2 °C until yellow or whitish embryogenic callus appears on the primary calli that have initially developed on the cut edges.

3.4.2 Establishment of Embryogenic Suspension Cultures

1. Weigh 250 mg of friable embryogenic callus and transfer to 25 ml liquid proliferation medium CP (van Boxtel and Berthouly 1996) in 250 ml Erlenmeyer flasks.

2. Incubate the Erlenmeyer flasks on a gyratory shaker at 100 rpm at 27 ± 2 °C in the dark.

3. Every 15 days, replace the old medium with 50 ml fresh liquid proliferation medium.

3.4.3 Regeneration of Somatic Embryos and Development into Plantlets

1. Culture 250 mg fresh weight of suspension cultures in Petri dishes (100 mm × 20 mm) containing 20 ml of regeneration medium (R) (van Boxtel and Berthouly 1996).
2. Incubate the petri dishes under a 16 h light photoperiod at 27 ± 2 °C for 8–10 weeks.
3. Transfer the somatic embryos to baby food jars containing 20 ml of germination medium (GER) with 16 h light photoperiod at 27 ± 2 °C for 6–8 weeks.
4. Transfer the plantlets to baby food jars containing 20 ml development medium (DEV) with 16 h light photoperiod at 27 ± 2 °C.

3.4.4 Hardening of In Vitro Plantlets in the Greenhouse

1. Do a transverse cut in the stem of the plants, below a node, and remove the leaves near the cut.
2. Place the freshly cut basal part of the stem in a solution of indoleacetic acid (IAA) (1 mg/mL) for 30 s.
3. Plant the seedlings in sterile substrate (peat moss with perlite) in plastic boxes (30 cm × 20 cm × 10 cm).
4. Cover the boxes with plastic and place in a growth room with controlled conditions (25 °C, photoperiod of 12 h).
5. Two weeks later, make small holes in the plastic covering of the boxes.
6. After five weeks, evaluate the rooting percentage and transfer the plants with emerged roots to the greenhouse. Keep the rest of the plants under controlled growth conditions for up to 8 weeks.
7. At the end of 8 weeks, determine the percentage of rooting and transfer those plants with roots to the greenhouse.
8. In the greenhouse, place the plants in polyethylene bags with a 3:1 mixture of substrate (peat moss with perlite: coconut fiber). Plant two plants per bag and identify according to the original numbering.
9. After 3 weeks, fertilize with granular slow-release fertilizer (e.g., Osmocote 14-14-14).
10. Irrigate the plants twice a week according to the climatic conditions of the greenhouse and the water requirement of the crop.

3.5 Determination of the Viability of the Embryogenic Calli

1. Weigh 100 mg of embryogenic calli sample.
2. Place the sample in a 2 ml reaction tube and add 1 ml of the TTC stock solution.
3. Incubate the samples for 24 h in the dark at 37 °C without shaking.

4. Remove the TTC solution by decanting or centrifugation and wash the sample with distilled water.
5. Add 1 ml of 95% (v/v) ethanol.
6. Extract the formazan by placing the samples in a water bath at 65 °C for 10 min with constant shaking.
7. Centrifuge the sample at 2,500 rpm and recover the supernatant.
8. Quantify absorbance at 490 nm in a spectrophotometer.

3.6 Sodium Azide Dose Determination

1. Review safety procedures of the chemical mutagenesis laboratory (see Notes 2 and 3).
2. Autoclave all non-disposable materials (e.g. sieves, forceps).
3. Prepare a fresh 500 mM NaN_3 stock solution by adding the required amount of NaN_3 to the phosphate buffer (100 mM, pH 3.0). Sterilize by filtering through a 0.2 μm filter using a sterile syringe.
4. Discard the syringe and filter in the hazardous waste.
5. Place 200 mg embryogenic calli in a 15 ml reaction tube with 10 ml of TEX medium (pH 3.0) (Teixeira et al. 2004).
6. In a laminar flow cabinet, add appropriate concentrations of NaN_3 solution. Shake the solution vigorously. Example dilutions and concentrations of NaN_3 are given in Table 1.
7. Label each tube with the appropriate treatment code (concentration and incubation time).
8. Place closed tubes in the dark at 27 ± 2 °C on a rotary shaker set at 100 rpm for 15 min.
9. After the incubation time, quickly but carefully decant each of the treatment batches and rinse the treated embryogenic suspension cultures with 10 ml of TEX medium (pH 5.6). This step and any subsequent steps must be carried out in a laminar flow cabinet.
10. Repeat washing step 3 times.
11. Collect all the liquid waste in a dedicated bucket labelled as hazardous waste. Dispose of toxic waste according to local regulations.
12. Add 10 ml of TEX medium and maintain the cultures for 24 h in the dark with constant shaking at 100 rpm and 27 ± 2 °C.
13. Record the absorbance and express the formazan content as a percentage of a positive control [(sample absorbance/absorbance of the control) $\times$ 100] as described in Sect. 3.5 and observe the cells using a stereoscope (see Note 5).
14. Calculate the survival rate as follows: [(the number of survival explant/the number of treated explant) * 100].
15. Alternatively, measure the absorbance values of the treated and non-treated embryogenic suspension cultures and determine survival rate.

16. Record the data for each treatment and enter it into a spreadsheet (e.g., Microsoft Excel).

17. Plot percentage of control against mutagenic treatment and estimate the mutagen concentration required to obtain 50% of viability compared to the control (LD_{50}) (see Fig. 2) (see Note 7).

18. Identify concentrations suitable for bulk mutagenesis of your material according to the LD_{50} previously estimated.

Table 1 NaN$_3$ concentrations chosen for the toxicity test in coffee embryogenic calli

NaN$_3$ concentration (mM)	TEX medium (ml)	500 mM NaN$_3$ (µl)[a]
0 (control; pH 5.6)	10	–
0 (control; pH 3.0)	10	–
2.5	9.95	50
5.0	9.90	100
7.5	9.85	150
10.0	9.8	200

[a] Prepare a fresh solution immediately before the experiment

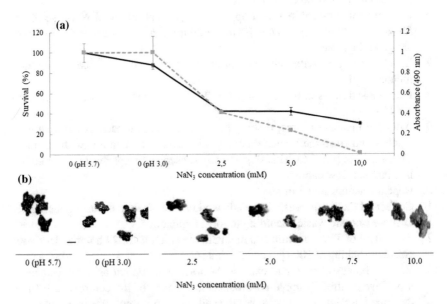

Fig. 2 Effect of NaN$_3$ concentration on survival and viability of coffee (*C. arabica* L. var. Catuaí) embryogenic calli. **a** Survival percentage (solid line) and absorbance (490 nm) (dotted line) versus NaN$_3$ concentrations. Each value represents the mean ± SD of two repetitions, **b** cell viability versus NaN$_3$ concentrations

3.7 *Ethyl Methanesulphonate Dose Determination*

1. Review safety procedures of the chemical mutagenesis laboratory (see Notes 2 and 3).
2. Autoclave all non-disposable materials (e.g. sieves, forceps).
3. Sterilize EMS by filtering through a 0.2 μm filter using a sterile syringe and 2 ml collection tube.
4. Discard the syringe, collection tube, and filter in the hazardous waste.
5. Place 200 mg embryogenic calli in a 15 ml reaction tube with 10 ml of TEX medium (Teixeira et al. 2004).
6. In a laminar flow cabinet, add appropriate concentrations of the EMS solution. Shake the solution vigorously. Example dilutions and concentrations for EMS are given in Table 2 (see Note 6).
7. Label each tube with the appropriate treatment code (concentration and incubation time).
8. Place closed tubes in the dark at 27 ± 2 °C on a rotary shaker set at 100 rpm and start incubation time.
9. After the incubation time, quickly but carefully decant each of the treatment batches and rinse the treated embryogenic calli with 10 ml of TEX medium (pH 5.6). This step and any subsequent steps must be carried out in a laminar flow cabinet.
10. Repeat washing step 3 times.
11. Collect all the liquid waste in a dedicated bucket labelled as hazardous waste. Dispose of toxic waste according to local regulations.
12. Add 10 ml of TEX medium and maintain the cultures for 24 h in the dark with constant shaking (100 rpm) at 27 ± 2°C.
13. Determine cell viability of embryogenic calli and observe the cells using a stereoscope as described in Sect. 3.5 (see Note 5).
14. Calculate the survival rate as follows: [(the number of survival explant/the number of treated explant) * 100]. Alternatively, measure the absorbance values of the treated and non-treated embryogenic suspension cultures and determine survival rate.
15. Record the data for each treatment and enter it into a spreadsheet (e.g. Microsoft Excel).
16. Plot percentage of control against mutagenic treatment and estimate the mutagen concentration required to obtain 50% of viability compared to the control (see Fig. 3) (see Note 7).
17. Identify concentrations suitable for bulk mutagenesis of your material according to the LD_{50}.

Table 2 EMS concentrations and incubation time chosen for the toxicity test in coffee embryogenic calli

EMS concentration (mM)	TEX medium (ml)	EMS (μl)	Incubation time (min)
0 (control)	10	–	20, 40, 60, 120, 150
5.0	10	5.4	20, 40, 60, 120, 150
10.0	9.99	10.8	20, 40, 60, 120, 150
15.0	9.99	16.3	20, 40, 60, 120, 150
20.0	9.98	21.7	20, 40, 60, 120, 150
80.0	9.91	87.0	60
100.0	9.89	108.5	60
120.0	9.87	130.0	60
140.0	9.85	152.0	60
185.2	9.8	200.0	60, 120
370.5	9.6	400.0	60, 120
555.7	9.4	600.0	60, 120
741.0	9.2	800.0	60, 120

[a] Prepare fresh immediately before experiment

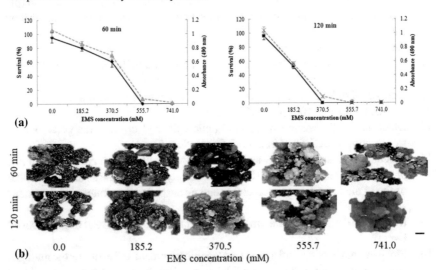

Fig. 3 Effect of EMS concentration on survival and viability of coffee (*C. arabica* L. var. Catuaí) embryogenic calli. **a** Survival percentage (solid line) and absorbance (490 nm) (dotted line) after 60 min and 120 min of exposure time to different EMS concentrations. Each value represents the mean ± SD of two repetitions. **b** Cell viability versus EMS concentrations. *Bar*, 0.5 cm

3.8 Bulk Mutagenesis

1. Autoclave all non-disposable materials (e.g. sieves, forceps).
2. Choose appropriate NaN_3 or EMS concentration and incubation time based on the results obtained from the chemical toxicity test.
3. See Fig. 4 for an overview of the bulk mutagenesis procedure.
4. Prepare embryogenic calli per each treatment chosen.
5. Place 200 mg one-month-old embryogenic suspension cultures in a 15 ml reaction tube with 10 ml of TEX medium (Teixeira et al. 2004).
6. Label each tube with NaN_3 and EMS concentration and incubation time.
7. Transfer closed tubes containing in vitro material into the chemical mutagenesis laboratory.
8. In a laminar flow cabinet, add appropriate concentrations of NaN_3 or EMS. Shake the solution vigorously.
9. Place closed tubes in the dark at 27 ± 2 °C on a rotary shaker set at 100 rpm for the chosen length of time.
10. After incubation, quickly but carefully decant each of the treatment batches and rinse the treated embryogenic suspension cultures with 10 ml of TEX medium (pH 5.6). This step and any subsequent steps must be carried out in a laminar flow cabinet.
11. Repeat washing step 3 times.
12. Collect all the liquid waste in a dedicated bucket labelled as hazardous waste. Dispose of toxic waste according to local regulations.
13. After the final wash, add 10 ml of liquid regeneration medium (R) (pH 5.6) and maintain the cultures for 48 h in the dark on a rotary shaker (100 rpm) at 26 ± 2 °C.
14. Carefully decant the treated embryogenic suspension cultures and plate them in Petri dishes containing 20 ml of R semisolid medium (see Note 8).
15. Incubate the Petri dishes under a 16 h light photoperiod at 27 ± 2 °C for 8–10 weeks.
16. Transfer the torpedo shape somatic embryos to baby food jars containing 20 ml of germination medium (GER) with 16 h light photoperiod at 27 ± 2 °C for 6–8 weeks (see Note 8).
17. Cut off cotyledons and roots of the developed plantlets.
18. Transfer the plantlets to baby food jars containing 20 ml of development medium (DEV) with 16 h light photoperiod at 27 ± 2 °C.
19. Acclimatize plantlets with 4 leaves and 3–4 cm tall in the greenhouse as described in Sect. 3.4.4.

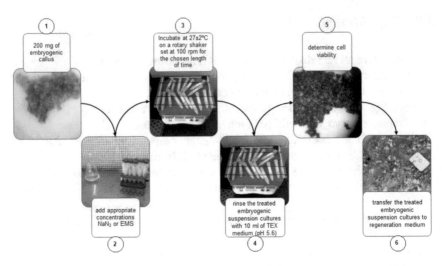

Fig. 4 Bulk mutagenesis process under in vitro conditions of embryogenic calli of coffee (*C. arabica* L. var. Catuaí)

4   Notes

1. This protocol has been established using *Coffea arabica* L. var. Catuaí seeds. It can be used as a reference for other Arabica coffee varieties, nevertheless, it is recommended to optimize the mutagenic parameter (NaN$_3$ and EMS concentration and incubation time) for each variety used.

2. All the mutagenesis experiments should be conducted in a dedicated chemical mutagenesis laboratory equipped with a ducted fume hood, toxic waste disposal and decontamination procedures.

3. Read the Materials Safety Data Sheet (MSDS) of materials being used and follow the recommendation of the manufacturer. Pay careful attention to the information on sodium azide and EMS and what to do in case of exposure. It is very important to wear personal protective equipment (gloves must be compatible with chemical mutagens, for instance PVC or neoprene gloves); safety glasses with side shields or chemical goggles; lab coat, closed-toe shoes, shoe protections, and long trousers. A double glove system is advised.

4. Coffee seeds loose viability rapidly if not properly stored. Therefore, it is recommended to use freshly harvested seeds or otherwise to store the seeds between 10 and 12% moisture at 15 °C for not longer than 3 months.

5. Tetrazolium chloride staining has been used to evaluate the cell viability of embryogenic suspension cultures of grapevine treated with EMS (Acanda et al. 2014). In our study, there was a robust correlation between the survival percentage and the absorbance measured (NaN$_3$, r^2: 0.97; EMS, r^2: 0.9) indicating that both methods could be used to determine cell viability.

6. In our study, embryogenic suspension cultures treated with 5, 10, 15, 20, 80, 100, 120, and 140 mM EMS, did not show a reduction in viability of more than 15% compared to the non-treated embryogenic suspension cultures.

7. Optimal dose determination of mutagenic treatment of in vitro cultures must be based on quantitative analysis of viability of cells exposed to the mutagen. When using embryogenic cultures, variables such as survival of cells and regeneration capacity after mutagenic treatment must be assessed to establish the dosage curve. Dual tests that allow qualitative and quantitative viability analysis are advised. One of the most widely used assays in in vitro cultures is the conversion of 2,3,5-triphenyltetrazolium (TTC) to its reduced form 1,3,5-triphenylformazan (TPF) in metabolically active (living) tissues. This process generates a red colored precipitate that can be easily detected and quantified.

8. Somatic embryos regenerated from mutagenized tissues can show germination and growth delay, germination inhibition can also appear.

Acknowledgements Funding for this work was provided by the University of Costa Rica, the Ministerio de Ciencia, Innovación, Tecnología y Telecomunicaciones (MICITT), the Consejo Nacional para Investigaciones Científicas y Tecnologicas (CONICIT) (project No. 111-B5-140; FI-030B-14) and the Cátedra Humboldt 2023. A. Gatica-Arias acknowledged the Cátedra Humboldt 2023 of the University of Costa Rica for supporting the dissemination of biotechnology for the conservation and sustainable use of biodiversity.

References

Acanda Y, Martínez O, Prado MJ, González MV, Rey M (2014) EMS mutagenesis and qPCR-HRM prescreening for point mutations in an embryogenic cell suspension of grapevine. Plant Cell Rep 33:471–481. https://doi.org/10.1007/s00299-013-1547-6

Anthony F, Combes MC, Astorga C, Bertrand B, Graziosi G, Lashermes P (2002) The origin of cultivated *Coffea arabica* L. varieties revealed by AFLP and SSR markers. Theor Appl Gent 104. http://doi.org/10.1007/s00122-001-0798-8

Campos NA, Panis B, Carpentier SC (2017) Somatic embryogenesis in coffee: the evolution of biotechnology and the integration of omics technologies offer great opportunities. Front Plant Sci 8. https://doi.org/10.3389/fpls.2017.01460

Food and Agriculture Organization of the United Nations (2018) FAOSTAT database. FAO, Rome, Italy. Retrieved 23 Feb 2018 from www.fao.org/faostat/en/#search/coffee

Gatica-Arias AM, Arrieta G, Espinoza AM (2008) Plant regeneration via indirect somatic embryogenesis and optimisation of genetic transformation in *Coffea arabica* L. cvs. Caturra and Catuaí. Electron J Biotechnol 11:1–12

Hendre P, Aggarwal R (2007) DNA markers: development and application for genetic improvement of coffee. In: Varshney R, Tuberosa R (eds) Genomics-assisted crop improvement, vol 2. Springer, Dordrecht, NL, pp 399–434

Quiroz-Figueroa FR, Fuentes CFJ, Rojas R, Loyola V (2002) Histological studies on the developmental stages and differentiation of two different somatic embryogenesis systems of *Coffea arabica*. Plant Cell Rep 20:1141–1149

Serrat X, Esteban R, Guibourt N, Moysset L, Nogués S, Lalanne E (2014) EMS mutagenesis in mature seed-derived rice calli as a new method for rapidly obtaining TILLING mutant populations. Plant Methods 10. https://doi.org/10.1186/1746-4811-10-5

Teixeira JB, Junqueira CS, Pereira JPC, Mello S, Silva PD, Mundim DA (2004) Multiplicação clonal de café (*Coffea arabica* L.) via embryogenesis somática. Embrapa Recursos Genéticos e Biotecnologia, Brasília (Embrapa Recursos Genéticos e Biotecnologia. Documentos, 121). Disponible en: http://www.cenargen.embrapa.br/publica/trabalhos/doc121.pdf

Van Boxtel J, Berthouly M (1996) High frequency somatic embryogenesis from coffee leaves: factors influencing embryogenesis, and subsequent proliferation and regeneration in liquid medium. Plant Cell Tissue Organ Cult 44(1):7–17

Xu C, Xiao J, He J, Hu G, Chen H (2011) The effect of ethyl methane sulphonate (EMS) and sodium azide (NaN_3) on plant regeneration capacity of an embryogenic cell suspension of 'Yueyoukang 1' (Musa, aaa), a banana cultivar resistant to fusarium wilt. Acta Hortic 897:301–302. https://doi.org/10.17660/ActaHortic.2011.897.41

Chemical Mutagenesis of *Coffea arabica* L. var. Venecia Cell Suspensions Using EMS

Joanna Jankowicz-Cieslak, Florian Goessnitzer, and Ivan L. W. Ingelbrecht

Abstract Arabica coffee is widely grown in Latin America where it is under threat of leaf rust, a fungal disease caused by *Hemileia vastatrix*. As a perennial crop, conventional breeding of Arabica coffee is challenged by its long juvenile period and narrow genetic base. Plant mutants are important resources for crop breeding and functional genomics studies. The ethylating agent ethyl methanesulfonate (EMS) is widely used for inducing random point mutations. In a wide range of species, treatment with EMS causes GC-to-AT transitions with great efficiency. These properties, combined with ease of use, make EMS a mutagen of choice for induced mutagenesis. In vitro cell and tissue culture integrated with mutation induction provide an attractive approach for broadening the genetic base and breeding purposes, especially for perennial crops such as Arabica coffee. Embryogenic cell cultures are suitable targets for mutation induction and can accelerate the development of chimera-free mutant plantlets. Here we describe a robust protocol for EMS mutagenesis of embryogenic cell suspensions of *Coffea arabica* var. Venecia. Dose-response curves were established within 3–4 weeks and showed LD_{30} and LD_{50} values in the range of 0.5% and 0.6% EMS respectively. Methods and media used for development of the treated cell suspensions and conversion to in vitro plantlets are also described.

Joanna Jankowicz-Cieslak, Florian Goessnitzer—Contributed equally.

J. Jankowicz-Cieslak (✉) · F. Goessnitzer (✉) · I. L. W. Ingelbrecht
Plant Breeding and Genetics Laboratory, Joint FAO/IAEA Centre of Nuclear Techniques in Food and Agriculture, IAEA Laboratories Seibersdorf, International Atomic Energy Agency, Vienna International Centre, Vienna, Austria
e-mail: j.jankowicz@iaea.org

F. Goessnitzer
e-mail: f.goessnitzer@iaea.org

© The Author(s) 2023
I. L. W. Ingelbrecht et al. (eds.), *Mutation Breeding in Coffee with Special Reference to Leaf Rust*, https://doi.org/10.1007/978-3-662-67273-0_8

113

1 Introduction

Coffee is one of the most valuable cash crops and provides employment for millions of people worldwide, especially in Latin America and parts of Africa and Asia (FAOSTAT 2021). Coffee belongs to the family *Rubiaceae* and the two main species of cultivated coffee are *Coffea arabica* L. and *Coffea canephora*. Coffee leaf rust (CLR) caused by the airborne fungus *Hemileia vastatrix* and coffee berry disease are among the most important diseases affecting coffee production. *C arabica* is the most severely affected by leaf rust. Leaf rust epidemic has hit countries in Mesoamerica, including Colombia, Peru, Ecuador, and Guatemala amongst others, in the past decade (Avelino et al. 2015).

Resistant varieties are perhaps the most appropriate means to manage CLR. Improvement of Arabica coffee using conventional cross breeding is challenged by its long juvenile phase and narrow genetic base (Wintgens 2012; Scalabrin et al. 2020). Induced mutagenesis is widely used an efficient method to induce genetic variability useful for genetic studies and breeding. Since the 1970s in vitro tissue culture technologies have been developed for coffee, including methods to regenerate plants from single cells through somatic embryogenesis (see Etienne et al. 2018 for a review). Both direct and indirect methods for somatic embryogenesis in Arabica coffee have been described (Quiroz-Figueroa et al. 2006; Murvanidze et al. 2021 and references therein). Single cells or cell clusters are attractive targets for mutagenesis given the high likelihood for directly yielding chimera-free, homohistont plants. In addition, in vitro systems could offer significant efficiency gains in terms of space and labour compared to greenhouse- or field-based experiments to establish large mutant populations for perennial crops and trees such as Arabica coffee. Here, a fast and reproducible protocol for EMS mutagenesis of embryogenic cell suspensions of Arabica coffee var. Venecia is presented. Protocols for converting the EMS treated somatic embryos into in vitro plantlets are also provided.

2 Materials

2.1 Culture Medium

1. Analytical balance.
2. Weighing trays.
3. Spatula.
4. Magnetic stir bar.
5. Hot plate.
6. pH meter.
7. Medium dispenser.
8. Forceps.
9. Surgical Blades.

Table 1 Media composition (mg/l) for somatic embryogenesis and plantlet regeneration of *Coffea arabica* var. Venecia

Component	CMA1	M5
	mg/L	mg/L
MS macronutrients	MS/2	(Sigma M6899) 4.4 g/l
$FeSO_4 \cdot 7H_2O$	27.8	–
Na_2-EDTA	37.3	–
B5 (Duchefa; G0415.0250)	56	–
Sucrose	15,000	30,000
2,4-D (1 mM)	5	–
IBA (1 mM)	5	–
2ip (1 mM)	10	–
BAP (1 mM)	0	0.5
Kinetin (1 mM)	5	–
NAA (1 mM)	–	0.1
Casein hydrolysate	100	–
Malt	200	–
L-Cystein	10	24
Gelrite	–	3,000
pH	**5.6**	**5.7**

10. Aluminium foil.
11. Sterile culture tubes (50 ml).
12. In vitro culture test tubes (30 ml).
13. 10 cm Petri-dish with vents.
14. Culture vessels for liquid media.
15. Laminar flow bench.
16. In vitro growth room.
17. Tissue culture grade water.
18. Gelling agent (e.g., Gelrite).
19. Coffee culture media components (Table 1).

2.2 Chemical Toxicity Test

1. Coffee cell suspensions (see Note 1).
2. Chemical mutagenesis laboratory equipped with fume hood and flow bench (see Note 2).
3. Labelled waste receptacle for dry hazardous material and collection vessels for EMS waste solution (see Note 3).

4. Ethyl-methanesulphonate (EMS) AR grade, M.W. 124.2 (see Note 4).
5. 10% (w/v) sodium thiosulfate ($Na_2S_2O_3.5H_2O$) (see Note 5).
6. Sterile deionized water.
7. 50 ml falcon tubes.
8. Syringe.
9. Needle.
10. Sterile membrane filter for filtering EMS solution: 25 mm diam., 0.2 μm pore size.
11. Pipette bulb.
12. Graduated cylinders.
13. Bottles (100 ml, 500 ml).
14. Beakers (500 ml and 1,000 ml).
15. Orbital shaker.
16. Disposable pipettes (5 ml, 25 ml).

2.3 Calculation of Lethal Dose (LD)

1. Pen.
2. Notebook.
3. Ruler
4. Standard spreadsheet software e.g., Microsoft Excel.
5. Camera.
6. Photobooth (optional).

3 Methods

3.1 Preparation of Liquid Culture Medium

1. Prepare MS, growth regulator and chemical stock solutions according to common procedures (concentrations shown in Table 1).
2. Filter sterilize all stock solutions.
3. Dispense into 50 ml batches and freeze for further use. Store the working solution at 4 °C.
4. Autoclave culture vessels for CMA1 medium before dispensing liquid medium.
5. Take desired amounts of solutions and chemicals (Table 1) and mix well.
6. Place the media on the mixer and let it mix properly.
7. Calibrate the pH meter as per manufacturer instructions.
8. While stirring, adjust medium to pH 5.8 using NaOH and HCl.
9. Autoclave for 20 min at 120 °C.
10. Allow the medium to cool.
11. Dispense liquid CMA1 medium into previously autoclaved culture vessels.

12. Dispense M5 medium in culture test tubes after autoclaving.
13. Store the medium in a cold room.

3.2 Mutagenesis of Coffee Cell Suspensions: Chemical Toxicity Test

1. Prepare sufficient suspension cultures to perform the mutagenic treatment. Procedures for obtaining cell suspensions can be found in Chap. "Somatic Embryogenesis and Temporary Immersion for Mass Propagation of Chimera-Free Mutant Arabica Coffee Plantlets".
2. Filter 5–6 weeks old cell suspension cultures through 0.5 × 0.5 mm mesh.
3. Transfer filtered suspension to fresh CMA1 medium with an end volume of 70 ml.
4. Review safety procedures of the chemical mutagenesis laboratory and Consult the Materials Safety Data Sheets for all chemicals used.
5. Prepare the laboratory (see Note 2).
6. Choose appropriate concentrations of EMS solution and incubation times for the mutagenesis of coffee cell cultures (see Note 6).
7. Prior to working with EMS, test the bottles and falcon tubes used for mixing EMS and for the mutagenic treatment of cell suspension cultures (see Note 7).
8. Prepare 50 ml falcon tubes containing cell suspensions in final volume of 15 ml. If necessary, dilute the cultures. Here, 3 ml of filtered cell cultures were added to 12 ml of CMA1 medium. Prepare at least 3 replicates per each treatment combination.
9. Transfer culture tubes to the chemical mutagenesis laboratory. Care should be taken not to expose the cultures to unfavorable conditions while transferring to the facility where the mutagenesis will be performed.
10. Prepare a bottle containing 100 mM sodium thiosulfate and place in the fume hood along with a role of paper towels.
11. In the fume hood, prepare fresh 10% EMS dilution by adding the required volume of EMS into the bottle containing autoclaved water (Fig. 1a). Use a sterile syringe and a 0.2 μm filter for this step. Place syringe and filter into a beaker containing 100 mM sodium thiosulfate to inactive EMS before placing in hazardous waste (see Note 8).
12. Seal the bottle of prepared 10% EMS dilution with a screw cap.
13. Wipe the outside of stock bottles with a paper towel soaked in sodium thiosulfate.
14. Ensure the sash is lowered on the fume hood and shake the solution vigorously for 15 s (see Note 9).
15. In the fume hood, prepare 50 ml Falcon tubes containing appropriate volume of CMA1 medium for each of the EMS concentration. The volume of the culture is calculated for all replicates.
16. Pipette the appropriate volume of 10% EMS (in water) into the tubes labelled with the respective concentration of EMS/treatment combination (Fig. 1b).

17. Mix the dilutions of EMS/culture medium properly.
18. Carefully distribute the determined volumes (here 5 ml) of EMS solution to every falcon tube containing cell suspensions (Fig. 1c). Ensure that all the volumes are being calculated properly, so that the final concentration of EMS corresponds to the one you wish to apply. In the chemical toxicity experiment described here following EMS concentrations were used: 0.2%, 0.5%, 0.8%, 1.5% and 2% EMS. In this experiment, the final reaction volume after adding the EMS dilution was 20 ml.
19. Prepare control batch in the same way by adding culture medium to the falcon tubes containing cell suspensions.
20. Wipe the outside of falcon tubes containing EMS with a paper towel soaked in sodium thiosulfate.
21. Place falcon tubes (including control) on a rotary shaker set at 60 rpm, record the time and temperature, and start incubation (Fig. 1d and see Note 11). One hour incubation was chosen for the experiment described in this protocol.
22. Pour 100 mM sodium thiosulfate into bottles used to prepare EMS dilutions.
23. Dispose of liquid and solid waste in appropriate toxic waste containers.
24. Wipe the fume hood with a paper towel soaked in sodium thiosulfate.
25. Fifteen minutes before the end of the incubation time, transfer falcon tubes into the fume hood and let the cells sediment. A clear pellet of mutagenized coffee cells should be visible at the bottom of the falcon tube (Fig. 1e).
26. Carefully decant each of the treatment batches into a waste beaker (Fig. 1f and see Note 12).
27. Wash each coffee cell pellet with 20 ml of CMA1 medium. Pour the EMS-medium solution off to the waste beaker (Fig. 1g, h).
28. Repeat the wash steps for a total of three washes (see Note 13).
29. After the final wash, add appropriate volume of culture medium to the treated cells. Here, 3 ml of CMA1 were added to each falcon tube containing treated, as well as non-treated cells.
30. Collect all the liquid waste in a dedicated bucket labelled as hazardous waste.
31. Detoxify the waste and all unused EMS solution by adding sodium thiosulfate in a 3:1 ratio by volume.
32. Dispose of toxic waste according to local regulations. Decontaminate all surfaces and equipment by wiping down with 100 mM sodium thiosulfate followed by a water rinse (see Note 14).
33. Move mutagenized cell cultures into the flow cabinet and immediately proceed with transferring 100 μl of treated suspension cultures to test tubes containing freshly prepared M5 medium (Fig. 1i and see Note 15). Here, 90 culture tubes containing 100 μl cells each were prepared per treatment combination.
34. Place tubes in the culture room with 16 h light and a temperature of 28 ± 1 °C. Observe daily for any changes in the color of cells or growth.

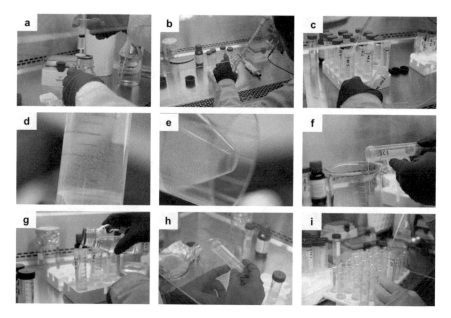

Fig. 1 Procedure for EMS chemical mutagenesis of in vitro coffee cell suspensions. **a** A 10% EMS stock mixture is prepared under the fume hood. **b** Final dilutions are prepared by adding appropriate volumes of the 10% EMS stock solution into the cell culture maintenance media. **c** The EMS dilutions are aliquoted into 50 ml falcon tubes containing 20 ml cell suspensions, replicates of 3 per treatment are prepared. **d** Cells are incubated for a specific time (here 1 h) under orbital rotation. **e** Shortly before the end of the incubation, falcon tubes containing mutagenized cells are removed from the shaker and put aside to allow cells to settle. **f** The supernatant is carefully decanted, not to lose the pellet. **g**, **h** Washing of mutagenized material with 40 ml maintenance liquid media, repeated at least 3 times. **i** After the washing step, a set volume of culture media is added to the mutagenized material. Tubes are transferred to the in vitro laboratory and 100 μl aliquots of the mutagenized cells are transferred to the regeneration media

3.3 Calculation of Lethal Dose (LD)

1. Monitor the cell growth daily.
2. Visible differences between mutagenic treatments can be observed approximately 2 weeks after mutagenesis.
3. Let the cells/cell clusters grow until 3–4 weeks post treatment when the scoring can be taken (Fig. 2).
4. Count tubes where cell growth is clearly visible.
5. Calculate survivability and graph data (Figs. 3 and 4).
6. Repeat the chemical toxicity test if the results and data obtained are not precise enough.
7. Choose one or more concentrations for bulk mutagenesis following this protocol. While making the selection, consider the values calculated for LD_{30} and LD_{50} (Fig. 4).

Fig. 2 Example data showing the response of Arabica coffee embryogenic cell suspension treated with different EMS concentrations, observed 3 weeks post EMS treatment. Growth inhibition of 100% was observed for cultures treated with 1.5% and 2% EMS

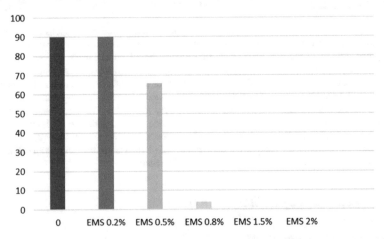

Fig. 3 Example of the survival count of cultured coffee cells taken 3 weeks after treatment with EMS. For the control material, 90/90 cultured tubes maintained growth. In the case of treated cultures, 0.2% EMS had similar growth rate to the control, a slight drop is being observed for 0.5% EMS treated cultures (66/90 tubes survived). A clear drop occurred for the material subjected to 0.8% EMS for which only 4 out of 90 cultured cell suspension tubes maintained the growth

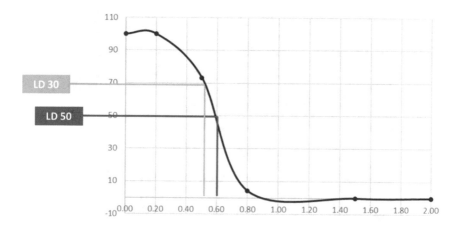

Fig. 4 Survival calculated as percentage of the control, whereby the control is 100%. The kill curve indicates the LD_{30} (the dose causing the death of 30% of the population) and LD_{50} (the dose causing the death of 50% of the population) values in the range of 0.5% and 0.6% EMS respectively

3.4 Development of Somatic Embryos and Conversion into Plantlets

1. After 4–6 weeks embryo development can be observed.
2. After 2–3-months transfer developed torpedo shaped embryos to new tubes containing M5 medium.
3. Embryo maturation begins after 1–2-months when first foliage leaf and root formation can be observed (Fig. 5).
4. Transfer all torpedo shape embryos into individual culture tubes containing M5 media.
5. Count the number of developed embryos for every treatment.
6. Place tubes in the culture room with 16 h light and a temperature of $28 \pm 1\ °C$.
7. Observe daily for any changes in the color or growth.
8. Take a count of developed plantlets.
9. Calculate the survival rate and compare with data obtained (Sect. 3.3; Figs. 3 and 4, Table 2).

Fig. 5 Conversion of *Coffee arabica* somatic embryos to plantlets, control cultures are shown. **a** The embryogenic cells are cultured on a solid medium for the induction of somatic embryos, **b** torpedo shape embryos are selected and transferred to individual culture tubes for **c, d** plantlet development

Table 2 The averages of torpedo shape embryos and regenerated plantlets per 100 μl cultured volume

EMS concentration	0	0.2	0.5	0.8	1.5	2
Torpedo-shape embryos	77	86	77	18	0	0
Torpedo-shape embryos %	100	112	100	23	0	0
Plantlets	36	22	16	7	0	0
Plantlets %	100	63	44	21	0	0

Percentage of the control has been also calculated

4   Notes

1. This protocol describes EMS chemical mutagenesis of *Coffea arabica* var. Venecia, a late maturing variety with excellent cup quality. The Venecia coffee variety has its origin in San Carlos, Alajuela where it was discovered on a coffee plantation of 100% Caturra. It was selected due to its increased productivity, larger fruit size and increased resistance to fruit drop in the rain. The procedures described here utilize coffee cell suspensions generated in the Plant Breeding and Genetics Laboratory, Seibersdorf, Austria. For details on establishing the cell suspension culture, see Chap. "Somatic Embryogenesis and Temporary Immersion for Mass Propagation of Chimera-Free Mutant Arabica Coffee Plantlets". Briefly, leaf discs served as the starting material to produce

embryogenic callus. The embryogenic callus was then transferred to a liquid medium to establish a homogenous embryogenic cell suspension culture. The cell suspension culture served to maintain and multiply embryogenic cell/cell clusters and was used for EMS treatments. The EMS-treated cultures were regenerated on semi-solid media described here.

2. Ideally, chemical mutagenesis experiments are conducted in a dedicated laboratory, using dedicated equipment, and equipped with a ducted fume hood, toxic waste disposal and decontamination procedures. It is advisable to work with another person during the steps handling EMS to assist with the provision of equipment (e.g., pipettes) and sodium thiosulfate to deactivate EMS. Personal protective equipment such as laboratory coats, gloves, goggles, and disposable shoe covers should be worn when working with hazardous chemicals. It is advised to wear double gloves so that contaminated gloves can be removed while avoiding contact of contaminated materials with skin. Consult the biosafety regulations on use and disposal of hazardous chemicals. In addition, in advance of performing the experiment it is advised to carefully plan out what will be done and to practice the critical steps. Ensure that enough empty, liquid, and dry waste buckets are placed in the laboratory. If using common space, inform co-workers of the experiment in advance to avoid accidental exposures.

3. Follow the waste disposal procedures established in your institute.

4. EMS is carcinogenic and thus extreme care should be taken. Pay careful attention to the information on the EMS Materials Safety Data Sheet (MSDS) and follow any recommendations in case of accidental exposure.

5. EMS can be inactivated by treatment with sodium thiosulfate. The half-life of EMS in a 10% sodium thiosulfate solution is 1.4 h at 20 °C and 1 h at 25 °C. Keep beakers of sodium thiosulfate (100 mM) on standby during the laboratory procedures to inactivate any spills and to clean tips and other consumables prior to disposing in hazardous waste.

6. Some publications use percentage (v/v) of EMS rather than molarity. Optimal concentrations of ~ 20–40 mM have been reported for many studies using cell cultures mutagenesis (Jankowicz-Cieslak et al. 2011; Jankowicz-Cieslak and Till 2016). It is advisable to include a no EMS control and to test concentrations below and above the concentration used in published studies of the same species. The timing and temperature of EMS treatment can also be tested, but it is recommended to start first with concentrations as optimal dosage can typically be determined by altering mutagen concentration alone. The current protocol uses five EMS concentrations (0.2%, 0.5%, 0.8%, 1.5%, 2%) plus control. Each concentration was performed in 3 replicates. The mutagenic treatment lasted 1 h.

7. EMS is not miscible in water. DMSO is usually added to improve miscibility (Ingelbrecht et al. 2018). In this experiment, we omitted the use of DMSO due to the very fragile nature of the material to be treated. The prepared mixture should be mixed thoroughly. The mixture is prepared in a bottle, sealed with a screw cap, and then shaken vigorously before adding to the coffee cell cultures.

It is important that the bottle does not leak. Test the bottle with water first and mimic the shaking procedure in the fume hood.

8. Prepare the concentration series of EMS commencing with the lowest concentration. Practicing this step allows for the proper placement of the stock bottle, dilution bottle, pipette, and waste container, so that the stock bottle containing the concentrated EMS doesn't spill when working in the confined space of a fume hood.

9. Lowering the sash is necessary for proper ventilation and provides some protection against leakage from the bottle.

10. Verify the exact volume to be added to ensure cells will be fully exposed to the chemical mutagen.

11. It is advisable to estimate the best ratio of cell cultures to the EMS mixture prior to the actual experiment. Add the amount of cell cultures used in the main experiment to the beaker, add water and place on the orbital shaker. The cell cultures should be completely immersed in the mutagen solution and fully exposed to the active ingredients of the mutagen. Adjust the speed of the orbital shaker so that all cell cultures can freely move, split cell cultures into multiple falcon tubes and reduce volumes if necessary. EMS is unstable in water solutions due to hydrolysis with a half-life of 26 h at 30 °C. At low temperatures, hydrolysis rate is decreased, implying that the mutagen remains stable for longer.

12. Arabica coffee var. Venecia cell suspensions were incubated for 1 h. Be very careful when pouring off the liquid, avoid splashing, use a mesh screen or a sieve to ensure that all cell cultures are captured.

13. The by-products of the incubation process and residual active ingredients should be promptly washed off the incubated cell cultures after treatment. This prevents continued absorption of the mutagen beyond the intended duration, so-called dry-back, which leads to lethality.

14. EMS is a toxic chemical and must be disposed of according to current safety regulations in the laboratory (check disposal procedures with personnel responsible for toxic materials or local health authority). All body parts or laboratory coats contaminated with EMS should be washed thoroughly with water and detergent and further neutralized with 100 mM sodium thiosulfate.

15. Care should be taken to ensure that any materials removed from the chemical mutagenesis laboratory are free from contamination with EMS.

Acknowledgements The authors wish to thank Dr. Noel Arrieta Espinoza, the Coffee Institute of Costa Rica (ICAFE), Costa Rica for providing the seed of *Coffea arabica* var. Venecia. Funding for this work was provided by the Food and Agriculture Organization of the United Nations and the International Atomic Energy Agency through their Joint FAO/IAEA Centre of Nuclear Techniques in Food and Agriculture.

References

Avelino J, Cristancho M, Georgiou S, Imbach P, Aguilar L, Bornemann G, Läderach P, Anzueto F, Hruska AJ, Morales C (2015) The coffee rust crises in Colombia and Central America (2008–2013): impacts, plausible causes and proposed solutions. Food Sec 7:303–321

Etienne H, Breton D, Breitler J-C, Bertrand B, Déchamp E, Awada R, Marraccini P, Léran S, Alpizar E, Campa C, Courtel P, Georget F, Ducos J-P (2018) Coffee somatic embryogenesis: how did research, experience gained and innovations promote the commercial propagation of elite clones from the two cultivated species? Front Plant Sci 9:1630

FAOSTAT (2021) FAOSTAT data. Available at: Coffee I FAO I Food and Agriculture Organization of the United Nations

Ingelbrecht I, Jankowicz-Cieslak, J, Szurman M, Till BJ, Szarejko I (2018) Chemical mutagenesis. In: Spencer-Lopes MM, Forster BP, Jankuloski L (eds) Manual on mutation breeding, 3rd edn, Chap 2. FAO/IAEA Publication. FAO, Rome, Italy, pp 51–65

Jankowicz-Cieslak J, Till BJ (2016) Chemical mutagenesis of seed and vegetatively propagated plants using EMS. Curr Protoc Plant Biol 1:617–635. https://doi.org/10.1002/cppb.20040

Jankowicz-Cieslak J, Huynh OA, Bado S, Matijevic M, Till BJ (2011) Reverse-genetics by TILLING expands through the plant kingdom. Emirates J Food Agric 23:290–300

Murvanidze N, Nisler J, Lerou O, Werbrouck SPO (2021) Cytokinin oxidase/dehydrogenase inhibitors stimulate 2iP to induce direct somatic embryogenesis in *Coffea arabica*. Plant Growth Regul 94:195–200. http://doi.org/10.1007/s10725-021-00708-6

Quiroz-Figueroa F, Monforte-González M, Galaz-Ávalos RM, Loyola-Vargas VM (2006) Direct somatic embryogenesis in *Coffea canephora*. In: Loyola-Vargas VM, Vázquez-Flota F (eds) Plant cell culture protocols. Methods in molecular biology™, vol 318. Humana Press. http://doi.org/10.1385/1-59259-959-1:111

Scalabrin S, Toniutti L, Di Gaspero G et al (2020) A single polyploidization event at the origin of the tetraploid genome of *Coffea arabica* is responsible for the extremely low genetic variation in wild and cultivated germplasm. Sci Rep 10:4642. https://doi.org/10.1038/s41598-020-61216-7

Wintgens JN (ed) (2012) Coffee: growing, processing, sustainable production. A guidebook for growers, processors, traders, and researchers. Wiley-VCH Verlag GmbH & Co, Germany

Chemical Mutagenesis of Zygotic Embryos of *Coffea arabica* L. var. Catuaí Using EMS and NaN₃

Andrés Gatica-Arias and Jorge Rodríguez-Matamoros

Abstract The genetic improvement of *C. arabica* L. is challenged by its low genetic diversity and autogamous reproductive biology. Induced mutagenesis offers an alternative approach to conventional cross-breeding to increase genetic variability in wild and cultivated Arabica coffee germplasm for further use in breeding programs and genetic studies. Here protocols are described for the preparation of zygotic embryos from *C. arabica* seed and for toxicity testing of zygotic embryos using two chemical mutagens, sodium azide (NaN_3) and ethyl methanesulfonate (EMS). Zygotic embryos were immersed for 10 min in a solution of NaN_3 (0, 2.5, 5.0, 7.5, 10.0, 12.5, 15.0 and 20. 0 mM) and for 2 h in a solution of EMS (0, 0.5, 1, 1.5, 2, 4 and 6 % v/v). The percentage survival was evaluated and the LD values for NaN_3 and EMS were determined at 12.5 mM (51.6%) and 1 % v/v (48.3%), respectively. Our protocols indicate that coffee zygotic embryos are suitable propagules for NaN_3 and EMS mutagenesis and expand the types of propagules suitable for induced mutagenesis, breeding and genetic studies in Arabica coffee.

1 Introduction

Coffee is one of the most important products around the world. Global coffee imports in 2021–2022 amounted to 133.59 million 60-kg bags with a global market value of US $107.93 billion in 2021, taking second place in international trade after crude oil. Coffee is cultivated in more than 80 countries in tropical and subtropical regions of the globe, especially in Africa, Asia, and Latin America. Coffee production generates directly or indirectly income to more than 100 million people around the world (Mishra and Slater 2012).

A. Gatica-Arias (✉) · J. Rodríguez-Matamoros
Laboratorio Biotecnología de Plantas, Escuela de Biología, Universidad de Costa Rica, 2060, San Pedro, Costa Rica
e-mail: andres.gatica@ucr.ac.cr

© The Author(s) 2023
I. L. W. Ingelbrecht et al. (eds.), *Mutation Breeding in Coffee with Special Reference to Leaf Rust*, https://doi.org/10.1007/978-3-662-67273-0_9

Approximately 60% of the world coffee production is derived from *C. arabica* L. because of its superior quality, aromatic characteristics, and low caffeine content compared to Robusta coffee (Mishra and Slater 2012; Alpízar 2014; Ahmed et al. 2013).

The cultivated varieties of *C. arabica* L. are derived from the "Typica" or "Bourbon" coffee lineages, resulting in low genetic diversity (Mishra and Slater 2012). The reproductive biology of Arabica coffee, being approximately 90% autogamous, along with historical and geographic data indicating the occurrence of genetic bottlenecks due to domestication and global spread from its center of origin in Ethiopia and a polyploidization event are responsible for the low genetic variation in Arabica coffee (Romero et al. 2010; Mendonça 2014; Scalabrin et al. 2020; Montagnon et al. 2021). Consequently, *C. arabica* varieties are often highly susceptible to different diseases and pests (Romero et al. 2010).

Induced mutagenesis offers a promising alternative to improve current coffee cultivars for enhanced tolerance to pathogens, as previously shown in other crops (Gressel and Levy 2006). Induced mutagenesis using chemical or physical mutagens typically introduces random changes throughout the genome and can thus generate a variety of mutations within a single plant. As opposed to cross-breeding, induced mutagenesis can be applied to a wide variety of plant propagules.

Ethylmethanesulfonate (EMS) is today the most widely used chemical mutagen. It selectively adds alkyl groups to guanine bases causing random point mutations; most of the changes (70–99%) are base-pair transitions from G/C to A/T (Jankowicz-Cieslak et al. 2016; Sikora et al. 2011). Another chemical mutagen is sodium azide (NaN_3), whose mutagenic effect is mediated through the production of an organic metabolite of the azide compound: L-azidoalanine. It creates point mutations, mostly transitions, of the type: G/C to A/T or vice versa (Prina et al. 2010; Srivastava et al. 2011).

This chapter describes protocols for the preparation and mutagenic treatment of zygotic embryos of *C. arabica* L. var. Catuaí using the chemical mutagens sodium azide and ethyl methanesulfonate, including methods for bulk treatment, germination, and development of mutant plantlets.

2   Materials

2.1   Plant Material

1. *Coffea arabica* L. var. Catuaí seeds (see Note 1).

2.2   Reagents

1. 10% (w/v) sodium thiosulfate (e.g., Sigma Cat Nr.: 217263).
2. 100 mM phosphate buffer (pH 3.0) (e.g., Sigma Cat Nr.: 217263).

3. 1 N KOH (e.g., Phytotechnology Cat Nr.: P682).
4. 1 N NaOH (e.g., Phytotechnology Cat Nr.: S835).
5. 2,3,5-Triphenyltetrazolium Chloride (TTC) stock solution (e.g., Phytotechnology Cat Nr.: T8164).
6. 6-benzylaminopurine (BAP) (e.g., Sigma Cat Nr.: D3408).
7. Biotin (e.g. Phytotechnology Cat Nr.: B140).
8. Citric acid (e.g., Phytotechnology Cat Nr.: C277).
9. Disodium phosphate (Na_2HPO_4).
10. D-Sucrose (e.g. Phytotechnology Cat Nr.: S829).
11. Ethanol 95° (v/v).
12. Ethyl methanesulphonate (EMS) (e.g., Sigma Cat Nr.: M0880) (see Notes 2, 3, and 4).
13. Gelling agent (e.g., Phytagel: Sigma Cat Nr.: P8169).
14. Gibberellic acid (GA_3) (e.g., Sigma Cat Nr.: G7645).
15. L-ascorbic acid (e.g., Phytotechnology Cat Nr.: M524).
16. Monopotassium phosphate (KH_2PO_4).
17. Murashige and Skoog (MS) basal salt mixture (e.g., Phytotechnology Cat Nr.: M524).
18. Nicotinic acid (e.g., Phytotechnology Cat Nr.: N765).
19. Phosphate buffer.
20. Phosphorus acid (H_3PO_3).
21. Pyridoxine HCl (e.g., Phytotechnology Cat Nr.: P866).
22. Sodium azide (NaN_3) (e.g., Sigma Cat Nr.: S2002) (see Notes 2, 3, and 4).
23. Sodium hypochlorite (3.5% v/v).
24. Sterile distilled water.
25. Thiamine HCl (e.g., Phytotechnology Cat Nr.: T390).
26. Tween 20 (e.g., Phytotechnology Cat Nr.: P720).

2.3 Glassware and Minor Equipment

1. 2 ml reaction tubes.
2. Beakers (250, 1,000 ml).
3. Bottles (100, 500 ml).
4. Box for dry hazardous material disposal.
5. Disposable pipettes (1 ml, 5 ml, 10 ml, 25 ml, or as needed according to calculations).
6. Erlenmeyer (250 ml).
7. Forceps.
8. Glass or disposable pipettes (1, 5, 10, 25 ml).
9. Graduated cylinder (50, 100, 500 and 1,000 ml).
10. Hazardous liquid waste receptacle (collection vessels for sodium azide and ethyl methanesulfonate waste solution).
11. Magnetic stir bar.

12. Personal protective equipment (disposable laboratory coat dedicated only to mutagenesis experiments, eye protection/goggles, shoe protection, nitrile gloves).
13. Petri dishes (100 mm × 20 mm).
14. Plastic boxes (88 × 42 × 16 cm).
15. Scalpel.
16. Spatula.
17. Volumetric flasks (1,000 ml).
18. Weighing trays.

2.4 Equipment

1. Analytical balance.
2. Autoclave.
3. Centrifuge.
4. Chemical mutagen laboratory equipped with fume hood and flow bench (see Notes 2, 3, and 4).
5. Fume Hood.
6. Hot plate shaker.
7. Orbital shaker.
8. pH meter.
9. Stereoscope.
10. Water bath at 65 °C.

2.5 Software

1. Standard spreadsheet software (e.g., Microsoft Excel or Open Office Excel).

2.6 Bulk Mutagenesis of Zygotic Embryos

All materials as listed in Sects. 2.1, 2.2, 2.3, 2.4, and 2.5.

3   Methods

3.1   Preparation of Stock Solutions

1. 100 mM phosphate buffer: add 13.6 g of KH_2PO_4 to 1,000 ml volumetric flask and dissolve by adding 250 ml of tissue culture grade water. Once completely dissolved, stir the solution while adding tissue culture grade water and bring to 1,000 ml. Adjust pH to 3.0 using phosphorus acid (H_3PO_3).

2. 6-Benzylaminopurine (BAP; 1 mg/ml stock solution): add 50 mg of the powder to a 100 ml volumetric flask and dissolve by adding 2–5 ml of 1 N NaOH. Once completely dissolved, stir the solution while adding tissue culture grade water and bring to 50 ml. Sterilize by filtering through a 0.2 μm filter and store aliquots (1 ml) at 4 °C.

3. Gibberellic acid (GA_3; 1 mg/ml stock solution): add 50 mg of the powder to a 100 ml volumetric flask and dissolve by adding tissue culture grade water. Once completely dissolved, stir the solution while adding tissue culture grade water and bring to 50 ml. Sterilize by filtering through a 0.2 μm filter and store aliquots (1 ml) at 4 °C.

4. KH_2PO_4 (0.07 M): add 907.8 mg of the powder to 250 ml volumetric flask and dissolve by adding 50 ml of tissue culture grade water. Once completely dissolved, stir the solution while adding tissue culture grade water and bring to 100 ml.

5. Na_2HPO_4 (0.08 M): add 1.1876 g of the powder to 250 ml volumetric flask and dissolve by adding 50 ml of tissue culture grade water. Once completely dissolved, stir the solution while adding tissue culture grade water and bring to 100 ml.

6. Sodium azide (500 mM): add 1.625 g of the powder to 250 ml volumetric flask and dissolve by adding 50 ml phosphate buffer (100 mM). Once completely dissolved, stir the solution while adding phosphate buffer (100 mM) and bring to 100 ml. Sterilize by filtering through a 0.2 μm filter and store at 4 °C in the dark.

7. 2,3,5-Triphenyltetrazolium Chloride (TTC) stock solution: add 4 ml of KH_2PO_4 (0.07 M) and 6 ml of Na_2HPO_4 (0.08 M). Adjust pH to 7. Add the amount of TTC needed to reach 1% (m/v) (100 mg for 10 ml of solution). Sterilize by filtering through a 0.2 μm filter and store aliquots in the dark at 4 °C.

8. Citric acid (100 mg/ml stock solution): add 1,000 mg of the powder to a 50 ml volumetric flask and dissolve by adding tissue culture grade water. Once completely dissolved, stir the solution while adding tissue culture grade water and bring to 10 ml. Sterilize by filtering through a 0.2 μm filter and store aliquots (1 ml) in the dark at 4 °C.

9. L-ascorbic acid (100 mg/ml stock solution): add 1,000 mg of the powder to a 50 ml volumetric flask and dissolve by adding tissue culture grade water. Once completely dissolved, stir the solution while adding tissue culture grade water and bring to 10 ml. Sterilize by filtering through a 0.2 μm filter and store aliquots (1 ml) in the dark at 4 °C.

3.2 Preparation of Tissue Culture Medium

1. Prepare stock solutions of BAP (1 mg/ml), GA_3 (1 mg/ml), Biotin (1 mg/ml), Calcium pantothenate (1 mg/ml), Nicotinic acid (1 mg/ml), Pyridoxine HCl (1 mg/ml), and Thiamine HCl (1 mg/ml).
2. Place a beaker containing 400 ml tissue culture grade water on a hot plate shaker and mix:
   - For 1 L of germination medium (EG): full-strength MS salts, 1.0 ml BAP, 1 ml GA_3, and 300 mg/l activated charcoal, and 2.5 g Phytagel™.
   - For 1 L of development medium (DEV): full-strength MS salts 1 ml thiamine HCl, 1 ml pyridoxine HCl, 1 ml nicotinic acid, 1 ml calcium pantothenate, 0.01 ml biotin, 100 mg myo-inositol, 0.3 ml BAP, 30 g sucrose, 2.5 g Phytagel™.
3. While stirring, add tissue culture grade water to a final volume of 1,000 ml.
4. Stir until the solution is homogenous and clear.
5. Calibrate the pH meter as per manufacturer instructions.
6. While stirring, adjust medium to pH 5.6 using 1 N NaOH and 1 N HCl.
7. For semi-solid medium, add the gelling agent and heat while stirring until the solution is homogenous and clear.
8. Dispense the culture medium in the respective culture vessels before or after autoclaving (depending on the application).
9. Sterilize all media in an autoclave at 0.1 kg/cm^2 for 21 min at 121 °C.
10. Allow the medium to cool prior to use.
11. Store the medium for up to a week in a cold room.

3.3 Seed Disinfection and Excision of Zygotic Embryos

1. Collect mature cherries from genetically homogenous mother plants maintained in the greenhouse or in the field (see Notes 1, 5 and 6 and Fig. 1).
2. Remove the pulp, the mucilage, and the parchment by hand.
3. Prepare a uniform seed stock by selecting disease-free seeds and removing any small, shriveled or damaged seeds.
4. Soak the seeds for 24 h in distilled water with two drops of Tween 20 with orbital rotation.
5. Disinfect the seeds with 3.5% (v/v) sodium hypochlorite for 1 h and finally rinse three times with sterile distilled water.
6. Soak seeds in sterile distilled water for 48 h.
7. In a laminar flow cabinet, excise zygotic embryos with the aid of tweezers and a scalpel and maintain them in a solution of citric acid and ascorbic acid (100 mg/ml each; pH 5.6) until the excision of all embryos prior to the germination experiments.

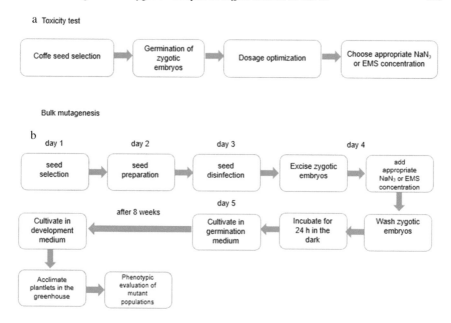

Fig. 1 Schematic representation of the mutagenesis of coffee (*Coffea arabica* L. var. Catuaí) zygotic embryos. **a** For chemical mutagenesis and toxicity testing, normal shaped, disease-free seeds are selected and NaN₃ and EMS dosage and incubation time are optimized using seed lots with a germination rate equal to or above 90%. These steps take approximately 8 weeks. **b** Bulk irradiation of zygotic embryos. For both **a** and **b**, the zygotic embryos are prepared by manually removing the pulp, the mucilage, and the parchment of the seed, disinfecting the seeds, excising the zygotic embryo, and incubating the zygotic embryos using the appropriate NaN₃ or EMS concentration followed by incubating, rinsing and planting the mutagenized zygotic embryos

8. Culture the zygotic embryos in petri dishes (100 mm × 20 mm) containing 20 ml of regeneration medium.
9. Incubate the petri dishes under a 16 h light photoperiod at $27 \pm 2\,°C$ for 4 weeks.
10. Record germination after 2–4 weeks based on visual scoring of the presence of leaves (see Fig. 2).

3.4 Determination of the Viability of the Zygotic Embryos

1. Place the zygotic embryos in a 2 ml reaction tube and add 1 ml of the TTC stock solution (see Sect. 2.5 for description of the TTC viability test).
2. Incubate the samples for 24 h in the dark at 37 °C without shaking.
3. Remove the TTC solution by decanting or centrifugation and wash the sample with distilled water.
4. Add 1 ml of 95° (v/v) ethanol.
5. The evaluation of the viability of the embryos was carried out by counting the embryos that stained red (see Fig. 2).

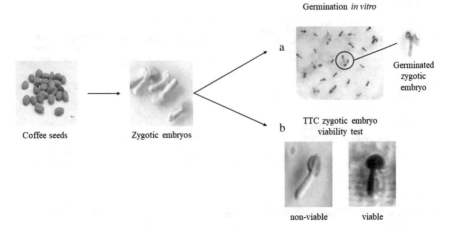

Fig. 2 Germination and viability test of zygotic embryos of coffee (*Coffea arabica* L. var. Catuaí).
a Excised zygotic embryos germinated under in vitro culture conditions. **b** TTC viability test

3.5   Sodium Azide Dose Determination

1. Review safety procedures of the chemical mutagenesis laboratory (see Notes 2 and 3).
2. Autoclave all non-disposable materials (e.g., sieves, forceps).
3. Prepare a fresh 0.5 M NaN_3 stock solution by adding the required amount of NaN_3 to the phosphate buffer (100 mM, pH 3.0).
4. Select similar and normal shaped seeds that are disease-free.
5. Disinfect the seed and excise the zygotic embryos beforehand according to the protocol described in Sect. 3.3.
6. Place 100 zygotic embryos in 250 ml labelled Erlenmeyer. The number of Erlenmeyer depends on the number of treatments. Label each tube with the appropriate treatment code (concentration and incubation time).
7. In a fume hood, add appropriate volumes of phosphate buffer solution (100 mM) (see Table 1).
8. Add NaN_3 to a final concentration of 2.5, 5.0, 7.5, 10.0, 12.5, 15.0, and 20.0 mM (see Table 1).
9. Shake the solution vigorously.
10. Incubate the mixture for 10 min in the dark at 27 ± 2 °C with gentle rotation (100 rpm).
11. After the incubation time, quickly but carefully decant each of the treatment batches and rinse the treated zygotic embryos with 100 ml of sterile distilled water. This step and any subsequent steps must be carried out in a fume hood.
12. Repeat washing step 3 times.
13. Collect all the liquid waste in a dedicated bucket labelled as hazardous waste. Dispose of toxic waste according to local regulations.

Table 1 Concentrations chosen for the toxicity test in coffee seeds

NaN$_3$ concentration (mM)	TEX medium (ml)	500 mM NaN$_3$ (μl)[a]
0 (control; pH 5.6)	10	–
0 (control; pH 3.0)	10	–
2.5	9.95	50
5.0	9.90	100
7.5	9.85	150
10.0	9.8	200
12.5	9.75	250
15.0	9.7	300
20.0	9.6	400

[a] Prepare a fresh solution immediately before the experiment

14. Twenty-four hour after the NaN$_3$ mutagenesis treatment, determine viability of zygotic embryos and observe them using a stereoscope as described in Sect. 3.4.
15. Calculate the survival rate as follows: [(the number of survival explant/the number of treated explant) * 100].
16. Record the data for each treatment and enter it into a spreadsheet (e.g., Microsoft Excel or Open Office Excel).
17. Plot percentage of control against mutagenesis treatment.
18. Estimate the mutagen concentration required to obtain 50% viability compared to the control (see Fig. 3).
19. Identify concentrations suitable for bulk mutagenesis of your material according to the LD$_{50}$ previously estimated.

3.6 Ethyl Methanesulphonate Dose Determination

1. Review safety procedures of the chemical mutagenesis laboratory (see Notes 2 and 3).
2. Autoclave all non-disposable materials (e.g., sieves, forceps).
3. Select similar and normal shaped seeds that are disease-free.
4. Disinfect the seed and excise the zygotic embryos according to the protocol described in Sect. 3.3.
5. Place 100 zygotic embryos in a 250 ml labelled Erlenmeyer. The number of Erlenmeyer depends on the number of treatments. Label each tube with the appropriate treatment code (concentration and incubation time).
6. In a fume hood, add appropriate volumes of sterile distilled water (see Table 2).
7. Add EMS to a final concentration of 0.5, 1, 2, 3, 4, 5, and 6% v/v (see Table 2). Shake the solution vigorously (see Note 7).

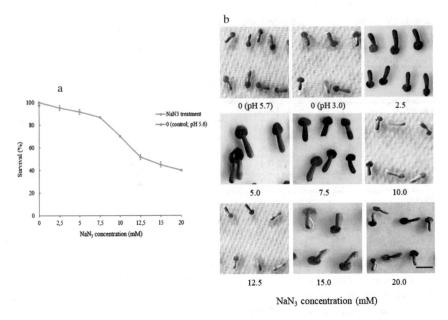

Fig. 3 Effect of NaN$_3$ concentration on survival and viability of coffee (*C. arabica* L. var. Catuaí) zygotic embryos. **a** Survival percentage (solid line) versus NaN$_3$ concentrations. Each value represents the mean ± SD of three repetitions. **b** Zygotic embryo viability versus NaN$_3$ concentrations. Zygotic embryos that stain red are considered viable. *Bar*, 1.0 cm

8. Incubate the mixture for 2 h in the dark at 27 ± 2 °C with gentle rotation (100 rpm).
9. After the incubation time, quickly but carefully decant each of the treatment batches and rinse the treated zygotic embryos with 100 ml of sterile distilled water. This step and any subsequent steps must be carried out in a fume hood.
10. Repeat washing step 3 times.
11. Collect all the liquid waste in a dedicated bucket labelled as hazardous waste. Dispose of toxic waste according to local regulations.
12. Twenty-four hour after the EMS mutagenesis, determine viability of zygotic embryos and observe them using a stereoscope as described in Sect. 3.4.
13. Calculate the survival percentage as follows:
    Survival percentage = Number of survived explant/the number of treated explants) * 100.
14. Record the data for each treatment and enter it into a spreadsheet (e.g., Microsoft Excel or Open Office Excel).
15. Plot percentage of control against mutagenesis treatment.
16. Estimate the mutagen concentration required to obtain 50% survival compared to the control (see Fig. 4).
17. Identify concentrations suitable for bulk mutagenesis of your material according to the LD$_{50}$ previously estimated.

Table 2 Concentrations chosen for the toxicity test in coffee zygotic embryos

EMS concentration (%)	Sterile distilled water (ml)	EMS (ml)
Negative control (−EMS)	100	–
0.5	99.5	0.5
1.0	99	1
1.5	98.5	1.5
2	98	2
4	96	4
6	94	6

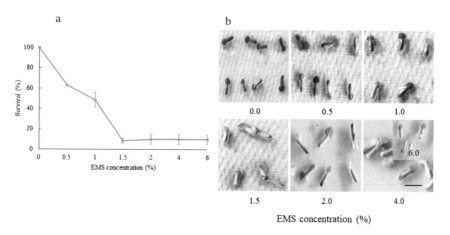

Fig. 4 Effect of EMS concentration on survival/viability of coffee (*C. arabica* L. var. Catuaí) zygotic embryos. **a** Survival percentage (solid line) versus EMS concentrations. Each value represents the mean ± SD of three repetitions. **b** Zygotic embryo viability versus EMS concentrations. Red stained zygotic embryos were considered as viable. *Bar*, 1.0 cm

3.7 Bulk Mutagenesis

1. Autoclave all non-disposable materials (e.g., sieves, forceps).
2. Choose appropriate NaN_3 or EMS concentration and incubation time based on the results obtained from the chemical toxicity test (see Sects. 3.4 and 3.5).
3. Select similar and normal shaped seeds that are disease-free.
4. Disinfect the seed and excise the zygotic embryos according to the protocol described in the Sect. 3.3.
5. Place 100 zygotic embryos in a 250 ml labelled Erlenmeyer. The number of Erlenmeyer depends on the number of repetitions.
6. Label each Erlenmeyer with the appropriate treatment code (NaN_3 and EMS concentration and incubation time).

7. Transfer Erlenmeyer containing the zygotic embryos into the chemical mutagenesis laboratory.
8. In a fume hood, add appropriate concentrations of NaN_3 or EMS.
9. Shake the solution vigorously.
10. Place Erlenmeyer in the dark at 27 ± 2 °C on a rotary shaker set at 100 rpm for
    the chosen length of time.
11. After incubation, quickly but carefully decant each of the treatment batches and
    rinse the treated zygotic embryos with 100 ml of sterile distilled water.
12. Repeat washing step 3 times.
13. Collect all the liquid waste in a dedicated bucket labelled as hazardous waste.
    Dispose of toxic waste according to local regulations.
14. After mutagenesis treatment, add 10 ml of liquid germination medium (pH 5.6)
    and maintain the cultures for 24 h in the dark on a rotary shaker (100 rpm) at
    26 ± 2 °C.
15. Perform a zygotic viability test as described in Sect. 3.4.
16. Carefully decant the treated zygotic embryos and plate them in Petri dishes
    containing 20 ml of semisolid germination medium.
17. Incubate the Petri dishes under a 16 h light photoperiod at 27 ± 2 °C for
    8–10 weeks.
18. Transfer the plantlets to baby food jars containing 20 ml of development medium
    (DEV) with 16 h light photoperiod at 27 ± 2 °C.

3.8   Acclimatization of M_1 Plantlets

1. After 8 weeks, do a transverse cut in the stem of the plantlets with 4 leaves and
   3–4 cm tall, below a node, and remove the leaves near the cut.
2. Place the freshly cut basal part of the stem in a solution of indoleacetic acid
   (IAA) (1 mg/mL) for 30 s.
3. Plant the plantlets in sterile substrate (peat moss with perlite) in plastic boxes
   (30 cm × 20 cm × 10 cm).
4. Cover boxes with plastic and place them under greenhouse conditions with 12 h
   light photoperiod at 27 ± 2 °C.
5. Two weeks later, make small holes in the plastic covering the boxes.
6. One week later, plant the plantlets in polyethylene bags with a 3:1 mixture of
   substrate (peat moss with perlite: coconut fiber). Two plants per bag were planted
   and identified according to the original numbering.
7. After 3 weeks, fertilize with granular slow-release fertilizer (e.g., Osmocote
   14-14-14).
8. Irrigate the plants twice a week according to the climatic conditions of the
   greenhouse and the water requirement of the crop.

4 Notes

1. This protocol has been established using *Coffea arabica* L. var. Catuaí zygotic embryos. It can be used as a reference for other Arabica coffee varieties, nevertheless, it is recommended to optimize the mutagenic parameter (NaN$_3$ and EMS concentration and incubation time) for each variety used.
2. All the mutagenesis experiments should be conducted in a dedicated chemical mutagenesis laboratory equipped with a ducted fume hood, toxic waste disposal and decontamination procedures.
3. Read the Materials Safety Data Sheet (MSDS) of materials being used and follow the recommendation of the manufacturer. Pay attention to the information on sodium azide and EMS and what to do in case of exposure. It is very important to wear personal protective equipment (gloves must be compatible with chemical mutagens, for instance PVC or neoprene gloves); safety glasses with side shields or chemical goggles; lab coat, closed-toe shoes, shoe protections, and full-length pants. A double glove system is advised.
4. Store the original EMS and NaN$_3$ always in an airtight colored bottle, preferably inside a sealed chamber containing a desiccant.
5. Coffee seeds loose viability rapidly if not properly stored. Therefore, it is recommended to use freshly harvested seeds or otherwise to store the seeds between 10 and 12% moisture at 15 °C not longer than 3 months.
6. Proceed to mutation induction protocol with seeds having germination rate close to or above 90%.
7. Some protocols use water-dimethyl sulfoxide (2% v/v DMSO) mixture in order to make the EMS mutagen more miscible.

Acknowledgements Funding for this work was provided by the University of Costa Rica, the Ministerio de Ciencia, Innovación, Tecnología y Telecomunicaciones (MICITT), the Consejo Nacional para Investigaciones Científicas y Tecnologicas (CONICIT) (project No. 111-B5-140; FI-030B-14) and the Cátedra Humboldt 2023. A. Gatica-Arias acknowledged the Cátedra Humboldt 2023 of the University of Costa Rica for supporting the dissemination of biotechnology for the conservation and sustainable use of biodiversity.

References

Ahmed W, Feyissa T, Disasa T (2013) Somatic embryogenesis of a coffee (*Coffea arabica* L.) hybrid using leaf explants. J Hortic Sci Biotechnol 88(4):469–475

Alpízar E (2014) Zonificación agroecológica del café (*Coffea arabica*) y el cacao (*Theobroma cacao*, Lin) en Costa Rica, mediante el sistema de zonas de vida. Master's thesis. Instituto Tecnológico de Costa Rica, Cartago, CR, 98 p

Gressel J, Levy A (2006) Agriculture: the selector of improbable mutations. Proc Natl Acad Sci 103:12215–12216

Jankowicz-Cieslak J, Till BJ (2016) Chemical mutagenesis of seed and vegetatively propagated plants using EMS. Curr Protoc Plant Biol 1:617–635. https://doi.org/10.1002/cppb.20040

Mendonça AP (2014) Desempenho de seleções clonais de *Coffea arabica*, em Campinas, SP. Master's thesis. Instituto Agronômico, Campinas, BR, 58 p

Mishra MK, Slater A (2012) Recent advances in the genetic transformation of coffee. Biotechnol Res Int 2012:1–17

Montagnon C, Mahyoub A, Solano W, Sheibani F (2021) Unveiling a unique genetic diversity of cultivated *Coffea arabica* L. in its main domestication center: Yemen. Genet Resour Crop Evol 68:2411–2422

Prina AR, Landau AM, Pacheco MG, Hopp EH (2010) Mutagénesis, TILLING y EcoTILLING. INTA-Fundación ARGENBIO, Argentina, 650 p

Romero JV, Camayo-Vélez GC, González-Martínez LF, Cortina-Guerrero HA, Herrera-Pinilla JC (2010) Caracterización citogenética y morfológica de híbridos interespecíficos entre *C. arabica* y las especies diploides *C. liberica* y *C. eugenioides*. Cenicafé 61(3):206–221

Scalabrin S, Toniutti L, Di Gaspero G et al (2020) A single polyploidization event at the origin of the tetraploid genome of *Coffea arabica* is responsible for the extremely low genetic variation in wild and cultivated germplasm. Sci Rep 10:4642

Sikora P, Chawade A, Larsson M, Olsson J, Olsson O (2011) Mutagenesis as a tool in plant genetics, functional genomics, and breeding. Int J Plant Genomics 2011:1–13

Srivastava P, Marker S, Pandey P, Tiwari DK (2011) Mutagenic effects of sodium azide on the growth and yield characteristics in wheat (*Triticum aestivum* L. em. Thell.). Asian J Plant Sci 10:190–201

Induced Mutagenesis of Seed and Vegetative Propagules of *Coffea arabica* L.

Physical Mutagenesis of Arabica Coffee Seeds and Seedlings

Abdelbagi Mukhtar Ali Ghanim, Souleymane Bado, and Keji Dada

Abstract Coffee, a perennial tropical crop, can be grown from seed or from cloned plants in the form of cuttings, grafts or tissue cultured plants. Arabica coffee is most commonly grown from seeds while Canephora is mostly grown vegetatively from cuttings and other propagules. Improving Arabica coffee through conventional breeding is seriously limited by the lack of genetic variation within the cultivated and wild species. Mutation breeding provides great potential to induce the novel genetic variation needed for Arabica coffee improvement. Here we present protocols for mutation induction of coffee seed and seedlings using the FAO/IAEA in-house gamma (^{60}Cobalt) and X-ray (RS2400 irradiator) sources. Methods for mutation induction using gamma- and X-ray mutagenesis techniques are described. Methods for the preparation, seed quality control and post-radiation treatment of materials are also provided along with example data for radio-sensitivity testing of *Coffea arabica* seed under laboratory conditions.

1   Introduction

Coffee is a perennial crop belonging to the genus *Coffea* in the family *Rubiacea*. There are about 125 species within this genus, with *Coffea arabica* and *Coffea canephora*, representing approximately 70% and 30% of coffee production, respectively (Lashermes et al. 2008). Arabica coffee is a tetraploid and self-pollinating (autogamous), while Robusta coffee is a diploid and allogamous (Wintgens 2004). The efficiency of traditional Arabica coffee breeding approach is greatly reduced by the lack of sufficient genetic diversity and the long time needed for coffee to flower and bear

A. M. A. Ghanim (✉) · S. Bado
Plant Breeding and Genetics Laboratory, Joint FAO/IAEA Centre of Nuclear Techniques in Food and Agriculture, IAEA Laboratories Seibersdorf, International Atomic Energy Agency, Vienna International Centre, Vienna, Austria
e-mail: abdmali@yahoo.com

K. Dada
Plant Breeding Unit, Cocoa Research Institute of Nigeria (CRIN), Ibadan, Oyo State, Nigeria

© The Author(s) 2023   143
I. L. W. Ingelbrecht et al. (eds.), *Mutation Breeding in Coffee with Special Reference to Leaf Rust*, https://doi.org/10.1007/978-3-662-67273-0_10

fruits (Scalabrin et al. 2020). Mutation breeding provides great potential to induce the novel genetic variation needed for Arabica coffee improvement.

Coffee can be grown from seed or from cloned plants in the form of cuttings, grafts or tissue cultured plants. Arabica coffee is most commonly grown from seeds while Canephora is mostly grown vegetatively from cuttings and other propagules. Despite widely reported spontaneous mutants of Arabica coffee, there are very few studies on induced mutagenesis in coffee. The first attempt to induced mutation breeding of *C. arabica* was reported by Carvalho et al. (1954) using X-ray irradiation. The process of optimizing dose involves dose-response experiments where the pattern of reduction in germination (Lethal Dose, LD) or growth rate (Growth Reduction, GR) is determined in relation to an increase of absorbed dose. From these experiments, the LD/GR$_{30}$ and LD/GR$_{50}$ is calculated. In case of coffee after adjustment trials of seed and vegetative part using our in-house gamma and X-ray irradiators, we came to a range of (0, 50, 100, 150, 200, 400 Gy) for *C. arabica* seeds and (0, 5, 10, 15, 20, 30 and 40 Gy) for seedlings and cuttings of *C. arabica* and *C. canephora*. The protocol for seed treatment follows the general procedure which starts with sorting clean and viable seeds, moisture equilibration in a desiccator with 60% glycerol, irradiation treatments, planting the treated material in suitable set-up such as moist filter papers in petri-dish, soil in trays or pots and incubate at appropriate condition under warm condition 28–30 °C. Germination or growth rate after 30 days is recorded and plotted relative to the untreated seeds over the series of the doses. From the plotted graph the doses for LD$_{50}$, GR$_{50}$ and LD$_{30}$, GR$_{30}$ are estimated. The same follows for vegetative propagules (cuttings, seedling, embryo etc.) except that the applied doses here are relatively low in the order of 0–40 Gy. The estimated dose can be used for bulk treatment. Induced mutations are random events, implying that even adherence to published irradiation conditions might not result in the same mutation events. A way of overcoming this is to work with large populations. In case of seed, it is generally recommended to target the production of an M$_2$ population of a minimum of 5,000–10,000 individuals.

In this chapter, step-by-step procedures for seed quality control, irradiation treatment, and radiosensitivity testing of seed and seedlings is described. Example LD/GR values resulting from radiosensitivity testing of seed germination and growth rate under laboratory conditions are presented.

2 Materials

1. High quality, disease-free seeds, clean and uniform in size (Note 1).
2. Sterilized soil mixture.
3. Trays or pots of appropriate size.
4. Glasshouse facility.
5. Growth incubator with temperature control.
6. Gamma radiation source.
7. X-ray irradiator RS-2400.

8. Paper envelopes (air- and water-permeable).
9. Vacuum desiccator.
10. Petri dishes (90 mm diameter).
11. Whatman filter papers for 90 mm Petri dishes.
12. 60% Glycerol (v/v).
13. Sterile and non-sterile deionised water.
14. Parafilm.

3 Methods

3.1 Seed Treatment

3.1.1 Preparation and Quality Control of Seed

(i) Use freshly harvested seeds as coffee seeds loose rapidly viability if not properly stored. Select seeds from genetically homogenous plants, after a few cycles of selfing from the targeted preferred variety. In case of freshly harvested cherries, de-pulp to remove the two seeds, separate seeds from mucilage, wash in running tap water and air dry at room temperature (under shade) for 3–7 days. Seeds are then ready for germination.

(ii) Sort out clean seeds and remove small, shriveled, and damaged seeds from the starting seed lots.

(iii) Perform viability test following appropriate germination test procedure; soak few seeds (50–60) from each genotype in warm water for 1–3 days and incubate on moist filter paper in petri dish (15–20 seeds/dish) or directly in light soil and keep under warm (27–29 °C) and moist condition until germination. Record germination based on visual scoring of full protrusion of the radicle and appearance of the coleoptile after about 2 weeks (Fig. 1). Proceed to mutation induction protocol with seeds having viability close to or above 90% germination.

(iv) Divide the seed lot with high viability seeds (> 90%) based on the amount available and the planned set-up for radio-sensitivity test (incubator vs. soil) into 20–50 seeds per treatment and place in appropriate paper bag (Fig. 1).

(v) Clearly label the bag with the genotype name, source of irradiation (Gamma vs. X-ray), treatment dose (0–400 Gy), replication number (R1–R3), date and the number of seeds (Fig. 1).

(vi) Pack the prepared seed bags, in step 5 above, for different genotypes in joint groups based on the planned source of irradiation treatment (Gamma vs. X-ray), similarity in treatment dose and replications. In total we will have 6 groups of treatments multiplied by the number of replications (2–3) (Fig. 1).

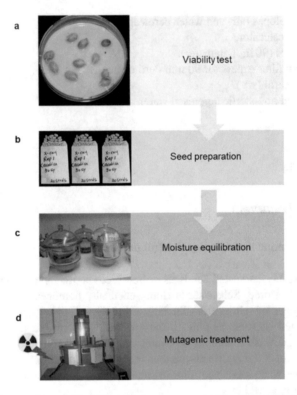

Fig. 1 Viability test of the seed lot by germination test on moist filter paper in petri dish (**a**), viable seed lot is divided into number of samples equal to the planned irradiation treatment, seed sample for each treatment is placed in small paper bag and labelled (**b**), bags with similar treatment are grouped in a larger bag and placed in a desiccator with 60% glycerol for moisture equilibration (**c**), after moisture equilibration seeds are ready for irradiation (**d**)

(vii)   Place the prepared seeds in step 6 above in a desiccator with 60% glycerol for moisture equilibration (Fig. 1). Keep the seeds in the desiccator for time sufficient to equilibrate the moisture of the seeds to (12–14%). In cereals and legumes this takes around 3–7 days. Longer stay in the desiccator is not expected to affect the treatment (Note 2).

(viii) Remove the seeds from the desiccator, vacuum pack (optional) and proceed to the irradiation facility for treatment. Up to this stage the seeds are known as M_0 seeds.

3.1.2   Gamma and X-ray Irradiation Treatment

The in-house Cobalt gamma-ray Cell and RAD-Source (RS 2400) was used for gamma-ray and X-ray irradiation treatment respectively, using previously described procedures (Spencer-Lopes et al. 2018). Briefly, the RS 2400 X-ray machine source

Post-treatment handling of coffee seed

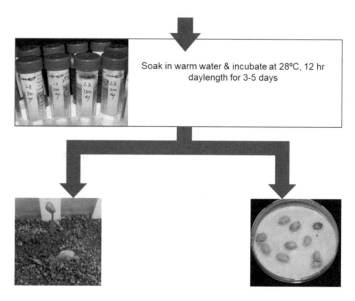

Fig. 2 Post treatment handling and radiosensitivity testing of irradiated seed. Steps of water soaking of treated seeds at 28 °C and 12 h photoperiod in incubator for 3–7 days, transfer to either petri dishes or soil for monitoring and recording germination (viability) and growth responses (growth reduction) of the irradiated seeds

is an upgraded irradiator for research and industry used by the FAO/IAEA Insect Pest Control Laboratory for the Sterile Insect Technique (Mastrangela et al. 2010; Mehta and Parker 2011) and adapted for mutation induction of plant propagules by the FAO/IAEA Plant Breeding and Genetics Laboratory. It is important in X-ray irradiation that samples are tightly packed to minimize air space and to maintain a near uniform field of X-rays through the entire sample. The dose rate of the RS2400 X-ray is about 12 Gy/min. After irradiation, the seeds and resulting plants are known as M_1 stage.

3.1.3 Post-treatment Handling and Radio-Sensitivity Testing

Viability and Growth Rate Testing of the M_1 Seeds in Petri Dish in an Incubator

1. Remove the paper bags containing the treated M_1 seeds from the packing set and rearrange them based on genotype, replication, and doses from 0 to 400 Gy.
2. Place the treated seeds in appropriate tube, add sufficient warm water (30–35°C), tightly close the tube, label for treatment, genotype, and replication, and leave in the incubator (27–29 °C) for 72 h for soaking of the M_1 seeds (Fig. 2).

3. Remove the soaked M_1 seeds and place them on moist filter paper in petri dish. Label the dish as in the soaking step above.
4. Place the petri-dish with soaked M_1 seeds in the incubator under the same condition above with 12 h day length (light) and incubate for germination (Fig. 2).
5. Monitor the germinating M_1 seeds and ensure sufficient moisture by wetting the filter papers every 2 days or whenever needed.
6. Record germination and measure root and coleoptile length after 2 weeks (Fig. 2).
7. Calculate the germination percentage and reduction in growth relative to the growth of the seedling from the untreated seeds.
8. Plot the germination percentage and reduction in growth rate to estimate the optimum dose based on lethal dose for 30 and 50% of the seeds (LD_{30} and LD_{50}, respectively) and growth reduction of 30 and 50% in the treated seeds compared to the control in root and coleoptile, and other vegetative parameters as appropriate (GR_{30} and GR_{50}, respectively).

Viability and Growth Rate Testing of the M_1 Seeds in Soil in the Glasshouse

Repeat steps 1 and 2 as described above under Sect. 3.1.3. Then, follow below steps:

1. Prepare soil mix and distribute in medium size trays using the recommended soil type for coffee seeds germination: light soil containing sand, clay and peat moss with acidic pH (5–6).
2. Prepare necessary labels (plastic) for genotype, replication and for each treatment.
3. Plant seeds in rows each with about 20 seeds from each treatment. Assign each replication in a separate tray. Sufficiently water the seeds (Fig 2).
4. Monitor the germinating M_1 seeds and ensure sufficient moisture by wetting every 2 days or when needed.
5. Record germination and measure hypocotyl length on regular basis (Fig. 2).
6. Calculate germination percent and reduction in growth relative to the growth of the seedling from the untreated seeds.
7. Plot germination percentage and reduction in growth rate to estimate the optimum dose based on lethal dose for 30 and 50% of the seeds (LD_{30} and LD_{50}, respectively) and growth reduction of 30 and 50% in the treated seeds compared to the control in root and coleoptile and other vegetative parameters as appropriate (GR_{30} and GR_{50}, respectively).

3.2   Coffee Seedlings Treatment

1. Grow coffee seedlings up to 56 days after germination (DAG) (Fig. 3).
2. Tie seedlings with a rope or tape into a pack to fit in the irradiation container.

Fig. 3 Key steps in the process of seedlings irradiation starting with preparation of seedlings at 56 days after germination (DAG) (**a**), transfer to gamma irradiation facility for treatment (**b**) and transplanting in pots in a glasshouse under appropriate conditions. Seedling at 14 (**c**) and 28 days after irradiation treatment (DAT) are shown (**d**)

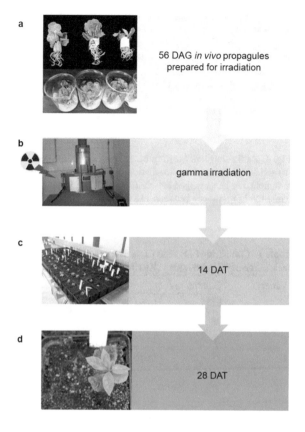

56 DAG *in vivo* propagules prepared for irradiation

gamma irradiation

14 DAT

28 DAT

3. Place the seedling pack in the center of an appropriate container for X or gamma-ray such that the material is tightly packed with minimum air space in case of X-ray.
4. Take the treated seedling for planting in glasshouse or field (Note 3).

3.3 Example Radio-Sensitivity Testing of Arabica Coffee Seed Under Laboratory Conditions

Seed from three *C. arabica* varieties (Kent, Geisha and Mundo Novo) from the Cocoa Research Institute of Nigeria were freshly harvested and sent within less than one month to the FAO/IAEA PBG Laboratory, Austria. Upon arrival, the seeds were immediately inspected, cleaned, and used for gamma and X-ray radiosensitivity testing as described above. Germination was scored using the laboratory-based procedure about 2 weeks after treatment and data analysed to estimate LD_{30} and LD_{50} values as well as measurement of seedling growth rate relative to the untreated seeds (control) to estimate GR_{30} and GR_{50} values (Note 4). The viability of the seeds of

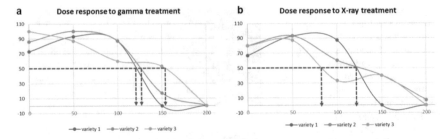

Fig. 4 Radio-sensitivity testing based on germination percent of the M_1 seeds using gamma-ray (**a**) and X-ray (**b**) in-house irradiation sources in three *C. arabica* varieties. As shown, for gamma treatment, the LD_{30} ranged between 38 and 75 Gy, compared to 16–52 Gy for the X-ray treatment, while the LD_{50} ranges between 118 and 150 Gy for gamma, and between 105 and 136 Gy using X-ray

Table 1 Comparing LD_{30} and LD_{50} using the FAO/IAEA in-house gamma and X-ray sources and their relative biological effect (RBE) in three *C. arabica* coffee varieties

Variety	Gamma-ray		X-ray		RBE	
	LD_{30}	LD_{50}	LD_{30}	LD_{50}	LD_{30}	LD_{50}
B	38	118	28	113	0.73	0.95
C	70	144	52	136	0.74	0.94
I	75	150	16	105	0.21	0.70

the three coffee varieties exceeded 95% of germination on moist filter paper after soaking in warm water for 3 days.

As show in Fig. 4, the LD_{30} ranged between 38 and 75 Gy among the three varieties for gamma treatment, compared to 16–52 for the X-ray, while the LD_{50} ranged between 118 and 150 Gy for Gamma and between 105 and 136 using X-ray (Table 1).

The relative biological effect (RBE) of gamma to X-ray ranged from 0.21 to 0.74 for LD_{30} and from 0.7 to 0.95 for LD_{50} indicating that Gamma and X-ray were relatively closer in their effect in LD_{50} (Table 1).

4 Notes

1. When performing a radio-sensitivity test 15–20 seeds are placed in paper bag per replication (3 replications). In case of X-ray irradiation, bags of the same dose are rolled together with the seed well distributed at the bottom and then placed in the center of the container to ensure uniform radiation. The remaining space is filled with instant rice for vacuum establishment. For bulk irradiation large amounts of seeds are used and these may be placed in an appropriate container for uniform irradiation without the use of a filler.

2. Moisture content. The moisture content of the plant propagule to be irradiated is a critical factor. In barley, for instance, it has been shown that at seed moisture content below 14%, there is marked increase in mutation frequencies as the moisture content decreases. It is therefore necessary to equilibrate the moisture content of seed prior to irradiation.

3. For radio-sensitivity test 10–20 propagules per replication are sufficient (3 replications). For bulk irradiation use large population sizes. In case of seed, it is recommended to target the production of an M_2 population of a minimum of 5,000–10,000 individuals.

4. It is strongly recommended to conduct a radiosensitivity testing of coffee seed also under greenhouse conditions (e.g., in pots or trays with soil) and monitor survival or growth for longer periods, e.g., until the cotyledons or first true leaves have appeared, to mimic as closely as possible the greenhouse or field conditions where the mutant population(s) will eventually be sown. Given the long germination time of coffee seed, such radiosensitivity testing may take up to 1 to 3 months, depending on the seed quality and growth conditions.

Acknowledgements The protocols and findings described here were developed at the FAO/IAEA PBG Laboratory in the context of an Internship for Mr. Keji Dada, Cocoa Research Institute of Nigeria (CRIN), Ibadan, Nigeria. Seeds of the different *C. arabica* varieties used in this study were provided by CRIN.

References

Carvalho A, Antunes Filho H, Nogueira RK (1954) Genética de coffea. XX. Resultados preliminares do tratamento de sementes de café com raios-X. Bragantia, Campinas, 13 XVII-XX (Nota, 7)

Lashermes P, Carvalho Andrade A, Etienne H (2008) Genomics of coffee, one of the world's largest traded commodities. In: Moore PH, Ming R (eds) Genomics of tropical crop plant, Chap 9

Mastrangela T, Parker AG, Jessup A, Pereira R, Orozco-Davila D, Islam A, Dammalage T, Walder JMM (2010) A new generation of X-ray irradiators for insect sterilisation. J Econ Entomol 103(1):85–94

Mehta K, Parker A (2011) Characterization and dosimetry of a practical X-ray alternative to self-shielded gamma irradiators. Radiat Phys Chem 80:107–113

Scalabrin S, Toniutti L, Di Gaspero G et al (2020) A single polyploidization event at the origin of the tetraploid genome of *Coffea arabica* is responsible for the extremely low genetic variation in wild and cultivated germplasm. Sci Rep 10:4642. https://doi.org/10.1038/s41598-020-61216-7

Spencer-Lopes MM, Forster BP, Jankuloski L (2018) Manual mutation breeding and biotechnology, 3rd edn

Wintgens JN (2004) Cofee bean quality assessment. In: Cofee: Growing, processing, sustainable production. A guidebook for growers, processors, traders, and researchers. Wiley-VCH Ver- lag Gmbh & Co., Weinheim, pp 810–819

Mutation Induction in *Coffea arabica* L. Using in Vivo Grafting and Cuttings

Weihuai Wu, Xuehui Bai, Kexian Yi, Xing Huang, Chunping He, Jinhong Li, Hongbo Zhang, Hua Zhou, Thomas Gbokie Jr, Tieying Guo, and Jingen Xi

Abstract Coffee leaf rust (CLR) caused by the obligate parasite, the biotrophic *Hemileia vastatrix* Berk. & Broome (Basidiomycetes: Pucciniales), is the most devastating disease of *Coffea arabica* L. Breeding resistant varieties is one of the most economic and environment friendly means to control the disease. However, this is challenged by the loss of resistance after a short period in commercial production. Catimor CIFC7963, an elite, leaf rust resistant *Coffea arabica* L. variety, has been cultivated in China for decades, which has resulted in the breakdown of its disease resistance. Due to the lengthy breeding process of coffee, the development of new resistant varieties is arduous. Physical and chemical mutagenesis offers an alternative means to more rapidly create novel and beneficial genetic variations. Bud grafting is a propagation technique frequently used for woody plants whereby a bud of one plant is attached to the rootstock of another plant. Likewise, cutting is a frequently used propagation technique. In coffee, physical irradiation of the bud followed by grafting or cutting can accelerate the mutation breeding process, as cutting or grafting increases the growth rate without affecting the major traits of the background varieties. Here, we present protocols to induce mutations on buds of the *C. arabica* variety Catimor CIFC7963 by gamma-ray irradiation and their subsequent propagation through cutting or bud grafting.

W. Wu · K. Yi (✉) · X. Huang · C. He · J. Xi
Hainan Key Laboratory for Monitoring and Control of Tropical Agricultural Pests, Environment and Plant Protection Institute, Chinese Academy of Tropical Agricultural Sciences, Haikou 571101, Hainan, China
e-mail: yikexian@126.com

X. Bai · J. Li · H. Zhang · H. Zhou · T. Guo
Dehong Institute of Tropical Agriculture, Ruili 678600, Yunnan, China

T. Gbokie Jr
College of Plant Protection, Nanjing Agricultural University, Nanjing 210095, China

I. L. W. Ingelbrecht et al. (eds.), *Mutation Breeding in Coffee with Special Reference to Leaf Rust*, https://doi.org/10.1007/978-3-662-67273-0_11

1   Introduction

Coffee is one of the most important beverages in the world. The genus *Coffea* has over 70 different species. Of these, only two species, namely *Coffea arabica* (Arabica) and *C. canephora* (Robusta) dominate global coffee production. *C. arabica* is a dominant species with a demand for genetic improvement, especially for disease resistance such as coffee leaf rust (CLR) (Tran et al. 2018). CLR caused by the fungus *Hemileia vastatrix* is one of the most important diseases of *C. arabica*. A few rust-resistant varieties have been developed, *e.g.* Oeiras MG 6851 (Pereira et al. 2000), however, the resistance was soon broken down by the newly emerged race XXXIII of *H. vastatrix* (Capucho et al. 2012). Thus, widening the genetic base would be key for developing sustainable disease resistance in coffee, given that Coffee-*H. vastatrix* rust interactions follow the gene-for-gene relationship (Flor 1942). Transcriptome and proteome methods have been used to identify leaf rust resistance genes (Guerra-Guimarães et al. 2015; Florez et al. 2017). While various *H. vastatrix* effector candidate genes (*HvECs*) have been reported, the knowledge of leaf rust resistance genes in coffee (S_H1-S_H9) is still limited (Maia et al. 2017). There is still insufficient knowledge to develop functional markers for marker-assisted selection for CLR resistance. Mutation breeding seems to be a promising approach for coffee improvement to help address the leaf rust resistance breakdown due to rapidly evolving *H. vastatrix* races.

Compared to conventional breeding techniques such as traditional crossing, induced mutagenesis can more efficiently generate novel variations and introduce new traits (Harten 1998). Gamma irradiation has proven to be effective in improving important agronomic traits such as yield, quality and disease resistance. The FAO/IAEA Joint Centre has significantly promoted the application of mutation breeding in agriculture, which generated more than 3400 varieties in 210 plant species and commercially planted in more than 70 countries (https://mvd.iaea.org/).

It takes over three years for *C. arabica* from planting until fruit production. This makes its propagation and breeding an extremely lengthy process. For tree crops, grafting offers an efficient way to propagate and maintain elite germplasm (Parlak 2017). Grafting technique has already been successfully applied in coffee for preventing damage by root-lesion nematodes (Villain et al. 2000). In addition, cutting is also widely used to propagate tree plants for maintaining outstanding traits. Cutting is also applied for clonal propagation of coffee plants, which is more stable than seed propagation for maintaining yield traits (Priyono et al. 2010). The survival rate of cutting is influenced by environmental factors such as exogenous phytohormones, humidity, temperature, etc. It results in a lower survival rate compared to grafting. On the other hand, cutting is a simpler method than grafting as it does not require a rootstock. Here, we present a protocol for gamma irradiation of coffee in combination with grafting and cutting techniques.

2 Materials

2.1 Grafting

1. Catimor CIFC7963 coffee plants.
2. Two-year-old Catimor CIFC7963 plant (rootstock).
3. Plastic strips.
4. Plastic label.
5. Grafting knife.
6. Tree pruner scissors.
7. Floral foam for flower arrangement.
8. Plastic storage box.

2.2 Cutting

9. Catimor CIFC7963 coffee plants.
10. Floral foam for flower arrangement.
11. Plastic bin.
12. River sand.
13. Flowerpot.
14. Difenoconazole (CAS: 119446-68-3).
15. 3000 mg/L of naphthaleneacetic acid (NAA) solution.
16. White plastic bag (keep moisture and transparency).
17. Newspaper.
18. Watering can.
19. Plastic greenhouses.

3 Methods

3.1 Mutagenesis and Propagation of Coffee Through Bud Grafting

3.1.1 Preparation of Straight Branches

1. Remove the lateral branches and leaves using tree pruner scissors, retain only the top leaves (*see* Note 1 and Fig. 1a).
2. Cut the bottom of the straight branches into a tongue-shape and put them into the floral foam for flower arrangement (*see* Note 2 and Fig. 1b).
3. Transfer floral foam with the coffee shoots into plastic containers for gamma-ray irradiation using Cobalt-60 source (*see* Note 3).

Fig. 1 Grafting procedure. **a** Straight shoot; **b** the floral foam for arrangement of the coffee shoots; **c** irradiation; **d** preparation of the rootstock; **e–g** grafting; **h** successful graft

3.1.2   Mutagenic Treatment

1. Take the straight branches out of the plastic container and divide into 6 sets, each containing 70 shoots, for irradiation using different dosages.
2. Perform mutagenic treatment using 0, 5, 10, 15, 20, and 25 Gy of gamma rays at the rate of 1 Gy/min (Fig. 1c).
3. Transfer the treated shoots into a greenhouse for grafting.

3.1.3   Preparation of the Rootstock

1. Remove the trunk of a 2-year-old Catimor CIFC7963 with only 10–15 cm stump retained (Fig. 1d).
2. Use a grafting knife to cut an incision on one side of the rootstock up to the xylem with a length of about 3 cm (Fig. 1e).

3.1.4   Preparation of the Scion

1. Take the irradiated shoots, retain the green unpinched cork with 1–2 buds on the top by cutting off the lower part.

2. Cut the scion with a grafting knife by two chops. The length of the two cutting faces should be about 3 cm, each side should be uneven with the other side in length as well as thickness. The width of the cutting side should be slightly narrower or equal to the diameter of the rootstock (Fig. 1f).

3.1.5   Grafting Method

1. Once the scion and rootstock have been cut, insert the graft in the rootstock, ensuring that the graft union is well aligned.
2. Immediately and tightly wrap the joint of the scion and stock with plastic strips (*see* Note 4 and Fig. 1g).

3.1.6   Post-grafting Management

1. Grow the grafted plants in a plastic greenhouse with temperature 20–25 °C and natural light.
2. Keep the rootstock clean by removing any emerging buds and shoots (*see* Note 5).

3.1.7   Survival Rate Statistics

1. One to two months after grafting, the successful graft is identified by the presence of emerged leaves on the scion and healing of the contact point between the scion and rootstock (*see* Note 6 and Fig. 1h).

3.1.8   Unbundling

1. Remove the wrapping tape when the scion tip is over 10 cm long and the interface is completely healed.

3.2   Mutagenesis and Propagation of Coffee Through Cuttings

3.2.1   Preparation of Straight Branches

1. Remove the lateral branches and leaves using tree pruners, retain only the top leaves on the shoots (*see* Note 1 and Fig. 2a).
2. Cut the bottom of the straight branches into a tongue-shape and arrange the cuttings in the floral foam (*see* Note 2 and Fig. 2b).
3. Transfer the floral foam with the coffee cuttings into plastic containers for gamma-ray irradiation using a Cobalt-60 source (*see* Note 3).

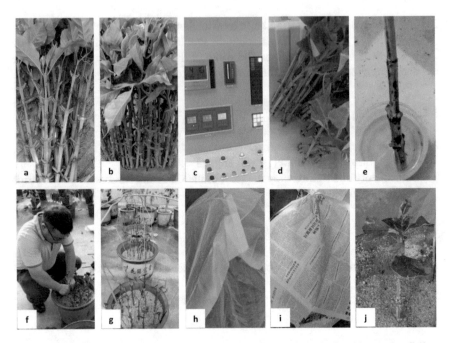

Fig. 2 Cutting procedures. **a** Straight shoot; **b** arrange coffee shoots in the floral foam; **c** irradiation; **d** 0.1% difenoconazole sterilizing solution; **e** NAA solution treatment; **f** stick cutting into the most rooting substrate; **g–h** set up wire support and cover with plastic bag to maintain moisture and humidity; **i** shading; **j** established cutting

3.2.2 Mutagenesis

1. Take the straight branches out of the plastic container and divide into six sets, each containing 70 shoots, for irradiation using different dosages.
2. Perform the mutagenic treatment using 0, 5, 10, 15, 20, and 25 Gy of gamma rays at the rate of 1 Gy/min (Fig. 2c).
3. Transfer the treated shoots to a greenhouse for cutting (*see* Note 4).

3.2.3 Preparation of the Cutting Substrate

1. Fill pots with river sand until it is 5–10 cm away from the top of the pot.
2. Level the sand bed and keep moist.
3. Spray 0.1% difenoconazole solution to sterilize the sand bed.
4. Perforate the sand bed at 10 cm ± 2 cm distance, and a depth of about 20 cm using bamboo skewers slightly thicker than the cutting strip.

3.2.4 Sterilization

1. Remove the irradiated shoots from the floral foam and soak them in 0.1% difenoconazole solution for 5–10 min (Fig. 2d).
2. Remove the corky part on the bottom of the straight branch and retain the upper, green part with 2–3 buds, *i.e.* the scion.

3.2.5 Cultivation

1. Cut the end of the cuttings at an angle of about 45° and dip in a solution of 3,000 mg/L NAA for 3–6 s (Fig. 2e).
2. Insert the cuttings vertically into prepared holes in the sand bed at a depth of 15–20 cm (Fig. 2f).
3. Wet the sand bed. Build a mini greenhouse environment with a height of 50–60 cm by using two wire brackets and plastic bags to maintain humidity and moisture (Fig. 2g, h).
4. Keep the cuttings in the dark for 2 days (Fig. 2i).
5. Maintain the temperature at 20–25 °C and observe the growth of the cuttings regularly.
6. Water every 3–5 days, and tightly wrap the plastic bags to maintain humidity and moisture.
7. The culture of the cuttings takes around 2–4 months.
9. Spray 0.1% difenoconazole for sterilization every 20–25 days.
10. Immediately remove any necrotic or dead cuttings.

3.2.6 Survival Rate Statistics

1. After 6 months, successful cuttings are identified by the newly emerging leaves and the presence of fibrous roots (Fig. 2j).

4 Notes

1. The shoots of Catimor CIFC7963 were collected from the Coffee Germplasm Resource Garden in Ruili, Ministry of Agriculture and Rural Areas. Coffee shoots containing 5–6 buds were used in this protocol.
2. Collect only straight branches, as side branches are not able to develop into upright trees.
3. In cases of long-distance transportation, try to avoid cold stress on the cuttings.
4. To ensure high survival rate, optimal timing for coffee grafting or cutting should be set during spring season, between March and April, when the coffee plants are not flowering, or in the fall, September to October, prior to the fruit harvest.

5. To ensure that the new leaves originate from the scions, newly emerging buds should be regularly removed from the rootstock once the grafting process is completed.
6. Dry scions or scions that do not grow or drop are typical features of failed grafting.

Acknowledgements Funding for this work was provided by the FAO/IAEA Coordinated Research Project (Contract No. 20380), and the International Exchange and Cooperation Project funded by the Agricultural Ministry Construction of Tropical Agriculture Foreign Cooperation Test Station and Training of Foreign Managers in Agricultural Going-Out Enterprises (SYZ2019-08) and the Central Public-interest Scientific Institution Basal Research Fund for Chinese Academy of Tropical Agricultural Sciences (No. 1630042017021).

References

Capucho AS, Zambolim EM, Freitas RL, Haddad F, Caixeta ET, Zambolim L (2012) Identification of race XXXIII of *Hemileia vastatrix* on *Coffea arabica* Catimor derivatives in Brazil. Australas Plant Dis Notes 7:189–191

Flor HH (1942) Inheritance of pathogenicity in *Melampsora lini*. Phytopath 32:653–669

Florez JC, Mofatto LS, Freitas-Lopes RL, Ferreira SS, Zambolim EM, Carazzolle MF, Zambolim L, Caixeta ET (2017) High throughput transcriptome analysis of coffee reveals prehaustorial resistance in response to *Hemileia vastatrix* infection. Plant Mol Biol 95:607–623

Guerra-Guimarães L, Tenente R, Pinheiro C, Chaves I, Silva MC, Cardoso FMH, Planchon S, Barros DR, Renaut J, Ricardo CP (2015) Proteomic analysis of apoplastic fluid of *Coffea arabica* leaves highlights novel biomarkers for resistance against *Hemileia vastatrix*. Front Plant Sci 6:478. https://doi.org/10.3389/fpls.2015.00478

Guo TY, Zhang HB, Xiao ZW, Li JH, Zhou H, Zhao MZ, Bai XH (2019) Practical cutting propagation technique of *Coffea arabica*. L. Trop Agricult Chin 2:75–76

Macovei A, Garg B, Raikwar S, Balestrazzi A, Carbonera D, Buttafava A, Bremont JFJ, Gill SS, Tuteja N (2014) Synergistic exposure of rice seeds to different doses of γ-ray and salinity stress resulted in increased antioxidant enzyme activities and gene-specific modulation of TC-NER pathway. BioMed Res Int 2014. https://doi.org/10.1155/2014/676934

Maia T, Badel JL, Marin-Ramirez G, Rocha CM, Fernandes MB, da Silva JCF, de Azevedo-Junior GM, Brommonschenkel SH (2017) The Hemileia vastatrix effector HvEC-016 suppresses bacterial blight symptoms in coffee genotypes with the S_H1 rust resistance gene. New Phytol 213:1315–1329

Parlak S (2017) Clonal propagation of mastic tree (*Pistacia lentiscus* var. chia Duham.) in outdoor beds using different rootstock and grafting techniques. J For Res 29:1061–1067

Pereira AA, Zambolim L, Chaves GM, Sakiyama NS (2000) Cultivar de café resistente à Ferrugem: Oeiras-MG 6851. Revista Ceres 47(269):121–124

Priyono FB, Rigoreau M, Ducos JP, Sumirat U, Mawardi S, Lambot C, Broun P, Pétiard V, Wahyudi T, Crouzillat D (2010) Somatic embryogenesis and vegetative cutting capacity are under distinct genetic control in *Coffea canephora* Pierre. Plant Cell Rep 29(4):343–357

Siegel CS, Stevenson FO, Zimmer EA (2017a) Evaluation and comparison of FTA card and CTAB DNA extraction methods for non-agricutural taxa. Appl Plant Sci 5(2):1600109

Siegel CS, Stevenson FO, Zimmer EA (2017b) Evaluation and comparison of FTA card and CTAB DNA extraction methods for non-agricultural taxa. Appl Plant Sci 5(2):1600109

Tran HT, Ramaraj T, Furtado A, Lee LS, Henry RJ (2018) Use of a draft genome of coffee (*Coffea arabica*) to identify SNPs associated with caffeine content. Plant Biotechnol J 16(10):1756–1766

van Harten AM (1998) Mutation breeding: theory and practical applications. Cambridge University Press, Cambridge. ISBN 9780521470742-353

Várzea VMP, Marques DV (2005) Resistance, population variability of *Hemileia vastatrix* vs. coffee durable. In: Zambolim L, Zambolim E, Várzea VMP (eds) Durable resistance to coffee leaf rust. UFV Press, Viçosa, pp 53–74

Villain L, Molina A, Sierra S, Decazy B, Sarah JL (2000) Effect of grafting and nematicide treatments on damage by root-lesion nematodes (*Pratylenchus* spp.) to *Coffea arabica* L. in Guatemala. Nematropica 30(1):87–100

Zambolim L (2016) Current status and management of coffee leaf rust in Brazil. Trop Plant Pathol 41:1–8

Chemical Mutagenesis of Mature Seed of *Coffea arabica* L. var. Venecia Using EMS

Joanna Jankowicz-Cieslak, Florian Goessnitzer, Stephan Nielen, and Ivan L. W. Ingelbrecht

Abstract Chemical mutagens are a major tool to generate novel genetic variation in crops, functional genomics and breeding. They are advantageous because they do not require any specialized equipment and can induce a high mutation frequency. Compared to physical methods, chemical mutagens cause point mutations rather than deletions or translocations. Point mutations can have varying effects on gene expression ranging from knockouts to changes in amino acids that may have subtle effects on protein function. Many important gene functions have been uncovered by evaluating the in vivo effect of mutated genes in a broad range of model and crop plants. Chemical mutagens have been successfully applied to induce tolerance to fungal diseases in cereals such as barley and wheat. Among the chemical mutagens used for plant mutagenesis, ethyl methanesulfonate (EMS) is the most widely applied. This protocol chapter describes the utilization of EMS for establishing kill curves and generating a mutagenized population of *Coffea arabica* var. Venecia via treatment of mature seed. The different steps of the mutagenesis process are described, including quality control and preparation of the seed batches, procedures for determining Lethal Dose (LD) and Growth Reduction (GR) values, and for post-treatment handling of the chemically mutagenized seed, specific for Arabica coffee.

1 Introduction

Mutation breeding has proven to be an efficient tool to develop improved crop varieties whereby various novel traits such as e.g., yield, growth, and disease resistance can be induced (Spencer-Lopes et al. 2018). Among various mutation induction techniques, mutagenesis using chemical agents has become an efficient and

J. Jankowicz-Cieslak (✉) · F. Goessnitzer · I. L. W. Ingelbrecht
Plant Breeding and Genetics Laboratory, Vienna International Centre, Joint FAO, IAEA Centre of Nuclear Techniques in Food and Agriculture, IAEA Laboratories Seibersdorf, International Atomic Energy Agency, Vienna, Austria
e-mail: j.jankowicz@iaea.org

S. Nielen
EMBRAPA Recursos Genéticos e Biotecnologia, Brasília, DF, Brazil

© The Author(s) 2023
I. L. W. Ingelbrecht et al. (eds.), *Mutation Breeding in Coffee with Special Reference to Leaf Rust*, https://doi.org/10.1007/978-3-662-67273-0_12

robust tool to induce point mutations (Ingelbrecht et al. 2018; Jankowicz-Cieslak and Till 2016). Mutations in coding regions of genes can be silent, missense, or nonsense, while mutations can equally occur in non-coding, regulatory sequences affecting gene expression, e.g., at intron splice sites. Among the different chemical mutagens, ethyl methanesulfonate (EMS) has been frequently used to induce random mutations because it is highly effective and relatively easy to handle. To date, mutation induction using EMS has been described for a wide range of plant species covering both seed and vegetatively propagated plants (Jankowicz-Cieslak et al. 2011; Jiang et al. 2022). Chemical mutagens have been successfully applied to induce tolerance to fungal diseases in cereals such as powdery mildew-resistance in barley (Molina-Cano et al. 2003) and resistance to leaf rust *Puccinia* sp. in wheat (Mago et al. 2017).

The workflow for chemical mutagenesis involves choosing the genotype and the appropriate tissue type to be mutated, the mutagen, and optimizing the treatment conditions and dosage. Once the optimal dose(s) have been determined, bulk mutagenesis can proceed. EMS mutagenesis of seed is widely used in diploid, self-pollinating cereals and legumes, amongst other plant species. Briefly, seed are pre-treated by soaking in water, EMS is added, and mutations are induced during a specific treatment period. Following this, seed are thoroughly washed and sown. This results in the M_1 generation which can harbor a high density of induced point mutations.

Growth and survival measurements remain the simplest route to dosage optimization and have the added appeal that they can be applied in almost any facility as little infrastructure or expertise are required. A reduction of growth rate in the early seedling stage, typically two weeks for small-seeded cereals, serves as an easy indicator (Jankowicz-Cieslak and Till 2016; Mba et al. 2010). A range between 30 and 50% growth reduction is typically chosen as optimal dose for bulk irradiation in cereals. As with any mutagenesis treatment, a compromise needs to be made between achieving a sufficiently high mutations frequency to have a reasonable chance to recover the desired mutations and suitable level of survival and fecundity (Jankowicz-Cieslak and Till 2015). It is advisable to use up to three doses of the mutagen, corresponding to $\pm$ 20% of the optimal dose found through the toxicity test.

For chemical mutagens such as EMS, both the concentration and duration of the treatment are evaluated during dose optimization. Different treatments are tested, and germination and survivability as well as growth reduction typically measured. Where possible, embryo lethality in the M_2 seed can be used as an indicator for the efficiency of mutagenesis (Till et al. 2003). There are additional issues with chemical mutagenesis that one needs to consider. Cytotoxicity may limit the efficacy of specific mutagens for certain plant species or genotypes, necessitating trials with different mutagens (Till et al. 2007). Following the mutagenesis procedure, M_2 mutant populations can be evaluated for phenotypic or genotypic variation distinct from the non-mutagenized parental line.

The advantages and limitations of different propagules for mutagenesis treatment of Arabica coffee are briefly described in Chap. Mutation Breeding in Arabica Coffee. Since Arabica coffee is self-fertilizing and mostly propagated through seeding, seed can serve as starting material for mutagenesis treatments. As with other seed crops, M_1 chimeric plants are expected which can be resolved through successive cycles of selfing (Mukhtar Ali Ghanim et al. 2018). In this chapter, the susceptibility of Arabica coffee seed var. Venecia to the chemical mutagen EMS was evaluated and optimal doses for EMS bulk irradiation were determined. Methods of evaluation of mutagenic effect at the seedling stage are presented. Further, example data on morphological and chlorophyl variegations observed at the M_1 stage are shown.

2 Materials

2.1 Chemical Toxicity Test

1. Chemical mutagen laboratory equipped with fume hood and flow bench (*see* Note 1).
2. Labelled waste receptacle for dry hazardous material and collection vessels for EMS waste solution.
3. High quality, disease-free coffee seeds (*see* Note 2).
4. Stainless steel tea-steeper.
5. Ethyl-methanesulphonate (EMS) AR grade, M.W. 124.2 (*see* Note 3).
6. 10% (w/v) sodium thiosulfate (Na2S2O3.5H2O) (*see* Note 4).
7. Dimethyl sulfoxide (DMSO).
8. Sterile water.
9. Deionized water.
10. 50 ml falcon tubes.
11. Syringe.
12. Needle.
13. Disposable pipettes.
14. Pipette bulb.
15. Graduated cylinders.
16. Bottles (100, 500 ml).
17. Beakers (500 and 1000 ml).
18. Sieves (metal, 70 mm diam., 10–100 μm pore size).
19. Orbital shaker.
20. Disposable pipettes (5, 25 ml).
21. Soil: light soil containing peat moss and sand (0.5–2 mm) in a ratio of 3:1 (peat: sand) with acidic pH (5–6).
22. Pots and multiwell trays.
23. Glasshouse with light and temperature control (*see* Note 5).

2.2 Calculation of Lethal Dose (LD) and Growth Reduction (GR)

1. Pen.
2. Notebook.
3. Ruler.
4. Standard spreadsheet software *e.g.,* Microsoft Excel.

2.3 Bulk Mutagenesis

1. All materials as listed in 2.1.
2. High quality, disease-free coffee seeds (*e.g.,* 1000).

2.4 Phenotyping

1. Pen.
2. Notebook.
3. Ruler.
4. Standard spreadsheet software *e.g.,* Microsoft Excel.
5. Coffee descriptors.
6. Camera.
7. Photobooth (optional).

3   Methods

3.1 Seed Quality Control and Pre-treatment

1. Perform a germination test of seed lots as part of quality control. The germination frequency should be minimum 90%.
2. Select seeds from genetically homogenous plants from the targeted preferred variety.
3. Sort out clean, similar in size, normal shaped and disease-free seeds. Discard small, shriveled, discolored and damaged seeds from the experimental lot (Fig. 1a and *see* Note 6).
4. Manually remove the outer parchment and the silverskin underneath from every seed.
5. Divide the seeds into batches, each about 50 seeds. Leave out one batch as a control (untreated). The remaining batches correspond to the possible combinations of concentrations of EMS (Fig. 1b and *see* Note 7).

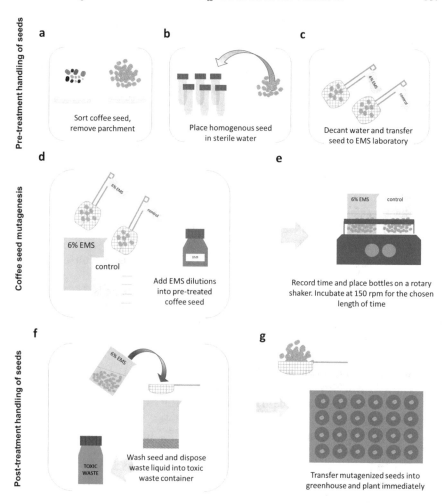

Fig. 1 Overview of EMS mutagenesis of mature coffee seed. The procedure involves three steps: (i) quality control and pre-treatment of the seeds, (ii) seed mutagenesis; (iii) post-treatment handling. **a** Ahead of chemical mutagenesis high quality coffee seeds are selected that are uniform, homozygous and isogenic, while discoloured, small or damaged seeds are discarded. **b** After removal of the parchment and the silverskin, the seeds are pre-soaked in sterile, distilled water for 48 h at room temperature. **c** After incubation, the water is decanted from the seed batches. **d** Seeds are placed into labelled beakers for treatment. A dilute mixture of EMS plus DMSO is prepared and stored under the fume hood. **e** The seeds are incubated for a specific time under orbital rotation. **f** The mutagen is removed, and seeds are thoroughly washed, minimum three times. **g** Coffee seeds are then transferred to the glasshouse and immediately planted in a light soil

6. Place seed batches in appropriately labelled 50 ml Falcon tubes filled with sterile distilled water.
7. Leave standing for 48 h at room temperature (*see* Note 8).
8. Carefully decant water and transfer coffee seed into either a labelled petri plate or a tea-steeper (Fig. 1c and *see* Note 9).
9. Transfer pre-soaked seed batches to the chemical mutagenesis facility.
10. Immediately proceed with the chemical mutagenesis experiment.

3.2 Coffee Seed Mutagenesis

1. Review safety procedures of the chemical mutagenesis laboratory and Consult the Materials Safety Data Sheets for all chemicals used.
2. Prepare the laboratory (*see* Note 10).
3. Choose appropriate concentrations of EMS solution and incubation times (Table 1 and *see* Note 11).
4. Prior to working with EMS, test the bottles used for mixing EMS/DMSO (*see* Note 12).
5. Prepare a bottle containing 100 mM sodium thiosulfate and place in the fume hood along with a role of paper towels.
6. In the fume hood, prepare fresh EMS dilutions by adding the required volume of EMS solution to the water/DMSO mixture (*see* Table 1). Use a sterile syringe and a 0.2 μm filter for this step. Place syringe and filter into a beaker containing 100 mM sodium thiosulfate to inactivate EMS before placing in hazardous waste (*see* Note 13).
7. Seal the bottle of each prepared EMS and DMSO solution with a screw cap.
8. Wipe the outside of stock bottles with a paper towel soaked in sodium thiosulfate.
9. Return EMS and DMSO stock bottles to their storage location.
10. Ensure the sash is lowered on the fume hood and shake the EMS/DMSO solution vigorously for 15 s (*see* Note 14).
11. In the fume hood, place each pre-soaked coffee seed batch into empty treatment beaker labelled with the appropriate treatment code (EMS concentration and replication number).
12. Carefully pour the determined volume of EMS solution to every beaker (Fig. 1d and *see* Note 15).
13. Prepare control batch in the same way by adding water to the beaker.

Table 1 Example calculations for solutions containing different EMS concentrations and 2% DMSO at a final volume of 1 L used for mutagenic treatment of var. Venecia

Final EMS concentration (%)	0	0.2	0.5	1	2	4	6
Volume EMS (ml)	0	2	5	10	20	40	60
Volume DMSO (ml)	20	20	20	20	20	20	20
Volume Water (ml)	980	978	975	970	960	940	920

14. Wipe the outside of beakers containing EMS mixture with a paper towel soaked in sodium thiosulfate.
15. Place beakers (including control) on a rotary shaker set at 80 rpm, record the time and temperature and start incubation (Fig. 1e and *see* Note 16).
16. Pour 100 mM sodium thiosulfate into bottles used to prepare EMS dilutions.
17. Wipe the outside of the bottle with a paper towel soaked in sodium thiosulfate.
18. Dispose of liquid and solid waste in appropriate toxic waste containers.
19. Wipe the fume hood with a paper towel soaked in sodium thiosulfate.
20. After the incubation time, transfer beakers into the fume hood.
21. Quickly but carefully decant each of the treatment batches into a waste beaker and rinse with sterile water (Fig. 1f and *see* Note 17).
22. Pour the EMS waste-solution into a hazardous waste container and repeat the wash steps.
23. Repeat steps 21 and 22 for a total of three water washes (*see* Note 18).
24. Collect all the liquid waste in a dedicated bucket labelled as hazardous waste.
25. Detoxify the waste and all unused EMS solution by adding sodium thiosulfate in a 3:1 ratio by volume.
26. Dispose of toxic waste according to your local Laboratory Safety Rules and Regulations (*see* Note 19).
27. Decontaminate all surfaces and equipment by wiping down with 100 mM sodium thiosulfate followed by a water rinse (*see* Note 20).
28. Transfer mutagenized seed into the greenhouse and immediately proceed with sowing (Fig. 1g and *see* Note 21).

3.3 Post-treatment Handling of Seeds

1. Prepare soil mix.
2. Distribute the soil uniformly in multiwell trays (Fig. 2a).
3. Irrigate the soil and let excess water drain.
4. Label the trays, include the genotype, replication, EMS concentration and treatment date.
5. To limit any detrimental effect of EMS on seed viability, sow the seeds immediately after treatment (Fig. 2a).
6. Water regularly under appropriate light and temperature conditions to maximize growth and survival.
7. Transplant to larger size pots approximately 7 months after treatment for further growth (Fig. 2b).

Fig. 2 Coffee plants grown in the FAO/IAEA Plant Breeding and Genetics Laboratory, Seibersdorf, Austria. **a** Mutagenized coffee seeds are sown in multiwell trays filled with light soil immediately after EMS treatment. **b** Seven months post-mutagenesis, plants are transplanted to bigger pots

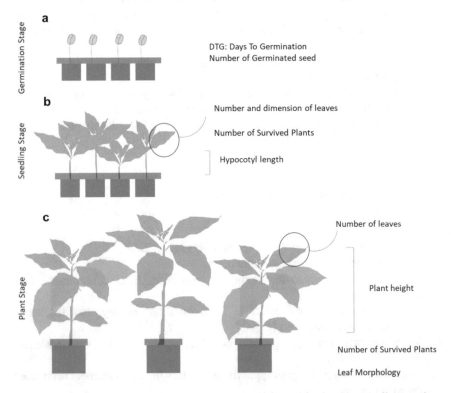

Fig. 3 **a** Coffee seed are germinated in the glasshouse and the germination date as well as germination rate is scored. **b** At the seedling stage (*e.g.,* 104 days post-mutagenesis) survival count is taken along with hypocotyl/seedling height and number/dimension of leaves. **c** At the plant stage (*e.g.,* 208 days post-mutagenesis) the survival rate, plant height, number of leaves and leaf morphology and other variegations are recorded

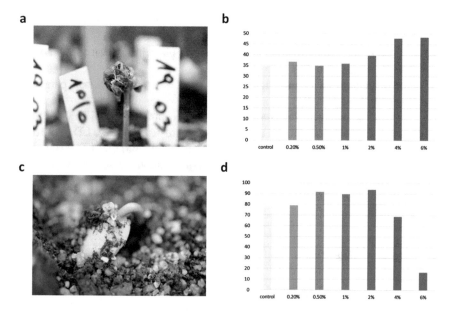

Fig. 4 Coffee seed germination scored for estimation of 50% lethality (LD_{50}): **a** Mutagenized materials are first grown, and the germination date is registered which serves as the basis for the calculation of the Days to Germination values (DTG). **b** The germination of EMS treated seed occurred between 34 and 63 days after sowing. Here the average of DTG values for each treatment is displayed which ranges between 35 days for the control and 0.5% EMS and 48 days for 4% and 6% EMS. **c** The number of germinated seed was recorded at the stage of the seedling emerging from the soil. **d** Nearly 80% of planted control seeds emerged. Germination was visibly enhanced for 4 EMS treatments (0.2% up to 2%) and then dropped to 69% at 4% EMS and to 17% at 6% EMS

3.4 Calculation of Lethal Dose (LD): Data Collection and Analyses

1. Maintain mutagenized coffee seeds in the greenhouse conditions.
2. Identify stages during which you will take measurements, starting from the germination stage, through seedling up to the plant stage (Figs. 3 and 4 c and *see* Note 22).
3. After ca 30 days of incubation start monitoring the germination of coffee seeds daily (the moment when the coffee plant emerges from the soil). Record the germination date of every seed. Mutagenic treatment might delay the seed germination process. In the current experiment, germination occurred between 34 and 63 days after sowing (Fig. 3a).
4. Count surviving coffee plants at the seedling and plant stage (Fig. 3b, c). Surviving plants are those which show continued growth. Plants which germinate but where growth is inhibited, and no development of true leaves is observed, are being counted as non-surviving plants.
5. Record other observations (*e.g.*, Fig. 3).

6. Enter data for each recorded trait into a spreadsheet (*e.g.,* Microsoft Excel).
7. Calculate the averages for each parameter and each treatment separately.
8. Calculate percentage of the parameter of the plant in relation to the control (Fig. 4 and *see* Note 23).
9. Plot percentage of control against mutagenic treatment for each parameter (Fig. 4 and *see* Note 24).
10. Estimate the mutagen concentration required to obtain 50% and 70% germination or survival relative to the untreated control to determine the LD_{50} (the dose expected to cause 50% death of an exposed population) and LD_{30} (the dose expected to cause 30% death of an exposed population) values (*see* Figs. 4 and 5).
11. Choose a concentration to perform larger scale mutagenesis, called bulk treatment.

3.5 Bulk Mutagenesis

1. Choose the appropriate EMS concentration(s) and incubation time based on the results obtained from the chemical toxicity test.
2. See Fig. 1 for an overview of the bulk mutagenesis procedure.
3. Prepare coffee seeds (*e.g.,* 1000) per each treatment chosen (*see* Note 25).
4. Follow procedures as outlined in Sects. 3.1, 3.2 and 3.3.

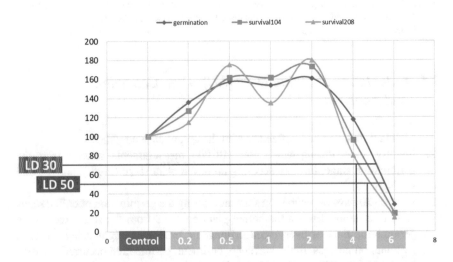

Fig. 5 Example of chemical toxicity test performed on the coffee seed. Coffee seed are exposed to 6 concentrations of EMS (0.2; 0.5; 1, 2, 4 and 6% EMS) plus the control and incubated for 2 h. Here, the percentage germination in relation to the control (100%) was plotted for each EMS concentration (blue line). The germination was recorded between 34 and 63 days after treatment. Note that germination is enhanced for 0.2, 0.5, 1 and 2% compared to the control, whereby a clear drop is visible for 4% and 6% EMS. Same for the survival scored 104- (orange line) and 208-days (grey line) post-mutagenesis, it drops as EMS concentration increases

Fig. 6 Examples of coffee leaf phenotypes scored 8 months post-mutagenesis for a non-mutagenized control (labelled as 'control') and mutant plants (no labels)

3.6 *Phenotyping*

1. Phenotyping protocols are provided in chapters "Use of Open-Source Tools for Imaging and Recording Phenotypic Traits of a Coffee (*Coffea arabica* L.) Mutant Population–A PCR-Based Assay for Early Diagnosis of the Coffee Leaf Rust Pathogen *Hemileia vastatrix*" of this book.
2. Example of morphological and chlorophyl variegations of 8-month-old M_1 coffee plants are illustrated in Fig. 6.

4 Notes

1. EMS mutagenesis must be conducted in a dedicated laboratory equipped with a ducted fume hood, toxic waste disposal and decontamination procedures. It is advisable to work with another person when handling EMS. Personal protective equipment such as laboratory coats, gloves, goggles and disposable shoe covers should be worn. It is advised to wear double gloves so that contaminated gloves can be removed while avoiding contact of contaminated materials with skin.

2. This protocol describes EMS mutagenesis of *Coffea arabica* var. Venecia, a late maturing variety with excellent cup quality. The var. Venecia originated in San Carlos, Costa Rica from a natural mutation of a Bourbon population, selected by the Instituto del Café de Costa Rica, and released in 2010 as a new variety for its increased productivity, larger fruit size and increased resistance to fruit drop in the rain. Venecia is susceptible to Coffee Leaf Rust, Coffee Berry Disease and nematodes (World Coffee Research 2023).

3. EMS is carcinogenic and thus extreme care needs to be taken. Carefully read the instructions on the Material Safety Data Sheet and follow the recommendation of the manufacturer.

4. EMS can be inactivated by treatment with sodium thiosulfate. The half-life of EMS in a 10% sodium thiosulfate solution is 1.4 h at 20 °C and 1 h at 25 °C. Keep beakers of sodium thiosulfate (100 mM) on hand during laboratory procedures to inactivate any spills, clean tips and other consumables prior to disposing in hazardous waste.

5. Coffee requires specific environmental conditions for optimal germination and growth (Wintgens 2012).

6. Use freshly harvested seed as Arabica coffee seed viability decreases rapidly after 6 months when stored at ambient temperature (Wintgens 2012). This is a critical point in case of coffee. Always perform a germination test to check seed viability.

7. This protocol describes treatment of Venecia seeds with 6 EMS concentrations (0.2; 0.5; 1; 2; 4; 6%); 50 seeds per treatment at one duration of 2 h. Additional concentrations can be tested, as well as various temperatures and treatment durations.

8. Soaking the seeds prior to the mutagenesis treatment enhances the total uptake, the rate of uptake and the distribution of EMS within the target tissue, maximizing infusion of the mutagen into the embryo tissue within the shortest possible time. The duration of pre-soaking depends primarily on the seed anatomy; hard and thick seedcoats require longer pre-soaking times than soft and thin ones. A pre-soaking test can be performed to estimate optimal pre-soaking conditions. In the case of Venecia seed, a pre-soaking duration of 48 h at room temperature was applied.

9. Tea-steepers were utilized for water decanting and at the same time for maintaining treatment batches. Each tea-steeper was labelled with an attached plastic indicating the EMS concentration.

10. It is strongly recommended to perform EMS chemical mutagenesis in a dedicated laboratory using dedicated equipment. Consult the regulations on use and disposal of hazardous chemicals. In addition, it is advised to plan and practice critical steps in advance of the actual experiment. Ensure that sufficient waste buckets and beakers are available. If using common space, inform co-workers of the experiment in advance, to avoid accidental exposures.

11. Concentrations of EMS for testing a specific plant species or variety can be estimated based on previously published studies, if available. Some publications use molarity rather than percentage (volume/volume) of EMS. Optimal concentrations of ~ 20–40 mM have been reported for many studies using seed mutagenesis (Jankowicz-Cieslak et al. 2011; Kurowska et al. 2011). Always include a control (no EMS) and concentrations below and above the concentration used in published studies. The duration of the exposure and temperature of EMS treatment can also be tested, but it is recommended to start first with concentrations as the optimal dosage can typically be determined by altering the mutagen concentration only.

12. EMS is not miscible in water. DMSO is added to 2% to improve miscibility. The EMS/DMSO solution is prepared in a bottle, sealed with a screw cap, and then shaken vigorously before adding to coffee seed. It is important that the bottle does not leak. Test the bottle with water first and mimic the shaking procedure in the fume hood.

13. Prepare the concentration series of EMS commencing with the lowest concentration. Practicing this step allows for the proper placement of the stock bottle, dilution bottle, pipette, and waste container, so that the EMS stock does not spill when working in the confined space of a fume hood.

14. Lowering the sash is necessary for proper ventilation and provides some protection against leakage from the bottle.

15. Avoid adding excess volume of EMS/DMSO to the beaker of seed as this may result in liquid spilling during orbital rotation.

16. One day before the experiment, set up a mock experiment with water to determine the best ratio of seed to liquid for the EMS incubation. Add the number of seed to be used in the main experiment to the beaker. Add water and place on the orbital shaker. The samples should be completely immersed in the mutagen solution to ensure that all seeds are equally and fully exposed to the active ingredients of the mutagen. Adjust the speed of the shaker so that all seed move freely in the water, avoid splashing, split seeds into multiple beakers if necessary. Take note of both the volume and the number of seeds and apply this during the main experiment. Generally, EMS is unstable in water with a half-life of 26 h at 30 °C at neutral to acidic pH. Hence, the temperature of the environment can influence the efficiency of the EMS mutagenic treatment due to hydrolysis of the EMS. At low temperatures, the hydrolysis rate is decreased, implying that the EMS mutagen remains stable for longer.

17. Be very careful when pouring off liquid. Avoid splashes. A mesh screen or a sieve should be used to capture the seed and to avoid that seed are poured out of the beaker.

18. The by-products of the incubation process and residual active ingredients should be promptly washed off the incubated seed after treatment. This prevents continued absorption of the mutagen beyond the intended duration, so-called dry-back, which leads to lethality.

19. EMS is mutagenic, teratogenic, and carcinogenic and hence must be disposed of according to the health and safety regulations of your laboratory or institution (check disposal procedures with personnel responsible for toxic materials or local health and safety authority).

20. All body parts or laboratory coats contaminated with EMS should be washed thoroughly with water and detergent and further neutralized with 100 mM sodium thiosulfate.

21. Care should be taken to avoid that any materials removed from the chemical mutagenesis laboratory are free from contamination with EMS.

22. Following parameters are usually scored in the process of kill curve establishment: days to germination (DTG), germination %, seedling height, seedling/plant survival %, chlorophyll mutation frequency, seed set. In the case of coffee seed, germination (as in seedling emergence from the soil, Fig. 4c) occurs approximately 30 days after planting. In the current experiment the first seeds germinated 34 days, and the last one 63 days after sowing. Average values were calculated for each treatment which resulted in DTG values of 34–63.

23. This calculation is made by dividing the number of germinated seeds of the mutated material (numerator) by the number of germinated seeds of the control, untreated seed measured at the same time (denominator) multiplied by 100. For example, if 12 coffee seeds germinated for a 2 h, 0.2% EMS treatment and 12 seeds germinated for the control, then the percentage is $12/12 \times 100 = 100\%$.

24. In the case of Venecia, the percentage of germination and survival were suitable parameters to estimate the damage due to the mutagenic treatment and determine the LD_{50} and LD_{30} values.

25. The total number of seed treated depends on the objectives of the experiment. An excess of seed should be treated as a percentage of seed will not germinate due to incubation with EMS. The optimal population size depends on the desired density of induced mutations, the available spatial and human resources, and the objectives of the study or breeding program.

Acknowledgements The authors wish to thank Dr Andrés Gatica-Arias, Universidad de Costa Rica, San Pedro, Costa Rica for guidance on coffee germination and growth, Dr Noel Arrieta Espinoza, The Coffee Institute of Costa Rica (ICAFE), Costa Rica for providing the coffee seed and Mr Islam Tazirul for technical support during these studies. Funding for this work was provided by the Food and Agriculture Organization of the United Nations and the International Atomic Energy Agency through their Joint FAO/IAEA Programme of Nuclear Techniques in Food and Agriculture.

References

Ingelbrecht I, Jankowicz-Cieslak, J, Szurman M, Till BJ and Szarejko I (2018) Chemical mutagenesis. In: Spencer-Lopes MM, Forster BP, Jankuloski L (eds) Manual on mutation breeding, 3rd edn. FAO/IAEA publication. FAO, Rome, pp 51–65 (Chapter 2)

Jankowicz-Cieslak J, Till BJ (2016) Chemical mutagenesis of seed and vegetatively propagated plants using EMS. Curr Protoc Plant Biol 1:617–635. https://doi.org/10.1002/cppb.20040

Jankowicz-Cieslak J, Huynh OA, Bado S, Matijevic M, Till BJ (2011) Reverse-genetics by TILLING expands through the plant kingdom. Emirates J Food Agric 23:290–300

Jankowicz-Cieslak J, Till BJ (2015) Forward and reverse genetics in crop breeding. In: Advances in plant breeding strategies: breeding, biotechnology and molecular tools, vol 1 (A.-K. J.M., J. S.M., and J. D.V., eds.) pp. 215–240. Springer International Publishing.

Jiang S, Wang M, Zhao C, Cui Y, Cai Z, Zhao J, Zheng Y, Xue L, Lei J (2022) Establishment of a mutant library of *Fragaria nilgerrensis* Schlechtendal ex J. Gay via EMS mutagenesis. Horticulturae 8:1061. https://doi.org/10.3390/horticulturae811106

Kurowska M, Daszkowska-Golec A, Gruszka D, Marzec M, Szurman M, Szarejko I, Maluszynski M (2011) TILLING—a shortcut in functional genomics. J Appl Genet 52(4):371–390

Mago R, Till B, Periyannan S, Yu G, Wulff BBH, Lagudah E (2017) Generation of loss-of-function mutants for wheat rust disease resistance gene cloning. In: Periyannan S (ed) Wheat rust diseases: methods in molecular biology, vol 1659. Humana Press, New York. https://doi.org/10.1007/978-1-4939-7249-4_17

Mba C, Afza R, Bado S, Jain SH (2010) Induced mutagenesis in plants using physical and chemical agents. In: Davey MR, Anthony P (eds) Plant cell culture: essential methods. Wiley, pp 111–130

Molina-Cano JL, Simiand JP, Sopena A, Perez-Vendrell AM, Dorsch S, Rubiales D, Swanston JS, Jahoor A (2003) Mildew-resistant mutants induced in North American two-and six-rowed malting barley cultivars. Theor Appl Genet 107:1278–1287. https://doi.org/10.1007/s00122-003-1362-5

Mukhtar Ali Ghanim A, Spencer-Lopes MM, Thomas W (2018) Mutation breeding in seed propagated crops: parental selection, mutant generation development, mutation detection, mutant evaluation and factors influencing success. In: Spencer-Lopes MM, Forster BP, Jankuloski L (eds) Manual on mutation breeding, 3rd edn. FAO/IAEA Publication, Rome, pp 119–156 (Chapter 5)

Spencer-Lopes MM, Jankuloski L, Mukhtar Ali Ghanim A, Matijevic M, Kodym A (2018) Physical mutagens. In: Spencer-Lopes, MM, Forster, BP, Jankuloski L (eds) Manual on mutation breeding, 3rd edn. FAO/IAEA Publication, Rome

Till BJ, Reynolds SH, Greene EA, Codomo CA, Enns LC, Johnson JE, Burtner C, Odden AR, Young K, Taylor NE, Henikoff JG, Comai L, Henikoff S (2003) Large-scale discovery of induced point mutations with high-throughput TILLING. Genome Res 13:524–530

Till BJ, Cooper J, Tai TH, Colowit P, Greene EA, Henikoff S, Comai L (2007) Discovery of chemically induced mutations in rice by TILLING. BMC Plant Biol 7:19

Wintgens JN (2012) Coffee: growing, processing, sustainable production: a guidebook for growers, processors, traders and researchers. Wiley, Weinheim, 1022pp

World Coffee Research https://varieties.worldcoffeeresearch.org/varieties/venecia. Accessed on 3 Jan 2023

Chemical Mutagenesis of Coffee Seeds (*Coffea arabica* L. var. Catuaí) Using NaN₃

Andrés Gatica-Arias and César Vargas-Segura

Abstract Coffee (Coffea arabica L.) is one of the most important crops in the world and one of the main export products in several developing countries. Coffee is a perennial crop threatened by multiple, serious diseases and pests. Induced mutagenesis of seeds is widely used for increasing the genetic diversity and improvement of annual seed crops and could equally be applied to Arabica coffee breeding and genetic studies. Here we describe protocols to induce genetic variability in Arabica coffee seeds through mutagenesis using sodium azide (NaN_3). Methods for NaN_3 chemical toxicity testing and bulk irradiation are described. Briefly, the coffee seeds were immersed for 4, 8 and 12 hours in a NaN_3 solution at different concentrations (0, 25, 50, 75, and 100mM). Two controls were used: one with distilled water and the other with the phosphate buffer (KH_2PO_4). Effects of the chemical mutagen on seed germination, seedling height, and root length were evaluated. As the concentration of applied NaN_3 increased, the germination, seedling height, and root length decreased. Eight hours exposure was determined as an adequate immersion time. The LD_{50} values for NaN_3 were between 50–75 mM. Our results indicate that NaN_3 is an effective mutagen for Arabica coffee seeds and can be applied to coffee breeding and to study gene function in coffee.

1 Introduction

Mutations can be induced in plants by exposure of seeds or meristematic cells, tissues, and organs, to both physical and chemical agents with mutagenic properties (Mba et al. 2010). Chemical mutagenesis is the exposure of plant material to chemical agents such as alkylating agents and azides under optimized doses. The mutagenic effect of sodium azide (NaN_3) is mediated through the creation of an organic intermediate (*L-azidoalanine*) which incorporates the azide group and interacts with DNA to mainly produce simple base substitutions (Gruszka et al. 2012). Sodium azide's

A. Gatica-Arias (✉) · C. Vargas-Segura
Laboratorio Biotecnología de Plantas, Escuela de Biología, Universidad de Costa Rica, 2060 San Pedro, Costa Rica
e-mail: andres.gatica@ucr.ac.cr

© The Author(s) 2023
I. L. W. Ingelbrecht et al. (eds.), *Mutation Breeding in Coffee with Special Reference to Leaf Rust*, https://doi.org/10.1007/978-3-662-67273-0_13

179

mutagenic effect greatly depends upon the acidity of the working solution, which must be prepared at low pH values (Gruszka et al. 2012).

Reducing the time required to develop improved plants through mutation breeding is a desirable characteristic, especially in long-life cycle plant species such as coffee. Coffee cultivation has great socio-economic impact; it is positioned as the world's second most exported commodity, only surpassed by oil (FAOSTAT 2018). The better cup quality and higher market value are associated to *Coffea arabica* L., the only allotetraploid ($2n = 4x = 44$) species among Coffea. Arabica coffee is an autogamous plant mostly incompatible with the remainder of Coffea species (Anthony et al. 2002). These characteristics, along with severe bottlenecks that happened during coffee domestication led to reduced genetic variability in *C. arabica* populations; this reduction enhances the general susceptibility of many *C. arabica* genotypes to diseases (Hendre and Aggarwal 2007). Conventional breeding faces limitations due to the long-life cycle of the coffee plant, requiring nearly 30 years to develop new cultivars.

This chapter describes the application of the chemical mutagen sodium azide for mutagenesis of *C. arabica* seeds as well as the germination, and development of mutant plantlets.

2 Materials

2.1 Plant Material

1. All experiments were conducted using seeds from *Coffea arabica* L. var. Catuaí (*see* Note 1).

2.2 Reagents

1. 10% (w/v) sodium thiosulfate (*e.g.* Sigma Cat Nr.: 217,263).
2. 100 mM phosphate buffer (pH 3.0).
3. Peat substrate.
4. Distilled water.
5. KH_2PO_4 (*e.g.* Sigma Cat Nr.: P5655).
6. Phosphorus acid (H_3PO_3) (*e.g.* Sigma Cat Nr.: 176,680).
7. Sodium azide (NaN_3) (*e.g.* Sigma Cat Nr.: S2002) (*see* Note 2).
8. Sodium hypochlorite (3.5% v/v).
9. Sterile distilled water.
10. Tween 20 (*e.g.* Phytotechnology Cat Nr.: P720).

2.3 Glassware and Minor Equipment

1. Beakers (250 ml).
2. Bottles (100, 500 ml).
3. Box for disposal of dry hazardous material.
4. Disposable pipettes (1, 5, 10, or 25 ml, as required).
5. Erlenmeyer (250 ml).
6. Graduated cylinders (100 ml or as needed according to calculations).
7. Hazardous liquid waste receptacle (collection vessels for the sodium azide waste solution).
8. Personal protective equipment (disposable laboratory coat dedicated only to mutagenesis experiments, eye protection/goggles, shoe protection, nitrile gloves).
9. Plastic boxes (88 × 42 × 16 cm).

2.4 Equipment

1. Analytical balance.
2. Chemical mutagen laboratory equipped with fume hood and flow bench (*see* Note 2 and 3).
3. Fume Hood.
4. Germination chamber (100% humidity and 30 °C temperature).
5. Orbital shaker.

2.5 Software

1. Standard spreadsheet software (*e.g.* Microsoft Excel or Open Office Excel).

3 Methods

3.1 Preparation of Stock Solutions

1. 100 mM Phosphate buffer: add 13.61 g of KH_2PO_4 to 1,000 ml volumetric flask and dissolve by adding 250 ml of tissue culture grade water. Once completely dissolved, stir the solution while adding tissue culture grade water and bring to 1000 ml. Adjust pH to 3.0 using phosphorus acid (H_3PO_3).
2. Sodium azide (0.1 M): add 6.48 g of the powder to 500 ml volumetric flask and dissolve in 100 ml phosphate buffer (100 mM). Once completely dissolved, stir the solution while adding phosphate buffer (100 mM) and bring to 200 ml. Make solution fresh as required.

3.2   Seed Germination Test

1. See Fig. 1 for an overview of the seed germination test.
2. Collect mature cherries from genetically homogenous mother plants maintained in the greenhouse or the field (*see* Note 3).
3. Remove the pulp, the mucilage, and the parchment by hand.
4. Prepare an uniform seed stock by selecting disease-free seeds and removing any small, shriveled or damaged material.
5. Disinfect the seeds with 3.5% (v/v) sodium hypochlorite for 1 h and rinse three times with sterile distilled water (*see* Note 4).
6. Soak 100 seeds in sterile distilled water for 48 h.
7. Sow the seeds in plastic boxes ($88 \times 42 \times 16$ cm) containing autoclaved peat substrate.
8. Place the boxes in the germination chamber (100% humidity and 30 °C temperature) for 8 weeks in darkness. Alternatively, the boxes could be placed under greenhouse conditions.
9. Record germination after 8 weeks based on visual scoring of full protrusion of the radicle and appearance of the coleoptile (*see* Note 5).

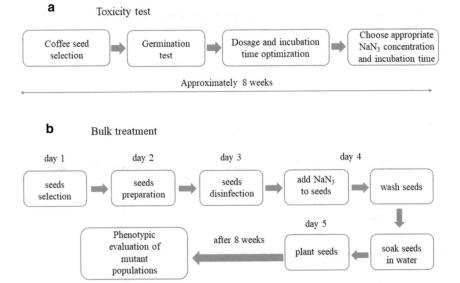

Fig. 1 Schematic representation of the process of mutagenesis of coffee seeds. **a** In advance of chemical mutagenesis, the NaN₃ concentration and incubation time are optimized using seeds having germination rate close to or above 90% (preferably normal shaped seeds that are disease-free). These steps take approximately 8 weeks. **b** bulk mutagenesis of coffee seed is a 6 steps procedure that consists of the seed sorting and selection, processing (manual removal of the pulp, the mucilage, and the parchment), disinfection of seeds, incubation of seeds with the appropriate NaN₃ concentration, post-treatment washing and planting of mutagenized seeds

3.3 Sodium Azide Toxicity Test and Dose Optimization

1. Review safety procedures of the chemical mutagenesis laboratory (*see* Note 6 and 7).
2. Autoclave all non-disposable materials (*e.g.*, sieves, forceps, etc.).
3. Prepare a fresh 0.1 M NaN_3 stock solution by adding the required amount of NaN_3 to the phosphate buffer (100 mM, pH 3.0).
4. Select normal shaped seeds that are disease-free. The seeds should have good germination (i.e. > 90%).
5. Disinfect the seed beforehand according to the protocol described in 3.2 (*see* Note 4).
6. Place 100 seeds in a 250 ml labelled Erlenmeyer. The number of Erlenmeyers depends on the number of treatments. Label each flask with the appropriate treatment code (concentration and incubation time).
7. In a fume hood, add appropriate volumes of phosphate buffer solution (100 mM) (*see* Table 1).
8. Add NaN_3 to a final concentration of 25, 50, 75, and 100 mM (*see* Table 1). Shake the solution vigorously.
9. Incubate the mixture for 4, 8, or 12 h in the dark at 27 ± 2 °C with gentle rotation (100 rpm).
10. After the incubation, quickly but carefully decant solution from each of the treatment batches and rinse the treated seeds with 100 ml of sterile distilled water. This step and any subsequent steps must be carried out in a fume hood.
11. Repeat washing step 3 times.
12. Collect all the liquid waste in a dedicated bucket labelled as hazardous waste. Dispose of toxic waste according to local biosafety regulations.
13. After NaN_3 mutagenesis, the seeds should be planted immediately in plastic boxes (88 × 42 × 16 cm) containing autoclaved peat substrate.

Table 1 NaN_3 concentrations and incubation time evaluated for the toxicity test of Arabica coffee seeds

NaN_3 concentration (mM)	Sterile distilled water (ml)	100 mM phosphate buffer (pH 3.0) (ml)	0.1 M NaN_3 (ml)*	Incubation time (h)
Positive control (water)	100	–	–	4, 8, 12
Negative control (−NaN_3)	–	100	–	4, 8, 12
25	–	95	5	4, 8, 12
50	–	90	10	4, 8, 12
75	–	85	15	4, 8, 12
100	–	80	20	4, 8, 12

* Prepare a fresh solution prior to each experiment

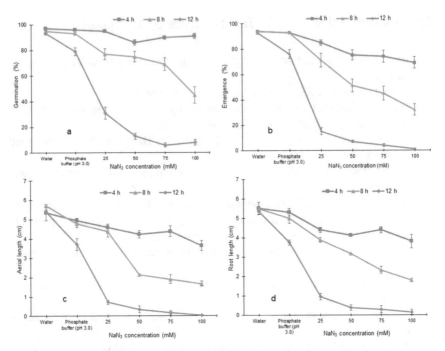

Fig. 2 a Germination percentage, **b** hypocotyl emergence percentage, **c** aerial length (cm), and **d** root length (cm) of the arabica coffee plantlets 8 weeks after incubation of seeds treated with different concentrations of NaN$_3$ and immersed for 4, 8, and 12 h. Each value represents the mean ± SD of four repetitions

14. Place the boxes in a germination chamber (100% humidity and 30 °C temperature) for 8 weeks and incubate in darkness. Alternatively, the boxes could be placed under greenhouse conditions.

15. Record the germination percentage, hypocotyl emergence percentage (seedlings with hypocotyl emerged from the substrate), aerial length (cm) and root length (cm) after 8 weeks.

16. Plot the germination percentage, hypocotyl emergence percentage, aerial length (cm), and root length of the control against each mutagenic treatment (*see* Fig. 2).

17. Estimate the mutagen concentration required to obtain 50 % germination compared to the control (see Fig. 3).

3.4 Bulk Mutagenesis

1. Autoclave all non-disposable materials (*e.g.*, sieves, forceps).
2. Choose appropriate concentration of NaN$_3$ and incubation time based on the results obtained from the chemical toxicity test.

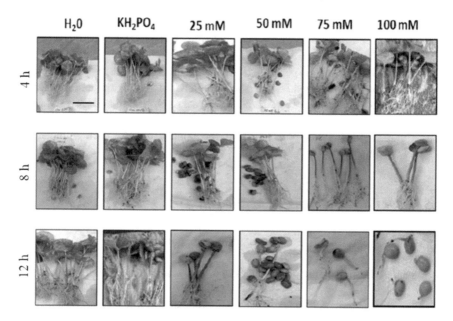

Fig. 3 Effect of NaN$_3$ treatment on germination of coffee seeds 8 weeks after incubation. Different concentrations of NaN$_3$ (25, 50, 75, and 100 mM) and incubation times (4, 8, and 12 h) were used. *Bar*, 1 cm

3. Select similar and normal shaped seeds that are disease-free. The seeds should have good germination (i.e. > 90%).
4. Disinfect the seed beforehand according to the protocol 3.2.
5. Place 100 seeds in a 250 ml labelled Erlenmeyer. The number of Erlenmeyer depends on the number of repetitions.
6. Label each Erlenmeyer with the appropriate treatment code (NaN$_3$ concentration and incubation time).
7. Transfer Erlenmeyers containing seed material into the chemical mutagenesis laboratory.
8. In a fume hood, add appropriate concentrations of NaN$_3$ and shake the solution vigorously.
9. Place the Erlenmeyer in the dark at 27 ± 2 °C on a rotary shaker at 100 rpm for the chosen length of time.
10. After incubation, quickly but carefully decant each treatment and rinse the treated seed batches with 100 ml of sterile distilled water.
11. Repeat washing step 3 times.
12. Collect all the liquid waste in a dedicated bucket labelled as hazardous waste. Dispose of toxic waste according to local regulations.
13. After mutagenic treatment, seeds should be planted immediately in plastic boxes ($88 \times 42 \times 16$ cm) containing autoclaved peat substrate.

14. Place the boxes in the germination chamber (100% humidity and 30 °C temperature) for 8 weeks in darkness. Alternatively, the boxes could be placed under greenhouse conditions.

15. Record the germination percentage, hypocotyl emergence percentage (seedlings with hypocotyl emerged from the substrate), aerial length (cm) and root length (cm) into a spreadsheet (*e.g.,* Microsoft Excel or Open Office Excel) (*see* Note 3).

16. Plot the germination percentage of the control against each mutagenesis treatment.

3.5 Planting Mutagenized M₁ Seedling

1. After 8 weeks, transfer coffee seedlings to pots containing autoclaved peat substrate.

2. Maintain the pots under greenhouse conditions with 12 h light photoperiod at 27 ± 2 °C.

3. Characterize the mutagenized M_1 population using following parameters: duration of seed germination, the germination percentage, hypocotyl emergence (seedlings with hypocotyl emerged from the substrate), aerial length (cm) and root length (cm).

4 Notes

1. This protocol has been established using *Coffea arabica* L. var. Catuaí seeds. It can be used as a reference for other arabica coffee varieties. Nevertheless, it is recommended to optimize the mutagenic parameters (NaN_3 concentration and incubation time) for each variety used.

2. Store NaN_3 in an airtight colored bottle, preferably inside a sealed chamber containing a desiccant.

3. Coffee seeds rapidly loose viability if not properly stored. Therefore, it is recommended to use freshly harvested seeds or otherwise to store the seeds between 10 and 12% moisture at 15 °C not longer than 3 month.

4. Disinfection is recommended only if stored seed is used, on the other hand, disinfection is not necessary for freshly harvested seeds.

5. Proceed to mutation induction protocol with seeds having a germination rate close to or above 90%.

6. All the mutagenesis experiments should be conducted in a dedicated chemical mutagenesis laboratory equipped with a ducted fume hood, toxic waste disposal and decontamination procedures.

7. Lab safety precautions: read the Materials Safety Data Sheet (MSDS) of materials being used and follow the recommendation of the manufacturer. Pay careful attention to the information on sodium azide and what to do in case of exposure. It is very important to wear personal protective clothing: gloves must be compatible with chemical mutagens, for instance PVC or neoprene gloves; safety glasses with side shields or chemical goggles; lab coat, closed-toe shoes, shoe protections, and full-length pants. A double glove system is advised.

Acknowledgements Funding for this work was provided by the University of Costa Rica, the Ministerio de Ciencia, Innovación, Tecnología y Telecomunicaciones (MICITT), the Consejo Nacional para Investigaciones Científicas y Tecnologicas (CONICIT) (project No. 111-B5-140; FI-030B-14). A. Gatica-Arias acknowledged the Cátedra Humboldt 2023 of the University of Costa Rica for supporting the dissemination of biotechnology for the conservation and sustainable use of biodiversity.

References

Anthony F, Combes MC, Astorga C, Bertrand B, Graziosi G, Lashermes P (2002) The origin of cultivated *Coffea arabica L.* varieties revealed by AFLP and SSR markers. Theor Appl Gent 104. https://doi.org/10.1007/s00122-001-0798-8

Food and Agriculture Organization of the United Nations (2018) FAOSTAT Database. Rome, Italy: FAO. Retrieved February 23, 2018 from www.fao.org/faostat/en/#search/coffee

Gruszka D, Szarejko I, Maluszynski M (2012) Sodium Azide as a Mutagen. In: Shu Q, Forster B, Nakagawa H (eds) Plant mutation breeding and biotechnology. ENG, CABI Publishing, Wallingford, pp 159–168

Hendre P, Aggarwal R (2007) DNA Markers: development and application for genetic improvement of coffee. In: Varshney R, Tuberosa R (eds) Genomics-assisted crop improvement, vol 2. Springer, Dordrecht, pp 399–434

Mba C, Afza R, Bado S, Jain SM (2010) Induced mutagenesis in plants using physical and chemical agents. In: Davey MR, Anthony P (eds) Plant cell culture: essential methods. Wiley. ISBN: 978-0-470-68648-5

Mutant Phenotyping and CLR Resistance Screening Methods

Use of Open-Source Tools for Imaging and Recording Phenotypic Traits of a Coffee (*Coffea arabica* L.) Mutant Population

Radisras Nkurunziza, Joanna Jankowicz-Cieslak (✉) · Stefaan P. O. Werbrouck, and Ivan L. W. Ingelbrecht

Abstract Mutation breeding in *Coffea arabica* offers a powerful tool to induce novel genetic variability for breeding and genetic studies. The success of a mutation breeding program depends largely on the ability to screen large populations for target traits. There is also a need to accurately record induced mutant traits at the individual plant level. Comprehensive phenotyping requires measuring and tracking traits of interest during the crop growth cycle and subsequent generations. Therefore, efficient and accurate data collection and recording of traits is essential, both at the individual plant level and populations. In recent years, various high-throughput plant phenotyping platforms have been developed. However, these are typically proprietary, and/or require costly infrastructures. In this chapter we illustrate the use of Field Book and ImageJ, two public domain software tools, for phenotyping and documenting growth and yield traits of a greenhouse-grown Arabica coffee mutant population. Example data of M_1 and M_2 mutant phenotypes induced through EMS and gamma-ray mutagenesis are presented. We further demonstrate the use of these tools for quantifying the canopy of mutants and non-mutagenized controls. These tools can be more widely applied to other visual phenotypes including plant or tissue responses to biotic or abiotic stresses. The use of free, open-access tools for integrating electronic data recording with image processing can greatly improve the efficiency, precision and speed of data collection for screening large mutant populations and is especially useful in resource-limiting settings.

R. Nkurunziza (✉) · J. Jankowicz-Cieslak (✉) · I. L. W. Ingelbrecht
Plant Breeding and Genetics Laboratory, Joint FAO/IAEA Centre of Nuclear Techniques in Food and Agriculture, IAEA Laboratories Seibersdorf, International Atomic Energy Agency, Vienna International Centre, Vienna, Austria
e-mail: radisras.nkurunziza@ugent.be

J. Jankowicz-Cieslak
e-mail: j.jankowicz@iaea.org

R. Nkurunziza · S. P. O. Werbrouck
Laboratory for Applied in Vitro Plant Biotechnology, Department of Plants and Crops, Faculty of Bioscience Engineering, Ghent University, Ghent, Belgium

© The Author(s) 2023
I. L. W. Ingelbrecht et al. (eds.), *Mutation Breeding in Coffee with Special Reference to Leaf Rust*, https://doi.org/10.1007/978-3-662-67273-0_14

1 Introduction

Coffee (*Coffea* spp.) is an indispensable source of income in Asia, Africa, South and Central America. It is ranked among the world's most valuable export commodities, on which more than 125 million people in the coffee growing areas derive their income directly or indirectly (FAO 2015). *C. arabica* contributes about 75% of the world coffee production due to its superior cup quality and aromatic characteristics. *C. arabica* has a unique biology compared to other species in the genus Coffea because it is a self-fertile, allotetraploid (2n = 4x = 44) species while almost all other coffee species are diploids (2n = 2x = 22) and generally self-incompatible (Carvalho et al. 1991). Due to its reproductive biology, *C. arabica* varieties tend to remain genetically stable. However, *C. arabica* varieties are typically low yielding and highly susceptible to a myriad pests and diseases and abiotic stresses, including Coffee Leaf Rust (CLR). Genetic improvement of *C. arabica* to withstand the afore mentioned constraints and with higher yield through classical breeding is laborious and can require up to 25–30 years to release an improved variety (Lashermes et al. 1996, 2009).

Mutation breeding offers a powerful tool to enhance genetic variation in the *C. arabica* gene pool which is very narrow. Random mutagenesis has been used in different crops to induce novel agronomic traits that were absent in their primary gene pool (Ulukapi and Nasircilar 2018). Induced mutagenesis can enhance stable, genetic variability not only in seed but also in vegetatively propagated plants (Jain 2005; Pathirana et al. 2009). Major traits improved through mutation-assisted breeding include plant architecture, early flowering and maturity, yield and quality traits, and tolerance to pests and diseases (Pathirana 2011; FAO/IAEA 2018). For a successful coffee mutation breeding program, phenotyping large mutant populations is paramount. Likewise, it is important to identify, and document induced mutant phenotypes at the individual plant level. Comprehensive phenotyping requires that traits of interest can be accurately and rapidly measured and documented during the crop growth cycle and subsequent generations (Sabina 2022). Recent technological advances have enabled accurate, high-throughput plant phenotyping. Commercial and open-source digital phenotyping technologies and methods have been developed to increase the precision and speed of data collection and analysis useful for plant breeders. However, high-throughput phenotyping methods typically require highly automated and sophisticated systems for image acquisition and analysis. Also, high-throughput technologies require significant infrastructural investment in the field or greenhouse facilities. The use of simple image capture tools such as manually operated cameras and downstream open-source image analysis tools such as ImageJ, Fiji and MATLAB provide affordable alternatives (Hartmann et al. 2011; Schindelin et al. 2012; Singh et al. 2017). In recent years several open-access applications for data recording have been developed. Examples include Field Book, an open-source application for taking phenotypic notes (Rife and Poland 2014), OneKK, an app designed to analyze seed lots, Coordinate, an open-source Android app used to collect and organize data into a predefined grid (Prasad et al. 2018) and Open Data Kit (ODK) for seed tracking (Ouma et al. 2019). Such public domain software tools facilitate a digital migration from manual methods of data capture and recording that are associated with unstandardised data that is difficult to process for analysis (Mechael 2009).

In this chapter, we illustrate the use of two public domain software tools, Field Book and ImageJ, for phenotyping and documenting growth and yield component traits of *C. arabica* M_1 and M_2 populations and plants developed and maintained in the greenhouse of the FAO/IAEA Plant Breeding and Genetics Laboratory in Seibersdorf, Austria. These tools are simple, user-friendly and especially useful for plant scientists, breeders or data collectors in resource-limiting environments where advanced, high-throughput phenotyping facilities and/or expertise is missing.

2 Materials

2.1 Establishment of the C. Arabica Mutant Populations

1. Acclimatised/hardened six-months old coffee mutagenized seedlings (*see* Note 1).
2. Greenhouse.
3. Peat soil.
4. Sand.
5. Potting mixer (shovel or spade).
6. 5-L pots.
7. Plastic labels/tags.
8. Marker pens/Pencils.
9. Water supply system.
10. Watering can, horse pipe or drip irrigation system.
11. Complex fertilizers for instance ENTEC® (NPK- $14 + 7 + 17(+ 2 + 9) + 0.02$ B $+ 0.01$ Zn; Eurochem).
12. Systemic pesticides (fungicides and insecticides).
13. Pesticide sprayer.
14. Chemical protective gear.

2.2 Phenotyping of the Mutant Populations

1. Electronic data collection tool (Android tablet/phone) or data sheet.
2. Field Book application.
3. Trait list and trait descriptors.
4. Rulers (30, 200 cm).
5. Vernier calliper.

6. Camera and camera stand.
7. Pencil.
8. Permanent markers.
9. Tags/labels.
10. Computer/workstation.

2.3 Data Analysis Tools

1. ImageJ software (digital image analysis).
2. Excel spreadsheet.
3. Statistical software (any software that you are familiar with).

3 Methods

3.1 Establishing the M_1 Mutant Population

1. Clean and disinfect the greenhouse at least one week before planting to eliminate potential pests and disease pathogens that may affect coffee plants.
2. Obtain well acclimatized/hardened M_1 seedlings (*see* Chap. "Chemical Mutagenesis of Mature Seed of Coffea arabica L. var. Venecia Using EMS", Note 1 and Fig. 1a).
3. Mix the potting medium containing peat soil and sand (3:1 v/v) at pH 5–6.
4. Distribute uniformly the potting medium into clean 5 L pots.
5. Carefully remove the hardened seedlings from the small container pots and transplant them in the 5 L pots (*see* Note 2 and Fig. 1b).
6. Do not fill the pots completely with the potting mix, leave about 3 cm depth.
7. During transplanting, handle treatments separately to avoid errors that may arise due to cross mixing.
8. Label the pots accordingly with the information about the genotype, mutagen concentration or dose and treatment date. Ensure that each seedling bears a unique identifier since it is an independent event.
9. Arrange the pots randomly into blocks in a complete randomised design (CRD) with uniform spacing of at least 30 cm within blocks and 100 cm between blocks (*see* Note 3).
10. During plant growth, monitor the pots and design an appropriate watering regime, usually three times a week, depending on climatic conditions.
11. Periodically apply fertilizers (*see* Note 4).
12. Inspect plants regularly for pests and diseases, apply appropriate and recommended (systemic) pesticides (*see* Note 5).

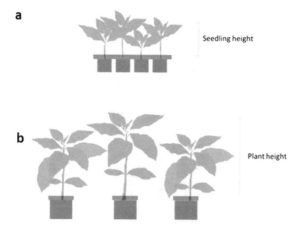

Fig. 1 Developmental stages of coffee mutants. **a** Coffee seedlings with 4–5 true leaves, ready for transplanting. Just before transplanting, data can be recorded on traits like number of leaves, size of canopy and leaf dimensions (leaf length and leaf width) (*see* Chap. "Chemical Mutagenesis of Mature Seed of Coffea arabica L. var. Venecia Using EMS"). **b** Mutant plants established in pots at about six months after transplanting. Growth traits including plant height can be measured to monitor plant growth

3.2 Phenotypic Characterisation of the M_1 Mutant Population

3.2.1 Electronic Phenotyping Tools

1. Use the Field Book application to record growth and yield traits (*see* Note 6 and Fig. 2).
2. Install the Field Book on your Android device, e.g., from Google Play Store (Fig. 2b).
3. Design an experimental layout in Excel on your laptop (Fig. 2a).
4. Import the experimental layout in the Field Book application into the import folder.
5. To activate the experimental layout file, open the application on your device and import the file as a field from local storage (Fig. 2c).
6. Add your traits of interest into the Field Book (Fig. 2d, e). The trait list and trait description should be prepared beforehand (IPGRI 1996).
7. Define your traits as numeric, categorical, percentage, date or text (in case you need to make notes/comments during data collection).
8. Data collection can start immediately (Fig. 2f). No internet connection is required.
9. Use forward/backward arrows to move to the next trait or plot and vice versa (Fig. 2g). If barcodes are used in the experiment, scanning barcodes (also provided in the application) is much better and faster than arrows.

Fig. 2 Schematic representation of the steps taken while using Field Book application for data collection on a coffee mutant population. **a** The overview of the layout of the field (import file) displaying information on coffee mutant labels only. No trait information is required. **b** Field Book logo. **c** Option for importing the experimental layout as a field into the application. **d** Option to input and define trait. **e** Appearance of the coffee traits as defined in the application. **f** Option to begin data collection. **g** An interface during data collection. It displays the trait (e.g., plant height), plant ID (e.g. Ca-2020–001 Gy 20) information as previously determined. Forward and backward arrows guide data collection. **h** Option to export data after collection for storage. **i** Selected procedure to export the data after collection

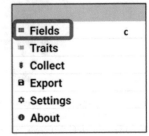

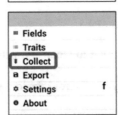

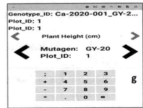

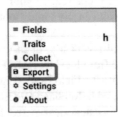

Genotype_ID	Genotype_Code	Mutagen	Block_ID	Plot_ID	ant Height (c	Plant Canopy	ber of primary bra	Angle of inse	Young Lea	Leaf Shap	Leaf Apex Sh	Leaf Lengt	Leaf Width (cm)
Ca-2020-001_GY-20_Plot-1	Ca-2020-001	GY-20	1	1	132	Intermediate	34	Sem-erect	Green	Elliptic	Acuminate	11.2	5.5
Ca-2020-003_EMS-4_Plot-2	Ca-2020-003	EMS-4	1	2	95	Compact	38	Sem-erect	Green	Elliptic	Acuminate	15	8
Ca-2020-025_GY-5_Plot-3	Ca-2020-025	GY-5	1	3	92	Compact	35	Sem-erect	Green	Elliptic	Acuminate	14.5	8.5
Ca-2020-006_EMS-6_Plot-4	Ca-2020-006	EMS-6	1	4	90	Compact	30	Sem-erect	Green	Elliptic	Acuminate	15	8
Ca-2020-023_GY-5_Plot-5	Ca-2020-023	GY-5	1	5	100	Intermediate	42	Sem-erect	Green	Elliptic	Apiculate	13	6.5
Ca-2020-008_EMS-6_Plot-6	Ca-2020-008	EMS-6	1	6	90	Compact	42	Sem-erect	Green	Elliptic	Acuminte	13	6
Ca-2020-020_EMS-6_Plot-7	Ca-2020-020	EMS-6	1	7	100	Compact	46	Sem-erect	Green	Elliptic	Acuminte	15	6.5
Ca-2020-012_Control_Plot-8	Ca-2020-012	Control	1	8	105	Compact	44	Sem-erect	Green	Elliptic	Acuminte	14	6.5
Ca-2020-014_EMS-4_Plot-9	Ca-2020-014	EMS-4	1	9	83	Intermediate	34	Sem-erect	Green	Elliptic	Acuminte	15	7.5
Ca-2020-021_EMS-2_Plot-10	Ca-2020-021	EMS-2	1	10	135	Open	32	Sem-erect	Green	Elliptic	Acuminte	41.5	7.2
Ca-2020-040_EMS-4_Plot-11	Ca-2020-040	EMS-4	1	11	95	Compact	36	Sem-erect	Green	Elliptic	Acuminte	15	7.8
Ca-2020-018_EMS-6_Plot-12	Ca-2020-018	EMS-6	1	12	80	Intermediate	32	Sem-erect	Green	Elliptic	Acuminte	16	8
Ca-2020-036_GY-50_Plot-13	Ca-2020-036	GY-50	1	13	90	Compact	42	Sem-erect	Green	Elliptic	Apiculate	15.5	7
Ca-2020-035_EMS-4_Plot-14	Ca-2020-035	EMS-4	1	14	95	Compact	38	Sem-erect	Green	Elliptic	Apiculate	17	8.5
Ca-2020-009_GY-20_Plot-15	Ca-2020-009	GY-20	2	15	105	Compact	42	Sem-erect	Green	Elliptic	Acuminte	14	7
Ca-2020-030_GY-50_Plot-16	Ca-2020-030	GY-50	2	16	105	Compact	38	Sem-erect	Yellowish	Elliptic	Acuminte	15	7
Ca-2020-010_EMS-4_Plot-17	Ca-2020-010	EMS-4	2	17	90	Compact	32	Sem-erect	Green	Elliptic	Acuminte	14.5	7.5
Ca-2020-026_GY-50_Plot-18	Ca-2020-026	GY-50	2	18	95	Compact	34	Sem-erect	Green	Elliptic	Acuminte	15	7.5
Ca-2020-005_Control_Plot-19	Ca-2020-005	Control	2	19	95	Compact	20	Sem-erect	Green	Elliptic	Acuminte	15.5	7.5
Ca-2020-027_GY-50_Plot-20	Ca-2020-027	GY-50	2	20	90	Compact	34	Sem-erect	Green	Elliptic	Acuminte	14	7
Ca-2020-013_Control_Plot-21	Ca-2020-013	Control	2	21	85	Compact	36	Sem-erect	Green	Elliptic	Acuminte	14	7.5
Ca-2020-042_EMS-2_Plot-22	Ca-2020-042	EMS-2	2	22	95	Compact	34	Sem-erect	Green	Elliptic	Acuminte	16	7.5
Ca-2020-022_EMS-2_Plot-23	Ca-2020-022	EMS-2	2	23	95	Compact	22	Sem-erect	Green	Ovate	Spatulate	14.5	9
Ca-2020-017_Control_Plot-24	Ca-2020-017	Control	2	24	95	Compact	32	Sem-erect	Green	Elliptic	Apiculate	15.5	7
Ca-2020-032_EMS-6_Plot-25	Ca-2020-032	EMS-6	2	25	95	Compact	36	Sem-erect	Green	Elliptic	Apiculate	14	7
Ca-2020-019_Control_Plot-26	Ca-2020-019	Control	2	26	105	Compact	34	Sem-erect	Green	Elliptic	Apiculate	16	7.5
Ca-2020-028_GY-20_Plot-27	Ca-2020-028	GY-20	2	27	70	Compact	24	Sem-erect	Green	Elliptic	Apiculate	16.5	8
Ca-2020-011_EMS-2_Plot-28	Ca-2020-011	EMS-2	2	28	95	Compact	36	Sem-erect	Green	Elliptic	Apiculate	16	7.5
Ca-2020-033_EMS-2_Plot-29	Ca-2020-033	EMS-2	3	29	90	Compact	30	Sem-erect	Green	Elliptic	Apiculate	15.5	7
Ca-2020-004_Control_Plot-30	Ca-2020-004	Control	3	30	165	Compact	34	Sem-erect	Green	Elliptic	Apiculate	17	8
Ca-2020-034_EMS-4_Plot-31	Ca-2020-034	EMS-4	3	31	85	Compact	36	Sem-erect	Green	Elliptic	Apiculate	16.5	7.5
Ca-2020-029_GY-20_Plot-32	Ca-2020-029	GY-20	3	32	95	Compact	36	Sem-erect	Green	Elliptic	Apiculate	15	7.5
Ca-2020-037_GY-50_Plot-33	Ca-2020-037	GY-50	3	33	90	Compact	34	Mosaic	Green	Elliptic	Apiculate	16	7.5
Ca-2020-015_GY-5_Plot-34	Ca-2020-015	GY-5	3	34	70	Compact	26	Sem-erect	Green	Elliptic	Apiculate	13.5	7
Ca-2020-038_GY-20_Plot-35	Ca-2020-038	GY-20	3	35	70	Open	36	Sem-erect	Green	Elliptic	Apiculate	15	7

Fig. 3 A screenshot of the exported data file from the Field Book application for the coffee mutant population. The file is retrieved as an Excel file (csv). Compared to the import file, the extra columns show traits with the corresponding data

10. In case of many traits, select or activate traits whose data is ready before you start data collection, since some traits appear before others, for instance flowers vs fruits.

11. After data collection, one can turn off the device. The entered data remains intact, you can resume from where you stopped upon the next data collection.

12. To archive the data, connect the device to the internet and export data (Fig. 2h, i).

13. The file name is automatically generated and can be changed as preferred. Save the data in the preferred destination.

14. The exported data is received as an Excel (csv) file containing all the traits as column headers, with the corresponding data records for every plant/plot/block etc. (Fig. 3).

3.2.2 Measurement of Growth Traits

1. Open the Field Book application on your device for data recording.
2. Use non-destructive methods to phenotype your mutant population.
3. At transplanting, assess the seedlings for any visible morphological characteristics, such as leaf colour and shape (Fig. 4).
4. If possible, fix the camera (Fig. 5) and take aerial pictures of the seedling (Chap. "Chemical Mutagenesis of Mature Seed of Coffea arabica L. var. Venecia Using EMS" and Fig. 6).
5. Using a ruler, measure the height of seedlings. Seedling height should be measured from the soil surface to the last apical node of the stem (Fig. 1a).

Fig. 4 M₂ seedlings five months after sowing. Abnormal phenotypes were identified at early stages of seedling development. The red-circled seedlings indicate aberrant leaf shapes. The seedlings within white circles show yellowing, an indicator of chlorophyll deficiencies

Fig. 5 Image capture of the seedlings at transplanting. **a** Camera stand with three support stands. **b** Camera fixed in one position to capture all images. **c** Acclimatized potted seedling ready for transplanting

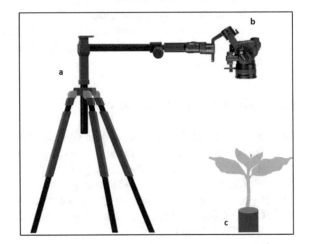

Fig. 6 Example images of M_1 seedlings captured using the fixed camera setting for ImageJ analysis (*see* Chap. "Chemical Mutagenesis of Mature Seed of Coffea arabica L. var. Venecia Using EMS")

6. Record the number of leaves per seedling. Leaves that are less than 50% green due to senescence should not be counted.

7. Using a Vernier caliper, measure and record stem diameter. Measure diameter at 3 cm above the soil level for seedlings and 5 cm for mature plants.

8. After transplanting, monitor growth traits with intervals of at least six months. As good practice for record tracking, always record the data collection dates.

9. Using a calibrated 200 cm ruler, measure plant height. Like for seedlings, plant height should be measured from the soil surface to the last apical node of the stem (*see* Note 7 and Fig. 1b).

10. Score the type of plant canopy qualitatively by visual assessment using predetermined descriptors such as compact, intermediate and open canopies. The size of the canopy can be determined quantitively by image analysis using tools like ImageJ.

11. Count the number of branches, starting from the oldest living branch to the highest (youngest) branch.

12. Score the angle of insertion of branches as drooping, horizontal or spreading, and as erect or semi-erect.

13. Score the colour of young leaves as green, dark green, yellow, etc.

14. Record leaf morphology as obovate, ovate, elliptic or lanceolate.

15. Record the morphology of leaf margin or apex shapes as round, obtuse, acute, acuminate, apiculate or spatulate.

16. Measure and record the average leaf length and width from five random mature leaves (> node 3 from the terminal bud). Leaf length is measured from leaf stalk to the apex while leaf width is measured from the widest part of the leaf.

17. Use leaf length and width to estimate leaf area that is an important measure of light interception and consequently plant productivity.

3.2.3 Measurement of Yield Component Traits

1. When the plants reach reproductive maturity, inspect and monitor plants daily and record yield component traits of interest.

2. Record the number of days to flower bud initiation for each plant.

3. Record the number of flower buds per axil. Determine the average number of buds from 10 axils, selected randomly from different nodes of different branches.

4. Record the number of days to flowering. This is determined from the planting date to the appearance of the first flowers.

5. Assess flower morphology and colour. Record any aberrations.

6. Record the average number of petals/corolla per flower, determined from the average of 10 flowers selected randomly from different nodes.

7. Measure and record the average length of petals per flower (mm), determined from 10 flowers selected randomly from different nodes.

8. Measure the average diameter (mm) of petals. Determine the petal diameter from the widest parts of the petals from 10 flowers selected randomly from different nodes.

9. Record the number of stamens/anthers per flower as the average of 10 flowers selected randomly from different nodes.

10. Record the length of stamens per flower (mm) as the average of 10 flowers selected randomly from different nodes.

11. Record the number of days or weeks to fruit/berry maturity (from bloom to harvest).

12. Record mature fruit colour as red, black, purple or orange.

13. Record fruit morphology as round, obovate or ovate.

14. Determine the weight of 20 mature fruits.

15. Measure and record fruit length (mm) as the average of five normal mature green fruits, measured using a Vernier caliper.

16. Measure fruit width (mm) as the average of five normal mature green fruits, measured at the widest part using a Vernier calliper.

17. Record the number of beans per berry, determined from five normal mature berries after de-pulping the berries.

18. Measure bean length (mm) as the average of five normal mature seeds.

19. Measure bean width (mm) as the average of five normal mature seeds, measured at the widest part of the beans.

20. Measure bean thickness (mm) as the average of five normal mature seeds, measured at the thickest (middle) part of the beans.

21. Screen the mutants for resistance to major pests and diseases or tolerance to abiotic stresses (for leaf rust resistance screening, *see* Chaps. "Screening for Resistance to Coffee Leaf Rust", "Inoculation and Evaluation of *Hemileia vastatrix* Under Laboratory Conditions" and "Evaluation of Coffee (*Coffea arabica* L. var. Catuaí) Tolerance to Leaf Rust (*Hemileia vastatrix*) Using Inoculation of Leaf Discs Under Controlled Conditions").

3.3 *Image Analysis to Estimate the Canopy Size*

1. Use ImageJ software to analyse images captured (Fig. 6).
2. ImageJ offers a non-destructive means to estimate e.g., the plant canopy as a measure of the area of spreading (Fig. 7).
3. Download and install ImageJ software on your computer (laptop or desktop) (Fig. 7b).
4. Choose between Windows or Mac depending on the operating system of your device (https://imagej.nih.gov/ij/download.html).
5. Open the software on your device to display the menu bar (Fig. 7a).
6. Import your image (png) into the software (File→Open→Select image) (Fig. 7c).
7. Calibrate your analysis by using the line tool to draw an imaginary line of the known distance, for instance internal diameter of the pot (60 mm) (Fig. 7d).
8. Set the scale for your analysis (Image→Set scale→Known distance→measured units e.g., cm or mm).
9. Adjust sliders to fully select the plant in the image (Image→Adjust→Colour threshold). Select red colour for the pixels (Fig. 7e).
10. Just before analysis, set parameters of interest (Analyse→Set measurement→Area, perimeter, etc.).
11. Invert the image output to black and white in the masks (Image→lookup tables→Invert LUT).
12. Perform image analysis to estimate area of spread (Analyse→Particles→Show masks→Display clear results→OK) (Fig. 7f).
13. Remove the noise from the mask (Process→Binary→Erode or Dilate) to generate an output image that is more similar to the input image (Fig. 7g).
14. Finally, repeat the analysis of particles (step 12) to obtain the measurement of the preferred parameter in the set units, for instance the area in square mm (mm^2) or perimeter (mm).
15. The data can be entered in the Field Book for the respective trait.

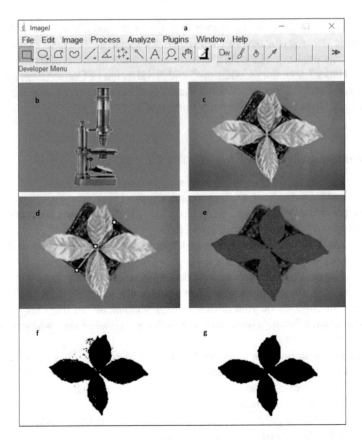

Fig. 7 Schematic representation of the steps taken to analyse images using ImageJ. **a** ImageJ software menu bar. **b** ImageJ software logo. **c** A picture of the seedling to be imported to the software for analysis. **d** Calibration line drawn from the internal diameter (100 mm) of the pot. **e** Red pixels adjusted using sliders to cover the seedling. **f** Output of the first image analysis with noise. **g** Output after noise removal

3.4 Statistical Analysis

1. Acquire the data from Field Book, or any other electronic data storage device.
2. Ensure internet connection on your Android device with the application containing data.
3. Export the data to your storage place (Fig. 2h, i).
4. Retrieve the data (Excel file, "csv" format) from the destination storage.
5. Save the latest Excel data file on your computer for statistical analysis. The file contains all the data from the entire data collection procedure.
6. Alternatively, connect the Android device to a computer with a USB cable, copy the data file directly from the export folder in the Field Book application.

7. Use any statistical software familiar to you to determine phenotypic variability within your population.
8. For every trait measured, calculate the average per treatment.
9. Perform pairwise comparisons for every mutagen treatment and non-treated control to determine an increase or reduction in the measured trait.
10. Likewise, perform multiple comparison analysis of every trait to determine whether phenotypic variation exists in the mutant population.
11. Explore the distribution of your data points per treatment. Pay attention to outliers, they could be of potential interest, for instance, high yielding.
12. If necessary, compute percentage increase or reduction in the trait with reference to control.
13. Determine levels of statistical significance ($p \leq 0.05$) between or among treatments.
14. The outputs can be represented graphically (Fig. 8) or in tabular formats.

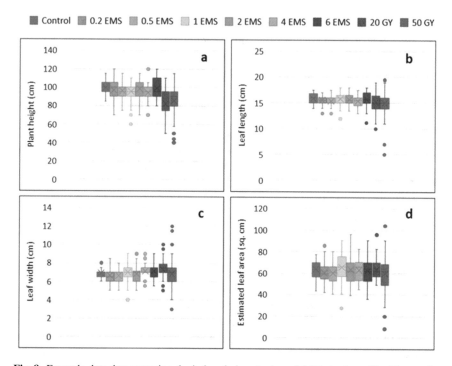

Fig. 8 Example data demonstrating the induced phenotypic variability in the coffee M_1 population treated with different doses of EMS (%EMS) and gamma-rays (gray—Gy). **a** Plant height measured three years after planting. **b** Leaf length measured from the petiole to the apex. **c** Leaf width measured at the widest part of the leaf. **d** Estimated Leaf Area (ELA) based on leaf length and leaf width measurements using allometric model, ELA $= 0.99927*(L*(-0.14757 + 0.60986*W))$ according to Unigarro-Muñoz et al. (2015)

3.5 Establishment of M_2 Population

1. To establish the M_2 population, harvest M_1 berries from each primary branch of each M_1 plant (*see* Note 8).
2. Mature berries can be sown/planted immediately after harvesting and depulping or kept for a period less than three months.
3. Prepare labels with all the necessary information including treatment, mother plant, date etc.
4. Remove the pulp from the berries.
5. Plant the beans in the previously prepared pots (*see* Note 8).
6. After sowing, water the pots periodically before and after emergence of the seedlings.
7. Record the number of days to emergence or germination.
8. At six months after sowing, the seedling will be ready for transplanting to 5 L pots.
9. Maintain M_2 plants following the procedures described in Sect. 3.1.
10. Collect data using the electronic phenotyping tools described in Sect. 3.2.1.
11. Monitor seedlings/plants for unique phenotypes (Fig. 4).
12. Prior to transplanting, take aerial images of the seedlings to determine the area of the seedlings canopies.
13. During plant growth, measure or score and record the growth traits as described in Sect. 3.2.2.
14. Measure and record yield traits as described in Sect. 3.2.3.
15. Assess the population for other traits including quality, resistance or tolerance to major pests and diseases, drought and other abiotic stresses (*see* Note 9).
16. Acquire data from the images using ImageJ as described in Fig. 7.
17. Likewise, export data from Field Book following procedures described in Fig. 2.
18. Analyse data using appropriate statistical analysis tools.

4 Notes

1. M_1 coffee seedlings were derived from *C. arabica* var. Venecia seed following EMS and gamma-ray mutagenesis (*see* Chap. "Chemical Mutagenesis of Mature Seed of Coffea arabica L. var. Venecia Using EMS"). The seeds were subjected to different doses of the respective mutagens after performing a chemotoxicity or radiosensitivity test.
2. Ensure that seedlings are transplanted with soil attached for optimal growth. Watering the seedlings one day before transplanting is recommended. The 5 L pots hold sufficient soil to support plant growth for at least three reproductive cycles under good management.

3. A completely randomised experimental design (CRD) is chosen since the experiment is set up in the greenhouse. The assumption is that a greenhouse provides a homogeneous micro-environment. The 100 cm space between blocks enables convenient working space during experimental management, data collection, etc.

4. Nutrients are important for plant growth and development. Some effects of induced mutagenesis are similar to those caused by nutrient deficiencies, such as chlorosis (yellowing, mosaics) and stunted growth. Therefore, apply complex fertilisers to support optimal plant growth and to have good and valid data on the mutant population. At transplanting, mixing fertilisers with the potting medium may not be necessary since the medium is still fresh. However, replenishing nutrients at three months intervals using nitrogen, phosphate and potassium (NPK) improves root and vegetative growth.

5. Pest and disease outbreaks in the mutant population should be prevented. Frequent monitoring of the plants is necessary. Like nutrient deficiencies, pests and disease symptoms can be similar to those resulting from mutations, such as chlorophyl aberrations, the curling of leaves and stunted growth. In case of any outbreaks, apply broad-spectrum systemic chemicals to provide an effective control strategy to diverse pests and diseases. Moreover, some insect pests attack the lower leaf surfaces or stems inside the canopy; hence, the pests will not be reached by contact pesticides administered by spraying. Systemic pesticides have less harmful effects on the natural biological agents such as ladybugs, wasps and ants as opposed to contact pesticides. For human safety reasons, follow the standard precautionary procedures for handling agricultural chemicals.

6. Field Book is an Android, open-access application that can be downloaded from the Google Play Store onto Android phones and tablets (https://github.com/PhenoApps/Field-Book). Using the application does not require connection to the internet. Prior to use, design your experimental layout in Excel on your computer. Connect your Android device to the computer, copy the Excel file and paste it into the import folder of the Fieldbook application. Finally, import the Excel file as a field in the application. Add your traits of interest in the application and define them as numerical, categorical, text, percentage, date, etc. After successful addition of the traits, data collection can start immediately. After completing data collection, it can be exported to a preferred data storage centre such as the cloud, Dropbox or shared via E-mail and Bluetooth, among others.

7. Measuring plant height is important to monitor differences in growth rate and identify dwarf or tall phenotypes in the mutant population.

8. Harvest berries from individual M_1 plants. Depending on the objective of the study, the position of the M_2 berries on individual M_1 plants can be mapped by recording the number and position of the branch and node for every harvested berry. Plant the beans in the previously prepared pots following the predetermined method, for instance, bulk method or single seed descent (FAO/IAEA 2018).

9. In most cases, the goal of a mutation breeding program is to improve one particular trait. However, one can simultaneously screen the mutant population for other traits of interest including quality, resistance or tolerance to major pests and diseases, drought, etc.

Acknowledgements The authors wish to thank the International Atomic Energy Agency and the Government of Belgium for their financial support through the CRP D22005 'Efficient Screening Techniques to Identify Mutants with Disease Resistance for Coffee and Banana' and the Peaceful Use Initiative project 'Enhancing Climate Change Adaptation and Disease Resilience in Banana-Coffee Cropping Systems in East Africa', respectively.

References

Carvalho A, Medina Filho HP, Fazuoli LC, Guerreiro Filho O, Lima MMA (1991) Aspectos genéticos do cafeeiro. Revista Brasileira de Genética l4:135–183

FAO (2015) Food and agriculture organisation, FAO statistical pocketbook. ISBN 978-108894-4

FAO/IAEA (2018) Manual on mutation breeding, 3rd edn. In: Spencer-Lopes MM, Forster BP, Jankuloski L (eds) Food and agriculture organization of the united nations. Rome, Italy, 301 p

Hartmann A, Czauderna T, Hoffmann R, Stein N, Schreiber F (2011) HTPheno: an image analysis pipeline for high-throughput plant phenotyping. BMC Bioinformatics 12(1):1–9

IPGRI (1996) Descriptors for Coffee (Coffea Spp. and Psilanthus Spp.). International Plant Genetic Resources Institute, Bioversity International p 36.

Jain MS (2005) Major mutation-assisted plant breeding programs supported by FAO/IAEA. Plant Cell Tissue Organ Cult 82:113–123

Lashermes P, Trouslot P, Anthony F, Combes MC, Charrier A (1996) Genetic diversity for RAPD markers between cultivated and wild accessions of *Coffea arabica*. Euphytica 87:59–64

Lashermes P, Bertrand B, Etienne H (2009) Breeding coffee (Coffea arabica) for sustainable production. In: Mohan JS, Priyadarshan PM (eds) Breeding plantation tree crops: tropical species. Springer, New York, pp 525–543. https://doi.org/10.1007/978-0-387-71201-7_14

Mechael PN (2009) The case for mHealth in developing countries. Innov Technol Gov Globalization 4(1):103–118

Ouma T, Kavoo A, Wainaina C, Ogunya B, Karanja M, Kumar PL, Shah T (2019) Open data kit (ODK) in crop farming: mobile data collection for seed yam tracking in Ibadan, Nigeria. J Crop Improv 33(5):605–619. https://doi.org/10.1080/15427528.2019.1643812

Pathirana R, Vitiyala T, Gunaratne NS (2009) Use of induced mutations to adopt aromatic rice to low country conditions of Sri Lanka. In: Induced plant mutations in the genomics era. Proceedings of an international joint FAO/IAEA symposium. International atomic energy agency, Vienna, Austria, 2009, pp 388–390

Pathirana R (2011) Plant mutation breeding in agriculture. Perspectives in agriculture, veterinary science, nutrition and natural resources. CAB International 2011 6, No 032

Prasad P, Agbona A, Kulakow P, Rabbi I, Egesi C, Parkes E, Mueller L (2018) Cassavabase, an advantage for IITA cassava breeding program. In: 4th International Cassava conference poster session presented at: the GCP21, at Cotonou Benin June 11–15; Cotonou, Benin

Rife TW, Poland JA (2014) Field book: an open-source application for field data collection on android. Crop Sci 54. https://doi.org/10.2135/cropsci2013.08.0579

Sabina L (2022) How data cross borders: globalizing plant knowledge through transnational data management and its epistemic economy. In: Krige J (ed) Knowledge flows in a global age: a transnational approach. University of Chicago Press, Chicago, pp 305–332. https://doi.org/10.7208/chicago/9780226820378-011

Schindelin J, Arganda-Carreras I, Frise E, Kaynig V, Longair M, Pietzsch T, Cardona A (2012) Fiji: an open-source platform for biological-image analysis. Nat Methods 9(7):676–682

Singh MK, Chetia S, Singh M (2017) Detection and classification of plant leaf diseases in image processing using MATLAB. Int J Life Sci Res 5(4):120–124

Ulukapi K, Nasircilar AG (2018) Induced mutation: creating genetic diversity in plants. In: Genetic diversity in plant species-characterization and conservation. IntechOpen

Unigarro-Muñoz CA, Hernández-Arredondo JD, Montoya-Restrepo EC, Medina-Rivera RD, Ibarra-Ruales LN, CarmonaGonzález CY, Flórez-Ramos CP (2015) Estimation of leaf area in coffee leaves (*Coffea arabica L.*) of the Castillo® variety. Bragantia 74:412–416. https://doi.org/10.1590/1678-4499.0026

Screening for Resistance to Coffee Leaf Rust

Vítor Várzea, Ana Paula Pereira, and Maria do Céu Lavado da Silva

Abstract Coffee leaf rust (CLR), caused by *Hemileia vastatrix* (*Hv*), is one of the main limiting factors of Arabica coffee production worldwide. Breeding for rust resistance is the most appropriate and sustainable strategy to control CLR. The characterization of coffee resistance to *Hv*, initiated in the 1930s in India, expanded with the creation of Coffee Rusts Research Center (CIFC) in 1955, in Portugal. Since then, the screening of coffee resistance to *Hv* races, from different geographical origins, has been carried out assisting breeding programmes of coffee growing countries and originating over 90% of the resistant varieties cultivated worldwide. However, the high adaptability of *Hv* has resulted in the gradual loss of resistance of some varieties. Thus, the characterization of new sources of resistance is crucial, also to face the recent epidemic resurgence of CLR across Latin America and the Caribbean.

Here, we provide a protocol for the screening of coffee resistance to *Hv* using different methods of inoculation on attached and detached leaves and on leaf disks. Information on environmental and pathogenicity factors that may affect the assessment of coffee resistance is also presented. This protocol allows the characterization of rust resistance on coffee mutants at laboratories, greenhouses, and field conditions.

1 Introduction

Coffee, the most important agricultural commodity, is crucial for the economy of more than 70 countries and is a livelihood source for between 12 and 25 million farmers worldwide (ICO 2019). The value of coffee exports amounted to USD 20 billion in 2017/18 being the revenue of the coffee industry estimated to surpass USD 200 billion (Samper et al. 2017; ICO 2019).

V. Várzea (✉) · A. P. Pereira · M. do C. L. da Silva
CIFC, Centro de Investigação das Ferrugens do Cafeeiro, Instituto Superior de Agronomia, Universidade de Lisboa, Quinta do Marquês, 2784-505 Oeiras, Portugal
e-mail: vitorvarzea@isa.ulisboa.pt

LEAF, Linking Landscape, Environment, Agriculture and Food Research Center, Associate Laboratory TERRA, Instituto Superior de Agronomia, Universidade de Lisboa, Tapada da Ajuda, 1349-017 Lisbon, Portugal

© The Author(s) 2023   209
I. L. W. Ingelbrecht et al. (eds.), *Mutation Breeding in Coffee with Special Reference to Leaf Rust*, https://doi.org/10.1007/978-3-662-67273-0_15

The two main cultivated coffee species, *Coffea arabica* (Arabica) and *C. canephora* (Robusta) account, on average, for 60% and 40%, respectively, of the world's coffee production (ICO 2020).

Coffee leaf rust (CLR), caused by the biotrophic fungus *Hemileia vastatrix* Berkeley and Broome, is considered the main disease of Arabica coffee. Since the historical first burst of CLR in the nineteenth century that caused the eradication of coffee cultivation in Sri Lanka, the disease gained a worldwide distribution, becoming practically endemic in all regions where coffee is grown (Wellman 1957; McCook 2006; Silva et al. 2006; Talhinhas et al. 2017; Keith et al. 2021). The disease produces economic losses over USD 1 billion annually (Kahn 2019).

H. vastatrix is a hemicyclic fungus producing urediniospores, teliospores and basidiospores, but only the dikaryotic urediospores, which form the asexual part of the cycle, reinfect successively the leaves whenever environmental conditions are favourable (Talhinhas et al. 2017 and references therein). After urediospore germination and appressorium differentiation over stomata, the fungus penetrates and colonizes the mesophyll tissues inter-and intracellularly giving rise to sporulation about 21 days after inoculation (Silva et al. 1999 and references therein; Silva et al. 2006; Talhinhas et al. 2017).

Breeding for resistance has been the most appropriate and sustainable strategy to control crop diseases.

Plant resistance to pathogens has been grouped into two different categories (Vanderplank 1968): complete resistance conditioned by single genes with major effects and incomplete resistance conditioned by multiple genes with minor and additive effects. A variety of terms have been used to refer to this perceived dichotomy, vertical versus horizontal, major-genes versus minor-genes, oligogenic versus polygenic, qualitative versus quantitative, race-specific versus race non-specific, hypersensitive versus non-hypersensitive, narrow-spectrum versus broad-spectrum (Parlevliet and Zadocks 1977; Roelfs et al. 1992). This diversity of terms reflects the assumptions made by the respective authors, but it also adds an element of confusion to the literature because some terms are used in different ways by different authors (Poland et al. 2009). Here the term incomplete resistance is considered as any form of resistance which allows for at least some reproduction of a given pathogen isolate on a given host plant (Eskes 1983).

Complete resistance results in phenotypes that fit into discrete classes of resistant and susceptible individuals according to Mendelian ratios (qualitative resistance). On the other hand, incomplete resistance cannot be easily categorized into distinct groups but in a continuous distribution of susceptible and resistant individuals (quantitative resistance) (Corwin and Kliebenstein 2017).

The traditional recording system for complete resistance on coffee to rust was developed at Coffee Rusts Research Center (CIFC) by d'Oliveira (1954–57) and consists of the identification of eight lesion types grouped into 4 classes of resistance and susceptibility. The incomplete resistance can be measured by its components, like infection frequency, lesion size, latent period and sporulation intensity (Browning et al. 1977; Parlevliet 1979; Eskes 1983).

The first effective characterization of coffee resistance to CLR, in experimental bases, was initiated in the 1930s in India (Mayne 1932, 1942). This work was greatly developed and broadened with the creation of CIFC in 1955, in Portugal. Inheritance studies have demonstrated that coffee-rust interactions follow the gene-for-gene relationship of Flor (1971) within a race-specific resistance system (complete resistance), being the resistance of coffee plants conditioned at least by nine major dominant genes (S_H1- S_H9) singly or associated. Reversely, it was possible to infer 9 genes of virulence (v_1-v_9) on *H. vastatrix* (Noronha-Wagner and Bettencourt 1967; Bettencourt and Rodrigues 1988). More than 55 *H. vastatrix* physiological races, from different geographic origins, were also identified over 60 years of world surveys carried out at CIFC (Rodrigues et al. 1975; Várzea and Marques 2005; Silva et al. 2006, 2022; Talhinhas et al. 2017; CIFC records), which allowed the characterization of coffee germplasm to support breeding programmes at coffee research institutions.

For many years, selection for *H. vastatrix* resistance has been based on highly specific complete resistance derived from major introgressed genes from *C. arabica* (S_H1, S_H2, S_H4 and S_H5) as well as from diploid species such as *C. canephora* (S_H6 - S_H9) or *C. liberica* (S_H3). To date, some of the most widely used sources of resistance to CLR are the Timor hybrids – HDTs (natural *C. arabica* x *C. canephora* hybrids) characterized and supplied by CIFC to research institutions of coffee growing countries (Rodrigues et al. 1975; Bettencourt and Rodrigues 1988).

The recent loss of resistance in some HDT-derived varieties, due to the appearance of more virulent rust races (Várzea and Marques 2005; Silva et al. 2006, 2022; Prakash et al. 2010; Talhinhas et al. 2017, CIFC records), as well as the current epidemics in Latin America and Caribbean, highlights the importance of the discovery and characterization of new sources of resistance.

Based on the CIFC's routine activities, we present a detailed protocol focused on the screening of complete resistance to CLR. A description of the qualitative scale used for the assessment of the reaction types and the environmental and pathogenicity factors that may affect this evaluation is also reported. The methods described here can be used in a greenhouse, laboratory or in field conditions and are useful for screening coffee mutants for leaf rust resistance.

2 Materials

2.1 To Collect and Store Inoculum

1. Urediniospores.
2. Gelatin capsules (8.5 mm).
3. Desiccator.
4. Vaseline (to close the desiccator).
5. Sulfuric acid solution.
6. Petri dishes.

7. Refrigerator (4 °C).
8. Cryogenic vials.
9. Minus 80 °C freezer.
10. Liquid nitrogen containers.
11. Cryo-gloves.
12. Flexible polyethylene tubing.
13. Laboratory lamp.
14. Scissors and tweezers.
15. Ethanol for surface and tools sterilization.

2.2 Spores Viability

1. Slides and coverslips.
2. Tweezers.
3. Micropipettes (100 μl).
4. Scalpel blades.
5. Watch glasses.
6. Formaldehyde solution.
7. Transparent nail polish.
8. Cotton blue lactophenol staining.
9. Light microscope.

2.3 Inoculation

1. Coffee leaves.
2. Urediniospores.
3. Distilled water.
4. Scalpels and soft brushes.
5. Test tubes.
6. Vortex mixer.
7. Micropipettes (10 μl, 20 μl).
8. Manual or electric sprayer.
9. Plastic bags.
10. Petri dishes.
11. Trays, nylon sponge and glass plates.
12. Analytical balance.
13. Ethanol for surface and tools sterilization.
14. Tween 80 solution.
15. Room with controlled light and temperature (Phytotron).
16. Incubation chambers.

2.4   Phenotyping of Coffee-Rust Interactions

1. Qualitative scale of reaction types.
2. Magnifying lens (if needed).

3   Methods

3.1   Procedures to Collect and Store Inoculum

For disease resistance screening, urediniospores collected with gelatin capsules are used as inoculum. The urediniospores must be collected from well sporulated young lesions. Note that spores from lesions in fallen leaves lose their viability very soon.

If enough spores cannot be harvested in the field for reliable screening tests, we can increase this amount with inoculations on vars. Caturra, Catuaí, Mundo Novo, Typica, Bourbon, etc. (carrying the resistance gene S_H5, i.e., with susceptibility to all the rust races infecting Arabicas) in greenhouse conditions.

Storage of rust samples must be done using recently collected spores thus to ensure high viability.

Rust samples can be stored for short and long term: (*i*) urediospores in gelatin capsules placed above sulfuric acid solution in a desiccator (50% relative humidity) and kept in refrigerator (4 °C) should retain good viability for about 180 days; (*ii*) in a freezer, at −80 °C, the spores keep the viability for more than 15 years; (*iii*) in liquid nitrogen, at −196 °C, spores can be stored for more than 20 years with high viability (CIFC records).

After storage at a negative temperature, a heat shock treatment (40 °C for 10 min) is required to break dormancy of the urediniospores and to recover their germination ability (CIFC records).

3.2   Spores Germination Tests

Before each experiment, laboratory germination tests are recommended to check the spores' viability. The urediniospore germination may be evaluated in vitro (glass slides) or in vivo (leaf pieces), being the last more accurate.

The germination in vitro is evaluated by placing aliquots of 100 μl of the urediniospore suspension (prepared as described in 3.3.1.2.) in glass slides which are kept in a moist chamber during 16 h at 23 °C. After this time, the germination is stopped with an aqueous solution of 3% formaldehyde. The glass slides are then covered with cover slips and observed under the microscope and the percentage of germinated spores counted on a minimum of ten fields of 100 urediniospores each (Silva et al. 1985).

The germination in vivo is evaluated on leaf pieces ($5cm^2$) cut from the previously inoculated leaves (*see* Sect. 3.3.), 16–24 h after inoculation (Silva et al. 1999). After let dried, the fragments are painted with transparent nail polish on the lower surface. About 24 h later, the dried nail polish (leaf replica) is removed with the help of tweezers and dipped into cotton blue lactophenol to stain the fungal structures (urediniospores, germ tubes and appressoria). The leaf replicas are placed in glass slides containing the same staining, covered with cover slips and observed under the microscope. With this technique it is possible to evaluate the rates of both the germinated urediniospores and the appressoria differentiated over stomata. Countings are made on a minimum of six microscope fields of 100 urediniospores each.

3.3   Inoculation Techniques

The screening of disease resistance is usually carried out by artificial inoculations by spreading fresh urediniospores on the lower surface of the coffee leaves with the help of a sterilized soft hairbrush or by using a urediniospore suspension on attached leaves in greenhouses or at field conditions, as well as on detached leaves and leaf disks at laboratory conditions.

Young, fully expanded leaves of the terminal node are used. In the day before inoculation, the plants are abundantly watered, and the turgid leaves are inoculated on the plant (leaves attached to the plant) or removed from the plant (detached leaves or leaf disks).

3.3.1   Attached Leaves

Brushing

This method can be used in attached leaves in the greenhouse or in the field. Following the routine procedure used at CIFC, fresh urediniospores of *H. vastatrix* (about 1 mg per pair of leaves) are placed with a scalpel on the lower surface of the leaf (*see* Fig. 1a) and then brushed gently with a camel's hairbrush (*see* Fig. 1b). The inoculated leaves are sprayed with distilled water (*see* Fig. 1c) and the plants are placed for 24 h under darkness at room temperature (18 °C to 24 °C) in moist chambers, (*see* Fig. 1d) after which they are placed in the greenhouses.

When the moist chambers are too small to allow the incubation of plants, the inoculated leaves after sprayed with distilled water are enveloped with a humid plastic bag during 24 h (*see* Fig. 1e). To avoid direct incidence of the sun rays, the plastic bags are covered with ordinary paper or newspaper sheets (*see* Fig. 1f) (D'Oliveira 1954–57). The same procedure can be used in field conditions.

Fig. 1 Inoculation of attached leaves using the brushing technique. *H. vastatrix* urediniospores on the scalpel (**a**) and then brushed on the lower surface of the leaf with a camel's hairbrush (**b**). Inoculated leaves are sprayed with distilled water (**c**) and placed in a moist chamber (**d**). For large plants, the inoculated leaves are enveloped in a humid plastic bag (**e**) and covered with newspaper sheets (**f**)

Urediniospore Suspension

The lower surface of the leaves is inoculated with an urediniospore suspension, with a concentration of 0.8 to 1.2 mg ml^{-1}, using a manual or electric sprayer (*see* Fig. 2), in a greenhouse or field conditions. These suspensions are prepared by suspending urediniospores in distilled water (to which 1–2 drops of Tween 0.02% is previously added). In the absence of Tween the following procedure is suggested: a spore mass is placed in a small test tube; one drop of distilled water is then added and kneaded into the spore mass with the help of a vortex mixer. This process is repeated by

Fig. 2 Inoculation of the lower surface of the coffee leaves with an urediniospore suspension using an electric sprayer

adding one drop of water at a time until the moistened spore mass has the pasty consistency of heavy cream. At this point, the bottom of the test tube is placed in a vortex mixer for several minutes and the remaining volume of water required for the final suspension is added during the stirring process. This step degasses the spore surfaces, which improves spore viability and yields almost complete dispersal of the spores in water. Good but incomplete suspensions leave a film of unwetted spores on the water surface if the stirring is omitted. Spores in suspensions prepared by this method germinate normally (CIFC records).

The upper inoculated leaves and part of the branch to which the leaves are attached are sprayed with distilled water and enveloped with a humid plastic bag. To avoid direct incidence of the sunrays, the plastic is covered with paper/newspaper sheets. The plastic bags are removed about 24 h after. Inoculations at field conditions are carried out in the late afternoon and the bags are removed early in the morning (Eskes 1989).

3.3.2 Detached Leaves

The leaves are placed with the abaxial surface upwards in trays lined on the bottom with a nylon sponge saturated with distilled water. Each leaf is inoculated with droplets (10–20 µl) from the urediniospores suspension (with a concentration of 250–500 spores/droplet). The droplets are deposited between the veins, using a micropipette (*see* Fig. 3). The trays are covered with glass plates and placed in the dark for 24 h at 22 ± 2 °C. After this time, the drops are dried out with small pieces

Fig. 3 Detached leaves
inoculated with droplets of
an urediniospore suspension

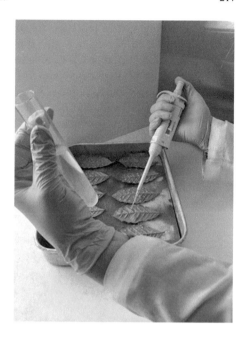

of filter paper, and the trays, covered with glass plates, are placed under moderate light conditions (fluorescent or indirect daylight of 500–1,000 lx) with a photoperiod of 12 h under similar temperatures (Eskes 1983).

3.3.3 Leaf Disks

Leaf disks are cut with cork borers from 1 to 2 cm in diameter and placed in Petri dishes or in trays with the upper leaf side down, on a sponge saturated with tap water. The disks are inoculated with droplets, from 10 to 20 μl of urediniospores suspension (with a concentration of 250–500 spores/droplet). After inoculation the boxes are closed with glass lids and incubated in the dark in the same conditions and environment described above for the detached leaves (Eskes 1982a).

3.4 Phenotypic Scoring Method for Disease Resistance

At greenhouse conditions, with a range of temperatures from 18 to 24 °C, the reading of the reaction types takes place usually 30–35 days after the inoculations, by a qualitative scale developed at CIFC by D'Oliveira (1954–57). However, the time to score the reactions can be extended to 45 days or more in the following situations: at higher or lower temperatures during the colonization process and with low aggressiveness of the fungal isolates.

This recording system has been followed at CIFC for more than 60 years to identify complete resistance on *Coffea* spp to CLR and to characterize rust races.

Qualitative scale used at CIFC to score the reaction types on attached leaves (D'Oliveira 1954–57; Bettencourt and Rodrigues Jr. 1988) (*see* Fig. 4).

i = immune (no visible symptoms).

fl = Flecks: small chlorotic flecks at the penetration sites, well visible with a pocket lens or when holding the leaf against the light.

; = Necrotic spots, visible macroscopically at the penetration site or dispersed over the infected area.

t = Punctiform tumefactions, often associated with flecks.

0 = Chlorotic spots, more or less intense, in the infected area, sometimes associated with small necrotic areas, but without spore production.

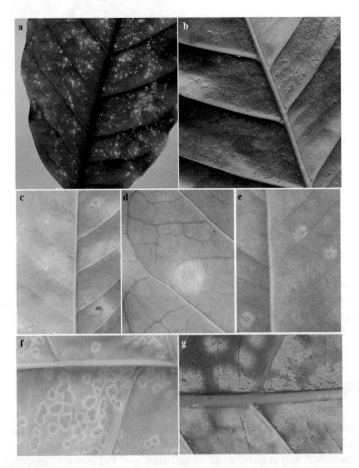

Fig. 4 Reaction types, according to the qualitative scale used at CIFC. Flecks visible when holding the leaf against the light (**a**); tumefactions (**b**); reaction 0 (**c**); reaction 1 (**d**); reaction 2 (**e**); reaction 3 (**f**); reaction 4 (**g**)

1 = Rare sporulating sori, always very small, sometimes only visible with a pocket-lens, in areas which are mainly chlorotic, sometimes associated with necrosis.

2 = Small or medium-sized pustules, diffused but visible macroscopically, in areas with intense chlorosis.

3 = Medium-sized or large pustules, surrounded by chlorosis.

4 = Large sporulating pustules, without true hypersensitivity, but sometimes surrounded by a slight chlorotic halo (highly susceptible or compatible).

X = Heterogeneous reaction with urediosporic pustules very variable in size associated with resistant reaction types.

The reaction types **i, fl, t** and **0** are jointly referred to as resistant (**R**), **1** as moderately resistant (**MR**), **2** as moderately susceptible (**MS**), and **3** and **4** as susceptible (**S**).

Detached leaves and leaf disks are useful to identify very susceptible genotypes to rust. However, intermediate levels of resistance expressed by a low reaction type (reactions 1 and 2) on attached leaves, at greenhouse or field conditions, may not be observable in leaf disks or detached leaves. In this way, whenever lesions without sporulation are found on detached leaves and leaf disks, we suggest to repeat the inoculations on attached leaves.

4 Notes

1. **To collect inoculum**

   When collecting rust samples in plants, either in the field or in the greenhouse it is important to avoid the presence of mycoparasites like the fungus *Lecanicillium lecanii* (Zimm.) Zare and W. Gams, with the ability of reducing spore viability and disease severity (Vandermeer et al. 2010; James et al. 2016; CIFC records). The first evidence of *L. lecanii* in rust lesions is in the form of small white spots at the center of the rust sori. The spots gradually enlarge; the cotton-like, white colored mycelium of the mycoparasite covered the rust sori. The development of this mycoparasite is restricted to the rust infected leaf parts, but never grow to over the entire width of the rust lesions (CIFC records).

2. **Inoculum**

   When the virulence of rust local populations is not known, the source of inoculum to be used to detect resistance in coffee mutants should be gathered from the same plants or similar genotypes where they come from. If resistance is found in the first inoculations, the screening on coffee mutants should continue with inoculum collected from different coffee genotypes in different regions to try to get rust samples with higher spectra of virulence. In general, the origin and distribution of rust races follow the resistance genes present in coffee populations.

3. **Factors influencing the infection process and the resistance symptoms**

   3.1. *Moisture*

The urediniospores do not germinate, even at high relative humidity, if the free water is absent. If the water dries off before penetration, then the process is inhibited (Nutman and Roberts 1963; Rayner 1972; CIFC records).

3.2. Temperature

(i) The temperature, while the leaf surface is wet, is one of the most important factors that determine the amount of spore germination and penetration. The optimum temperatures for germination are 20 to 25 °C (CIFC records).

(ii) Extreme temperatures after inoculation causes some depressive effect on fungal colonization and sporulation, with the slight lower reaction types on susceptible plants and in extreme may kill the fungus inside the leaves (Montoya and Chaves 1974; Ribeiro et al. 1978; Silva et al. 1992). The small chlorotic lesions, developed in these conditions, are likely to be confused with resistant reactions (CIFC records).

(iii) The enlargement of lesions on leaves and the sporulation are limited by temperatures over 35 °C and lower than 10 °C.

3.3. Light intensity and/or leaf age

(i) Leaves exposed to higher light intensities before inoculation show more lesions than those exposed to lower intensities (Eskes 1982b, 1983, 1989).

(ii) In screening tests, the light intensity should preferably be kept at medium levels before inoculation and medium to low levels after inoculation (Eskes 1982b, 1983, 1989, CIFC records).

(iii) Some derivatives of interspecific tetraploid hybrids (*C. arabica* x *C. canephora*) like Icatu and Timor Hybrid (HDT) show lower and even resistant reaction type lesions at a lower light intensity under greenhouse conditions (Marques and Bettencourt 1979; CIFC records).

(iv) Studies on the effect of leaf age and light intensity on CLR found higher resistance on young leaves growing in the shade, and lower resistance for old leaves exposed to sunlight (Eskes 1983).

4. Phenotypic scoring for disease resistance

4.1 In the assessment of complete resistance, the use of susceptible controls is needed to exclude the possibility of escapes (inadequate exposure to the pathogen and/or extreme temperatures during the initial infection process).

4.2 The reaction type "i" (Immunity = no macroscopically visible symptoms), which may appear to be very desirable, is rarely observed under CIFC greenhouse conditions. Care should be taken not to confuse this reaction type with escape. Each time immunity occurs confirmation with a new test with the same pathotype is needed.

4.3 Coffee plants irradiated at a dose of 100 Gy exhibited several morphologic changes in leaves like shape, length and width, the color of young leaves, number of leaves per plant, etc., and distance from the cotyledon to the first node (Quintana et al. 2019). These changes may influence the expression of resistance to leaf rust. Irradiated coffee plants should be inoculated, if possible, on leaves of different ages.

4.4 "Resistance" should be distinguished from "Tolerance" which is defined as the ability of a crop to maintain a high yield in the presence of disease, being a difficult characteristic to measure, and its component traits are generally undefined (Newton 2016). Note that tolerance should not be confused with incomplete resistance.

4.5 Partial resistance characterized by a reduced rate of epidemic development despite a high- or susceptible-infection type (Parlevliet 1975) was never detected at CIFC greenhouse conditions.

5. Incomplete resistance

The contribution of Albertus B. Eskes (1989 and references therein) for the characterization of incomplete resistance on coffee to CLR, using leaf disks and detached leaves, was of paramount importance and his works are a reference to those who intend to develop studies on this kind of resistance.

5.1 The quantitative or incomplete resistance of a host genotype cannot be assessed in absolute terms; it is always a relative measure compared with that of a well-known standard genotype. The latter is often the most susceptible genotype available (Parlevliet 1989).

5.2 The degree of incomplete resistance evaluated in a particular coffee genotype may be masked by different levels of *H. vastatrix* aggressiveness. Intermediate compatibility in host/pathogen interaction (low reaction type and long latent period) can be due to incomplete resistance of coffee plants/ or lower aggressiveness (fitness) of rust.

5.3 Components of incomplete resistance to CLR

Infection frequency: Number of lesions per leaf or leaf area unit, or the percentage of disks with lesions.

Latent period: The time from inoculation to spore production. Normally it is calculated as the time taken for 50% of the lesions to sporulate or the time between the inoculation and the formation of the first spores.

Incubation period: The number of days between the inoculation and the appearance of the first chlorotic lesions per leaf or disk

Proportion of sporulating lesions: Percentage of sporulated lesions in relation to the total number of lesions by leaf or disk.

Sporulation intensity: The number of spores produced per sporulating lesion or per infected leaf area, over a certain time interval.

Lesion size: Normally evaluated at the end of the experiment

5.4 Relations amongst reaction types (RT) and components of incomplete resistance

The majority of the components of incomplete resistance are a quantitative extension of the scale used for RT. These components, as well as the RT's are related to the same basic criteria, like lesion size, sporulation intensity, and the occurrence of chlorosis or necrosis. The latent period is related to lesion size when fungal growth is slow, the sporulation will generally be delayed, and the lesions will be smaller. The reaction types "0" (chlorosis is without sporulation) or necrotic spots will reduce the sporulation intensity and/or the duration of sporulation (Eskes 1981).

Acknowledgements Funding for this work was provided by the Food and Agriculture Organization of the United Nations and the International Atomic Energy Agency through their Joint FAO/IAEA Research Contract n° 20902/R0 of IAEA Coordinated Research Project D22005 and Portuguese funds through FCT—Fundação para a Ciência e a Tecnologia, I.P., under the project UIDB/04129/2020 of LEAF-Linking Landscape, Environment, Agriculture and Food, Research Unit.

References

Bettencourt AJ, Rodrigues Jr CJ (1988) Principles and practice of coffee breeding for resistance to rust and other diseases. In: Clarke RJ, Macrae R (eds), Coffee agronomy, vol. IV, Elsevier Applied Science Publishers LTD, London, pp 199–234

Browning JA, Simons MD, Torres E (1977) Managing host genes: epidemiologic and genetic concepts. In: Horsfall JG, Cowling EB (eds) Plant disease—advanced treatise, vol I. Academic Press, New York, p 448

Corwin JA, Kliebenstein DJ (2017) Quantitative resistance: more than just perception of a pathogen. Plant Cell 29:655–665. https://doi.org/10.1105/tpc.16.00915

D'Oliveira B (1954–57) As ferrugens do cafeeiro. *Revista do Café Português* 1(4):5–13; 2(5):5–12; 2(6):5–13; 2(7):9–17; 2(8):5–22; 4(16):5–15

Eskes AB (1981) Incomplete resistance to coffee leaf rust In: Lamberti F, Waller JM, van de Graff NA (eds), Durable resistance in crops, vol. 55, NATO ASI Series, pp 291–316

Eskes AB (1982a) The effect of light intensity on incomplete resistance of coffee to *Hemileia vastatrix*. Neth J Pl Path 88:191–202. https://doi.org/10.1007/BF02140882

Eskes AB (1982b) The use of leaf disk inoculations in assessing resistance to coffee leaf rust (*Hemileia vastatrix*). Neth J Pl Path 88:127–141. https://doi.org/10.1007/BF01977270

Eskes AB (1983) Incomplete resistance to coffee leaf rust (*Hemileia vastatrix*). Ph.D. thesis, Agriculture University, Wageningen, The Netherlands

Eskes AB (1989) Resistance. In: Kushalappa AC, Eskes AB (eds) Coffee rust: epidemiology, resistance and management. CRC Press, Boca Raton, pp 171–277

Flor HH (1971) Current status of the gene-for-gene concept. Annu Rev Phytopathol 9:275–296

ICO (2019) Coffee development report 2019. Growing for prosperity economic viability as the catalyst for a sustainable coffee sector. ISBN: 978-1-5272-4994-3. 80pp

ICO (2020) ICO coffee production 2020. https://www.ico.org/prices/poproduction.pdf. Accessed 4th Aug 2022

James TY, Marino JA, Perfecto I et al (2016) Identification of putative coffee rust mycoparasites via single-molecule DNA sequencing of infected pustules. Appl Environ Microbiol 82:631–639

Kahn LH (2019) Quantitative framework for coffee leaf rust (*Hemileia vastatrix*), production and futures. Int J Agric Ext 7:77–87

Keith L, Sugiyama L, Brill E et al (2021) First report of coffee leaf rust caused by *Hemileia vastatrix* on coffee (*Coffea arabica*) in Hawaii. Plant Dis 34:2–4

Marques DV, Bettencourt AJ (1979) Resistência a *Hemileia vastatrix* numa população de Icatú. Garcia De Orta, Sér Est Agron 6:19

Mayne WW (1932) Annual report of coffee scientific officer, 1931–32. Mysore Coffee Exp Stn Bull N° 7

Mayne WW (1942) Annual report of the coffee scientific officer, 1941–42. Mysore Coffee Exp Stn Bull N° 24

McCook S (2006) Global rust belt: *Hemileia vastatrix* and the ecological integration of world coffee production since 1850. J Glob Hist 1:177–195

Montoya RH, Chaves GM (1974) Influência da temperature e da luz na germinação, infectividade e periodo de geração de *Hemileia vastatrix*. Experientiae 18:239

Newton AC (2016) Exploitation of diversity within crops –the key to disease tolerance? Front Plant Sci 7:665. https://doi.org/10.3389/fpls.2016.00665

Noronha-Wagner M, Bettencourt AJ (1967) Genetic study of the resistance of *Coffea* spp. to leaf rust I. Identification and behavior of four factors conditioning disease reaction in *Coffea arabica* to twelve physiologic races of *Hemileia vastatrix*. Can J Bot 45:2021–2031

Nutman FJ, Roberts FM (1963) Studies on the biology of *Hemileia vastatrix* Berk. &Br. Trans Br Mycol Soc 46:27

Parlevliet JE (1975) Partial resistance of barley to leaf rust, *Puccinia hordei*. I. Effect of cultivar and development stage on latent period. Euphytica 24:21–27

Parlevliet JE (1979) Components of resistance that reduce the rate of epidemic development. Annu Rev Phytopathol 17:203–224

Parlevliet JE, Zadoks JC (1977) The integrated concept of diseases resistance, a new view including horizontal and vertical resistance in plants. Euphytica 26:5–21

Parlevliet JE (1989) Identification and evaluation of quantitative resistance. In: Leonard KJ, Fry WE (eds) Plant disease epidemiology—genetics, resistance and management, vol II, McGraw-Hill Publishing Company, New York, pp 215–248

Poland JA, Balint-Kurti PJ, Wisser RJ et al (2009) Shades of gray: the world of quantitative disease resistance. Trends Plant Sci 14:21–29

Prakash NS, Mishra MK, Padamajyothi D et al (2010) Evaluation of coffee varieties derived from diverse genetic sources of resistance for prospective exploitation—an international cooperative effort. In: Proceedings of the 23rd international conference on coffee science (ASIC), Bali, Indonesia. Abstract PB739, p 176

Quintana V, Alvarado L, Saravia D et al (2019) Gamma radiosensitivity of coffee (*Coffea arabica* L. var. typica) Estudio de la radiosensibilidad gamma del café (*Coffea arabica* L. var. typica). Peruvian J Agron 3:74–80

Rayner RW (1972) Micologia, historia y biologia de la roya del cafeto. Publicacion Misceffma no. 94. Instituto Interamericano de Ciencias Agricolas de la Organizacion de Estados Americanos, Centro Tropical de Enseranza e Investigacion, Turrialba, Costa Rica, p 67

Ribeiro IJA, Mónaco LC, Filho OT et al (1978) Efeito de altas temperaturas no desenvolvimento de *Hemileia vastatrix* em cafeeiro susceptível. Bragantia 37:1

Rodrigues Jr CJ, Bettencourt AJ, Rijo L (1975) Races of the pathogen and resistance to coffee rust. Annu Rev Phytopathol 13:49–70

Roelfs AP, Singh RP, Saari EE (1992) Rust diseases of wheat: concepts and methods of disease management. DF Cimmyt, Mexico, 81 pp

Samper L, Giovannucci D, Marques Vieira L (2017) The powerful role of intangibles in the coffee value chain. In: Economic research working paper N° 39:1–78

Silva MC, Rijo L, Rodrigues Jr CJ, Vasconcelos MI (1992) Estudos histológicos da acção de tratamentos pelo calor nas expressões de susceptibilidade e de resistência na interacção *Coffea arabica - Hemileia vastatrix*. Turrialba I:200–206

Silva MC, Nicole M, Rijo L et al (1999) Cytochemical aspects of the plant-rust fungus interface during the compatible interaction *Coffea arabica* (cv. Caturra)-*Hemileia vastatrix* (race III). Int J Plant Sci 160:79–91

Silva MC, Várzea V, Guerra-Guimarães L, Azinheira HG et al (2006) Coffee resistance to the main diseases: leaf rust and coffee berry disease. Braz J Plant Physiol 18:119–147

Silva MC, Guerra-Guimarães L, Diniz I et al (2022) An overview of the mechanisms involved in coffee-*Hemileia vastatrix* interactions: plant and pathogen perspectives. Agronomy 12:326. https://doi.org/10.3390/agronomy12020326

Silva MC, Rijo L, Rodrigues Jr CJ (1985) Differences in aggressiveness of two isolates of Race III of *Hemileia vastatrix* on cultivar Caturra of *Coffea arabica*. In: Proceedings of the 11th international scientific colloquium on coffee, Lomé, Togo, pp 635–645

Talhinhas P, Batista D, Diniz I et al (2017) The Coffee Leaf Rust pathogen *Hemileia vastatrix*: one and a half centuries around the tropics. Mol Plant Pathol 18:1039–1051. https://doi.org/10.1111/mpp.12512

Vandermeer J, Perfecto I, Philpott S (2010) Ecological complexity and pest control in organic coffee production: uncovering an autonomous ecosystem service. Bioscience 60:527–537

Vanderplank JE (1968) Disease resistance in plants. Academic Press, New York, p 194

Várzea VMP, Marques DV (2005) Population variability of *Hemileia vastatrix* vs coffee durable resistance. In: Zambolim L, Zambolim E, Várzea VMP (eds) Durable resistance to coffee leaf rust. Universidade Federal de Viçosa, Viçosa, Brasil, pp 53–74

Wellman FL (1957) *Hemileia vastatrix*. Federation Cafetalera de America, San Salvador

Inoculation and Evaluation of *Hemileia vastatrix* Under Laboratory Conditions

Miguel Barquero-Miranda, María José Cordero-Vega, and Kimberly Ureña-Ureña

Abstract The coffee leaf rust, a disease caused by the biotrophic fungus *Hemileia vastatrix*, is one of the main limitations in coffee production today as it causes significant economic losses to the coffee production sector. Genetic improvement is an option to solve these problems. The Arabica varieties have a very narrow genetic base therefore the induction of mutations, through e.g. physical methods such as gamma rays, could be an efficient tool to increase the genetic diversity of the crop. This would allow to obtain desirable agronomic characteristics such as resistance to pests and diseases. To determine the effect of irradiation on the plants, protocols enabling evaluation of improved traits must be applied. In the case of the assessment of plant resistance to pests and diseases, screening protocols that take into account their biology should be considered. This chapter provides a detailed protocol for the inoculation and evaluation of *Hemileia vastatrix* under laboratory conditions.

1 Introduction

Coffee is the second most commercialized product worldwide. It is produced in over 50 countries and secures livelihoods for millions of farmers (Vega et al. 2003; ICAFE 2017). *Hemileia vastatrix* Berk. & Broome, the causal agent of coffee leaf rust, is one of the biotic factors that affects coffee, causing significant economic losses due to the defoliation, subsequent harvest losses, and renewal needs due to severe damage caused in plants (Barquero 2013; ICAFE 2013).

This disease caused a widespread impact in 2012 in Central America, mainly due to the susceptibility of planted varieties of *Coffea arabica* such as Caturra and Catuaí (Avelino and Rivas 2013). Due to the origin, domestication process, reproduction and evolution of the genome, Arabica varieties are characterized by a low genetic diversity (Hendre et al. 2008; Prakash et al. 2002). As a result of a natural hybridization process between *C. eugenioides* and *C. canephora*, *C. arabica* is the only tetraploid

---

M. Barquero-Miranda (✉) · M. J. Cordero-Vega · K. Ureña-Ureña
Phytoprotection Laboratory, Costa Rican Coffee Institute, Coffee Research Center, 37-1000, San Pedro, Heredia, Costa Rica
e-mail: mbarquero@icafe.cr

© The Author(s) 2023
I. L. W. Ingelbrecht et al. (eds.), *Mutation Breeding in Coffee with Special Reference to Leaf Rust*, https://doi.org/10.1007/978-3-662-67273-0_16

Fig. 1 Coffee leaf rust symptoms and signs, chlorotic spots and urediniosporic sori on the lower leaf surface

species of the genus *Coffea* (2n = 4x = 44). This generates a limitation for genetic improvement of resistance genes due to the homogeneity of the varieties (Jefuka et al. 2010; Naranjo Zúñiga 2018).

Hemileia vastatrix is a biotrophic fungus that penetrates the plant through the stomata located on the abaxial side of the leaf. Its taxonomic classification is as follows: Phylum: *Basidiomycota*, Class:Pucciniomycetes, Order: Pucciniales, Family: Zaghouaniaceae, Genus: *Hemileia* (gbif 2022).

H. vastatrix is a hemicyclic fungus producing urediniospores, teliospores and basidiospores, but only the dikaryotic urediniospores, which form the asexual part of the cycle, reinfect coffee leaves successively and are responsible for the disease (as revised by Talhinhas et al. 2017). As a first symptom, small yellow chlorotic spots are observed in the foliage that subsequently, as the infection progresses, produce masses of orange urediniosporic sori (*see* Fig. 1) (Arauz Cavallini 2011). An epidemic of coffee leaf rust can be divided into two stages: stage of the production of the initial inoculum, whose main source is the residual inoculum, and a second stage that comprises the production of the secondary inoculum, which is the result of the successive repetition of the infection process on the same leaf (Avelino and Rivas 2013, Naranjo Zúñiga 2018).

The disease cycle of coffee leaf rust consists of five stages which can be affected by factors such as fruit load, plant resistance, microclimate and plant nutrition (Avelino 2004; Rhiney et al. 2021).

These stages include:

- **Dissemination**: It occurs in three stages, the release of urediniospores, dispersion through factors such as rain, wind and people and deposition in plant tissue.
- **Germination**: Once the urediniospores are deposited in the leaves, 4–6 germinative tubes are emitted with an appressorium necessary to force stomatal entry. The optimal conditions are 22 °C temperature, 24 h of darkness and free water until the penetration stage (Silva et al. 1999; Naranjo Zúñiga 2018).
- **Penetration**: The presence of well-formed stomata is necessary to be able to enter the leaf, so that the age of the leaf influences the receptivity to infection.

- *Colonization*: This phase requires the growth of hyphae of the fungus in the intercellular spaces of the spongy parenchyma and haustoria within the cells of the palisade parenchyma and even the upper epidermis to give rise to the first symptoms, a macroscopic chlorosis (McCain and Hennen, 1984).
- *Sporulation*: Once hyphae invade the substomatal chamber, they differentiate to form protosori. Later urediniosporic sori protrude through the stomata (Silva et al. 1999 and references therein).

Genetic improvement is an attractive approach that enables solving production constraints caused by pests and diseases. Induced mutagenesis is one of the tools that can be used to increase genetic diversity (ICAFE 2011; Novak and Brunner 1992; Shu et al. 2011). Gamma rays have proven to be an efficient tool to improve traits of agronomic importance such as resistance to pests and diseases (Borzouei et al. 2010; Yadav and Singh 2013; Shu et al. 2011). An efficient screening protocol is therefore required for evaluation of mutant populations developed via induced mutagenesis.

This protocol describes the procedures for the inoculation and evaluation of the defense response of the coffee genetic material infected with *Hemileia vastatrix* under laboratory conditions, or the biological efficacy of molecules for the control of the disease. This protocol is based on Eskes and Toma-Braghini (1982), with modifications made by the Costa Rican Coffee Institute-Coffee Research Center, Phytoprotection Laboratory.

2 Materials

2.1 Preparation of Rust Inoculum

1. Sterile scalpel.
2. 50 μL centrifuge tubes.
3. Falcon tubes of the necessary size.
4. Distilled water.
5. Neubauer chamber.
6. Microscope.

2.2 Rust Inoculation

1. Scissors.
2. Plastic boxes.
3. Foams.
4. Plastic grid.
5. Water.
6. Micropipette (50 μl graduation).

7. Micropipette tips.
8. Ceramic spoon.
9. Adhesive plastic.

2.3 Rust Evaluation

1. Magnifying glass.
2. Light source.

3 Methods

3.1 Preparation of Rust Inocula

1. Collect coffee leaves with abundant rust spores (race II, or the most important for the region or country) from the field or greenhouse-grown susceptible varieties (Caturra or Villa Sarchí) (*see* Note 1).
2. Scrape the spores with a sterile scalpel and store in 50 μL centrifuge tubes.
3. Prepare a suspension of spores in distilled water and determine the concentration of urediniospores using a Neubauer chamber. Count the spores with orange coloration located in the corners and the center of the central sub-chamber (*see* Fig. 2).
4. Determine the concentration of urediniospores applying the formula (*see* Note 2):

$$Urediniospore\ concentration/ml = urediniospores\ counted\ \times 5 \times 1e10^4$$

Fig. 2 Urediniospores count points in Neubauer chamber

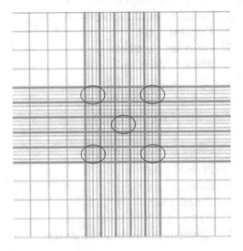

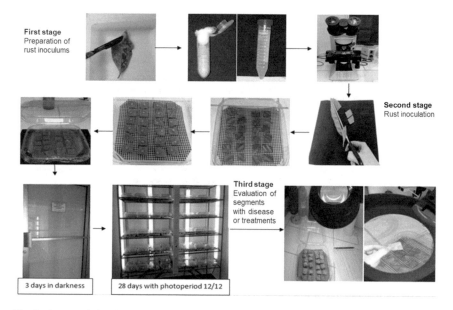

Fig. 3 A general description of the preparation stage of the inoculum and of the plant material, of the process for the germination and incubation of the pathogen and of the quantification of the presence or absence of the disease

3.2 Rust Inoculation

1. Collect healthy leaves from the second node of branches from the middle stratum of the plants selected for rust resistance evaluation (*see* Fig. 3 and Note 3).
2. Cut 2 × 2 cm square segments from collected leaves.
3. Prepare humid chambers that consist of a plastic box with a wet foam in the background and a plastic grid on the foam.
4. Place the leaf segments on the foam and inoculate with 50 μl of suspension by placing a drop in the center and spreading it with a ceramic spoon.
5. Cover the wet chamber with clear adhesive plastic, ensuring it is airtight.
6. Incubate in darkness at room temperature for 3 days (*see* Note 4).
7. Transfer into a room with a photoperiod of 12 h light /12 h dark and 22 ± 1 °C for 28 days.

3.3 Rust Evaluation

1. Uncover the humid chambers and place in a location with enough light.
2. Remove necrotic segments.
3. Count number of segments inoculated and number of segments with rust.

Table 1 Scale used to measure the severity of rust in coffee segments (Eskes and Toma-Braghini 1982) (*see* Note 5)

Level of severity	Description
0	Without injury
1	10% chlorotic lesions without sporulation
2	25% chlorotic lesions without sporulation
3	50% chlorotic lesions without sporulation
4	75% chlorotic lesions without sporulation
5	100% chlorotic lesions without sporulation
6	Less than 10% sporulation lesions
7	25% sporulated lesions
8	50% sporulated lesions
9	75% sporulated lesions
10	100% sporulated lesions

4. The incidence rate of the disease is determined by the formula:

$$\% \text{ incidence} = \frac{\text{Number of segments with presence of uredospores} \times 100}{\text{total segments inoculated}}$$

5. Determine the presence and abundance of signs and symptoms (Table 1).

   This inoculation technique can also be used to evaluate the response of different natural or chemically synthesized molecules for the defense of plants susceptible to the disease. To do this, the use of Table 1 allows us to understand the mechanism of action of the products according to the incubation and latency periods of the pathogen.

4 Notes

1. When selecting leaves to collect rust spores, it is necessary to check whether the fungus *Lecanicillium lecanii*, a hyperparasite of *H. vastatrix*, is not present. Application of fungicides should be avoided two months prior to collection of leaves.
2. The concentration of the spore suspension should be approximately 1×10^5 urediniospores/ml.
3. It may be important to evaluate different leaf ages of individual mutant plants with putative resistance to the disease.
4. Some protocols (in greenhouse and laboratory conditions) indicate that 24 h of incubation is enough. At this time, the fungus concludes the germination, appressoria differentiation, and penetrates the host tissues. In our laboratory,

the sporulation is more abundant and successful when dark conditions remain for three days. Therefore, it is recommended that this step is being tested and adjusted to the conditions of the laboratory in which the evaluation is going to be performed.

5. In this protocol, in addition to the rust incidence the scale described in Table 1 is being used. The inoculation technique presented in this chapter can also be used to evaluate the response of different natural or chemically synthesized molecules for the defense of plants susceptible to the disease. To do this, the use of Table 1 allows us to understand the mechanism of action of the products according to the incubation and latency periods of the pathogen.

Acknowledgements Funding for this work was provided by the Costa Rican Coffee Institute-Coffee Research Center and the FAO/IAEA Joint Center. This work is part of the Coordinated Research Project D22005 titled "Efficient Screening Techniques to Identify Mutants with Disease Resistance for Coffee and Banana", Contract Number 20475.

References

Arauz Cavallini LF (2011) Fitopatología: un enfoque agroecológico, 2 edn. UCR, San José, Costa Rica, 514 p

Avelino et al (2004) Effects of crop management patterns on coffee rust epidemics. Plant Pathol 53:541–547. https://doi.org/10.1111/j.1365-3059.2004.01067.x

Avelino J, Rivas G (2013) La roya anaranjada del cafeto. hal-01071036, pp 1–47. Available in https://hal.archives-ouvertes.fr/hal-01071036 Accessed 19 Nov 2019

Barquero M (2013) Recomendaciones para el combate de la roya del cafeto, 3 edn. Heredia, CR, ICAFE, 72 p. Available in https://www.researchgate.net/publication/281625030_Recomendaciones_para_el_combate_de_la_roya_del_cafeto

Borzouei A, Kafi M, Khazaei H, Naseriyan B, Majdabadi A (2010) Effects of gamma radiation on germination and physiological aspects of wheat (*Triticum aestivum* L.) seedlings. Pakistan J Botany 42(4):2281–2290

Eskes AB, Toma-Braghini M (1982) The effect of leaf age on incomplete resistance of coffee to Hemileia vastatrix. Netherlands J Plant Pathol 88(6):219–230. https://doi.org/10.1007/BF02000128Gbif. Global Biodiversity Information Facility—https://www.gbif.org/search?q=Hemileia%20vastatrix. Accessed 21 Dec 2022

GBIF (2022) Gobal Biodiversity Information Facility. www.gbig.org

Hendre PS, Phanindranath R, Annapurna V, Lalremruata A, Aggarwal RK (2008) Development of new genomic microsatellite markers from robusta coffee (*Coffea canephora* Pierre ex A. Froehner) showing broad cross-species transferability and utility in genetic studies. BMC Plant Biol 8(1):51

ICAFE (2011) Guía técnica para el cultivo del café. 1 ed. Heredia, CR, ICAFE, 72 p

ICAFE (2013) Informe sobre la actividad cafetalera de Costa Rica. Heredia, CR, ICAFE, 69 p

ICAFE (2017) Informe sobre la actividad cafetalera de Costa Rica. Heredia, CR, ICAFE, 59 p

Jefuka C, Fininsa C, Adugna G, Hindort H (2010) Coffee leaf rust epidemics (*Hemileia vastatrix*) in Montane Coffee (*Coffea arabica* L.) Forests in Southwestern Ethiopia. East. Afr J Sci 4(2):86–95

Naranjo Zúñiga VR (2018) Evaluación del efecto de diferentes manejos de nutrición y sombra sobre la resistencia fisiológica de la planta de café (*Coffea arabica*) a la roya (*Hemileia vastatrix*), en discos de hoja en condiciones controladas de laboratorio (Doctoral dissertation, CATIE, Turrialba, Costa Rica), 68 p

Novak FJ, Brunner H (1992) Tecnología de mutación inducida para el mejoramiento genético de los cultivos. IAEA Newsletter 4:25–33

McCain JW, Hennen JF (1984) Development of the uredinial thallus and sorus in the orange coffee rust fungus, Hemileia vastatrix. Phytopathology 74:714–721

Prakash NS, Combes MC, Somanna N, Lashermes P (2002) AFLP analysis of introgression in coffee cultivars (*Coffea arabica* L.) derived from a natural interspecific hybrid. Euphytica 124:265–271

Rivillas Osorio CA, Serna Giraldo CA, Cristancho Ardila MA, Gaitan Bustamante AL (2011) La roya del cafeto en Colombia—Impacto, manejo y costos del control. Caldas, Colombia, Cenicafé, 51 p

Rhiney K, Guido Z, Knudson C, Avelino J, Bacon CM, Leclerc G, Aime MC, Bebber DP (2021) Epidemics and the future of coffee production. Proc Natl Acad Sci USA 118:e2023212118

Shu QY, Forster BP, Nakagawa H (2011) Plant mutation breeding and biotechonology. FAO/IAEA, Viena, 589 p

Silva MC, Nicole M, Rijo L, Geiger JP, Rodrigues CJ (1999) Cytochemistry of plant-rust fungus interface during the compatible interaction *Coffea arabica* (cv. Caturra)-*Hemileia vastatrix* (race III). Int J Plant Sci 160:79–91

Talhinhas P, Batista D, Diniz I, Vieira A, Silva DN, Loureiro A, Tavares S, Pereira AP, Azinheira HG, Guerra-Guimarães L, Várzea V, Silva MC (2017) The Coffee Leaf Rust pathogen Hemileia vastatrix: one and a half centuries around the tropics. Mol Plant Pathol 18:1039–1051. https://doi.org/10.1111/mpp.12512

Vega FE, Rosenquist E, Collins W (2003) Global project needed to tackle coffee crisis. Nature 425:343

Yadav A, Singh B (2013) Effects of gamma irradiation on germination and physiological aspect of maize genotypes. Int J Biotechnol Bioeng Res 4(6):519–520

Evaluation of Coffee (*Coffea arabica* L. var. Catuaí) Tolerance to Leaf Rust (*Hemileia vastatrix*) Using Inoculation of Leaf Discs Under Controlled Conditions

José Andrés Rojas-Chacón, Fabián Echeverría-Beirute, and Andrés Gatica-Arias

Abstract   Coffee leaf rust (CLR), caused by the obligate biotrophic fungus *Hemileia vastatrix*, is considered one of the most devastating diseases of Arabica coffee. The use of leaf rust resistant or tolerant coffee varieties is a critical component for effective management of this disease at the farm level. Conventional breeding of Arabica coffee for leaf rust resistance requires many years of breeding and field-testing. Induced mutagenesis is an effective tool to increase genetic variability and generate new alleles with potential benefit for addressing abiotic and biotic stresses such as leaf rust in Arabica coffee. Efficient screening methods are required to evaluate coffee germplasm or mutant populations for resistance to *H. vastatrix*. Here, we present a screening method that uses inoculation of leaf discs in a controlled environment. The method was evaluated using M_1V_1 and M_2 plants derived from chemically mutagenized Arabica coffee cell suspensions. In this method, the first rust symptoms appear on the leaf discs approximately 29 days after inoculation while the disease severity and incidence can be scored about 47 days after inoculation. Our results show that the methodology is simple, efficient and suitable to rapidly screen large mutant populations in a small area.

J. A. Rojas-Chacón · F. Echeverría-Beirute
Escuela de Agronomía, Instituto Tecnológico de Costa Rica-San Carlos, 21002 San Carlos, Costa Rica

A. Gatica-Arias (✉)
Laboratorio Biotecnología de Plantas, Escuela de Biología, Universidad de Costa Rica, 2060 San Pedro, Costa Rica
e-mail: andres.gatica@ucr.ac.cr

I. L. W. Ingelbrecht et al. (eds.), *Mutation Breeding in Coffee with Special Reference to Leaf Rust*, https://doi.org/10.1007/978-3-662-67273-0_17

1 Introduction

Coffee (*Coffea arabica* L.) is one of the most important beverages in the world and the second most important commercial product exported by developing countries (Alemayehu 2017). Coffee leaf rust (CLR), caused by the biotrophic fungus *Hemileia vastatrix* Berk. and Broome, is one of the main limiting factors of Arabica coffee production worldwide (Waller et al. 2007). The disease can reduce global coffee production by 20 to 25%, with losses of over $ 1 billion annually (McCook 2006; Talhinhas et al. 2017).

The application of fungicides has been the most widely used method to control CLR, even when the development of varieties with genetic resistance is the best alternative (Zambolim 2016). The quest for natural resistance to CLR by traditional breeding has been the focus of research for decades (Melese Ashebre 2016; Mishra and Slater 2012). However, conventional genetic control of CLR has been hampered by the prodigious pathological diversity and rapid genetic evolution of the fungus overcoming the plant resistance genes deployed so far (Cabral et al. 2016; Lima et al. 2020). The induction of genetic variability in Arabica coffee through mutagenesis provides an important complementary tool for crop improvement programs, since a range of variants can be generated (Dhumal and Bolbhat 2012; Vargas-Segura et al. 2019).

Chemical mutagens such as sodium azide (NaN_3) and ethyl methanesulfonate (EMS), have been used in crop breeding for developing mutants (Bolívar-González et al. 2018; Laskar et al. 2018). These chemical mutagens induce a broad variation of morphological and yield-related traits. Other authors reported cases of crops treated with chemical mutagens and improved for fungal resistance or tolerance, for example, powdery mildew-resistant barley (Khan et al. 2010) and wheat resistant to leaf rust *Puccinia* sp. (Mago et al. 2017).

Genetic studies related to *H. vastatrix* and coffee genotypes pursuing resistance, require periodic inoculation of different uredospores of the fungus into the host. A safe and efficient way to evaluate the resistance to different *H. vastatrix* races is carried out by infection *in situ*, using detached leaves or leaf discs under controlled conditions of humidity, light, and temperature that stimulate the development of the pathogen (Cabral et al. 2016; Eskes 1982).

This chapter presents a Coffee leaf rust resistance screening method based on inoculation of leaf discs under controlled conditions. The method was evaluated using $M_1 V_1$ mutant plants obtained from M_0 embryogenic callus treated with NaN_3 and EMS and the resulting M_2 population. The method proved suitable to rapidly screen large coffee populations for CLR resistance.

2 Materials

2.1 Plant Material

1. *Coffea arabica.* var. Catuaí plants M_1V_1 (*see* Note 1).
2. *Coffea arabica* var. Catuaí seeds M_2 (*see* Note 2).
3. *C. arabica* var. Obatá (or any other CLR resistant variety).
4. *C. arabica* var. Caturra (or any other CLR susceptible variety).
5. *C. canephora* (or any other CLR resistant species).

2.2 Other Biological Materials

1. Coffee leaves with rust spores.
2. Healthy coffee leaves (in greenhouse).

2.3 Consumables and Minor Equipment

1. Black polyethylene bags (6 × 8 in).
2. Calibrated scoops for fertilizer application.
3. Commercial potting soil.
4. Cylindrical punch (10 mm-diameter) (*e.g.*, Korff model 06940-5/16).
5. Falcon tube (50 ml).
6. Latex gloves.
7. Microcentrifuge tubes (1.5 ml).
8. Mix of screened compost.
9. Napkins.
10. Permanent markers.
11. Plant labels (with progenies number, plant number).
12. Plastic boxes (12 × 10 × 3.5 cm).
13. Plastic dropper.
14. Plastic grid.
15. Plastic pots (1–2 L capacity).
16. Rice husk.
17. Scalpels blades.
18. Scalpels.
19. Shovels.
20. Slow-release fertilizer.
21. Soil.
22. Wheelbarrow.

2.4   Reagents and Agrochemicals

1. Bayfolan forte (10 ml/L) (Bayer S.A, Amatitlán, Guatemala).
2. Distilled water.
3. Fungicide Vitavax 40 WP (Chemtura Corporation, Middlebury, USA).
4. Osmocote Pro, 19:9:10 + 2MgO + TE.
5. Tween 20 (Research Products International, Illinois, USA).

2.5   Equipment

1. LED light lamps (Heliospectra, model LX 602).
2. Microscope.
3. Neubauer chamber or Haemocytometer slide.
4. Stereo microscope.
5. Incubator with light, humidity, and temperature control (*e.g.*, BIOBASE brand, model BJPX-L400.Shandong, China).
6. Electronic Digital Vernier Caliper (*e.g.*, TOTAL, model TMT322001).

3   Methods

3.1   Germination of M_2 Seeds

1. Collect ripe cherries from M_1V_1 plants in the field and place the fruits in labeled paper bags.
2. Remove the pulp and the mucilage, wash and let dry for 12 days without full sun exposure (*see* Note 3).
3. Select normal-shaped seeds that are free of visible disease and insects.
4. Treat and cure the seeds with the fungicide Vitavax 40 WP (1 g/Kg) (*see* Note 4).
5. Label the plastic pots, keeping the same code obtained from their original fruit.
6. Place 10 cm of a substrate in plastic pots and sow the seeds.
7. Add a 1 cm layer of substrate over the sown seeds (approximately 20 and 30 seeds per pot).
8. Place the pots under controlled conditions; humidity greater than 90%; 25–30 °C, with a photoperiod of 12 h.
9. Record the date of planting.
10. Estimate the duration of seed germination and the percentage of germination per progeny.

Fig. 1 M$_2$ plant development stages. **a** Seedbed preparation and seed germination. **b** Transplanting and planting of seedlings. **c** Establishment of plants under controlled conditions in a greenhouse or growth chamber

3.2 Planting Seedlings M$_2$

1. Once the plants have their cotyledonary leaves (*see* Fig. 1a), transfer the plantlets to polyethylene bags (6 × 8 in) containing substrate (soil, mix of screened compost, and rice husk at 2:1:1).
2. Sow 2 seedlings of the same height and tap root per bag (*see* Fig. 1b and Note 5).
3. Give a permanent and unique identification code to each plant after planting.
4. Prepare a field map to indicate full details of plant identification and location.
5. Maintain the pots under controlled conditions in a greenhouse or growth chamber with 12 h light LED photoperiod at 28 ± 2 °C, and (*see* Fig. 1c).
6. Fertilize with slow-release fertilizer (5 g/plant) (*e.g.*, Osmocote Pro, 19:9:10 + 2MgO + TE) and weekly applications of Bayfolan forte (10 ml/L) to support growth and development.

3.3 Preparation of Coffee Leaf Rust Inoculum

1. Select coffee leaves with abundant *Hemileia vastatrix* spores, without the presence of *Lecanicillium lecanii* (*see* Fig. 2a).
2. Place the leaves in plastic bags and label (sample number, location, and coffee variety from which it was collected).
3. In the laboratory, using a stereo microscope and a sterile scalpel, scrape the sporulated lesions (only intense orange lesions) (*see* Fig. 2b, c).
4. Collect spores in sterile 1.5 ml microcentrifuge tubes.
5. Pour 30 ml of distilled water, 0.1 ml of Tween 20, and the uredospores into a sterile tube and shake the suspension of spores (*see* Fig. 3).

Fig. 2 Sample collection for *Hemileia vastatrix* inoculum, **a** leaves with CLR spores, **b** sporulated lesions, **c** uredospores

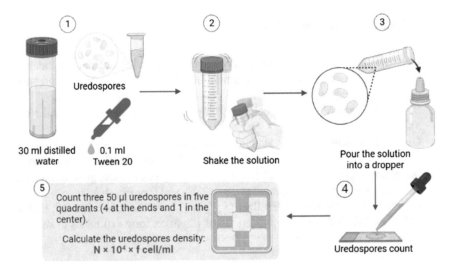

Fig. 3 Preparation of the CLR inoculum and counting cells

6. Determine the uredospore concentration of the spore suspension using a haemocytometer and a microscope at 10× magnification. Examine five quadrants (4 at the ends and 1 in the center) (*see* Fig. 3).

7. Count three 50 μl drops of the spore suspension (count only orange-colored spores).

8. Calculate the uredospore density as following: $N \times 10^4 \times f$ cell/ml, where "N" is the total counted cells, and "f" is the dilution factor (*see* Note 6).

9. Adjust inoculum to a concentration of approximately 2.3×10^5 spores/ml.

3.4 Inoculation of the Coffee Leaf Discs with CLR

1. Collect healthy full-grown M_1V_1 leaves in the greenhouse and keep in a plastic bag on sterilized foam moistened with water (*see* Note 7).
2. In the laboratory, carefully wash the leaves with water and dry them for 1 h at 24 °C (*see* Note 8).
3. Clearly mark the humidity chambers (each plastic box) with the plant accession number.
4. Using a 10 mm diameter cylindrical punch, cut out circular leaf discs (*see* Fig. 4a) without midribs that do not contain stomata (CLR entry point) (Eskes 1982).
5. Ten leaf discs per plant are needed for the detection of resistance. Include susceptible (*e.g.*, *C. arabica* L. var. Caturra) and resistant plants (e.g., *C. canephora*, *C. arabica* var. Obatá) as controls.
6. The discs are moistened and placed upside down in the moist chambers (*see* Fig. 4b and Note 9).
7. Inoculate each leaf disc with one droplet of approximately 50 μl of the spore suspensions (1 mg spores per mL) (see Fig. 4c).
8. Close the boxes with a transparent lid and keep them at 22.5 ± 1.5 °C in the incubator in the darkness for 36 hours and, a relative humidity greater than 90%, to allow rust germination.
9. After this period, the humidity chambers are uncovered to allow the suspension to dry for 3 h, allowing the evaporation of the inoculation droplets.
10. Incubate at approximately 2,000 lux intensity of artificial light, with 12 h light period and temperature 22.5 ± 1.5° C, with a relative humidity greater than 90% for 15 days.
11. After this period, keep at 23 ± 1.5 °C and relative humidity of 85% under natural light for approximately 10 h and 14 h of darkness for a total of 22 days.

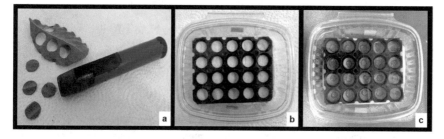

Fig. 4 Preparation of the humidity chambers, **a** cylindrical punch to cut circular (10mm) leaf discs, **b** moist chambers, and **c** inoculum in each leaf disc

Table 1 Disease
severity-rating scale used to
record symptoms caused by
Hemileia vastatrix in coffee
plants (Rozo-Peña and
Cristancho-Ardila 2011).

Scale	Disease symptoms
0	Absence of injury
1	Appearance of chlorosis
2	Increase in disease area
3	Tendency of lesions to coalesce
4	Appearance of the first signs of sporulation
5	Sporulation of less than 25% of the lesion
6	Sporulation between 25–50% of the lesion
7	Sporulation greater than 50% of the lesion

3.5 Evaluation of Plant Resistance Against CLR

1. Record weekly until the appearance of symptoms (chlorotic lesions according to the scale degree 1; *see* Table 1).
2. After observing the first symptoms, the incidence and severity are evaluated every 72 hr for 26 days.
3. Record symptoms and disease severity rate from 29 to 47 days after inoculation on a scale from 0 to 7 using the disease severity rating shown in Table 1 (0 = resistant and 7 = highly susceptible).
4. Calculate disease incidence as follows: [(no. of diseased discs/no. of inoculated discs) * 100] (Rozo-Peña and Cristancho-Ardila 2011).
5. The disease severity is calculated based on the scale shown in Table 1, as follows: [Σ (scale grade * frequency) / (total units observed) * 100] (Rozo-Peña and Cristancho-Ardila 2011; Leguizamón et al. 1998).
6. From infection, calculate the Incubation Period (IP: number of days from inoculation to appearance of chlorosis) and the Latency Period (LP: number of days to appearance of sporulated lesions i.e., uredospores) (Rozo-Peña and Cristancho-Ardila 2011; Leguizamón et al. 1998).
7. Select the plantlets that show CLR resistance (*see* Note 10).

4 Notes

1. M_1V_1 coffee plants (*Coffea arabica* var. Catuaí) were obtained after treatment of embryogenic callus (M_0) with the mutagenic agents NaN_3 (5 mM for 15 min) and EMS (185.2 mM for 120 min) according to Bolívar-González et al. (2018).
2. M_2 coffee seeds of 81 different progenies were obtained after treatment of seeds (M_0) with NaN_3 (50 mM for 8h) according to Vargas-Segura et al. (2019).
3. Coffee beans rapidly lose their viability when dried (their humidity content cannot be less than 10%).

4. Curing the seed before sowing is the first step to obtain healthy plants; it allows to eliminate pathogens and prevents possible diseases originating from the soil.
5. Young plantlets are highly susceptible to diseases. The best size of the plantlet is 8–10 cm with a tap root of 6–8 cm.
6. To count cells using a haemocytometer or a Neubauer chamber add 15–20 μl of the cell suspension between the haemocytometer and a cover glass. Count the number of cells in all five quadrants (4 at the ends and 1 in the center) and divide by five (*see* Fig. 3 step 5). The number of cells per square $\times 10^5 =$ the number of spores/ml of the suspension.
7. It is convenient to collect samples of leaves with rust from different locations and coffee varieties to have greater variability of the pathogen.
8. It is recommended to carefully clean the coffee leaves before cutting the discs.
9. The chambers consist of clear sterile plastic boxes (12 $\times$ 10 $\times$ 3.5 cm) with foam on the bottom. Place a plastic tray with 20 1.2 cm diameter depressions to hold the leaf discs (*see* Fig. 4b).
10. The procedure proves to be reliable and very sensitive; it is not time-consuming, requiring small amounts of inoculum and plant tissue.

Acknowledgments This work was funded by the Consejo Nacional de Rectores (CONARE), project "Evaluation of alternative sources of genetic resistance to coffee rust (*Hemileia vastatrix*)", (project No. 5401-1701-6140). A-Gatica-Arias acknowledged the Cátedra Humboldt 2023 of the University of Costa Rica for supporting the dissemination of biotechnology for the conservation and sustainable use of biodiversity.

References

Alemayehu D (2017) Review on genetic diversity of coffee (*Coffea arabica* L.) in Ethiopia. Int J For Hortic (IJFH) 3(2):18–27. https://doi.org/10.20431/2454-9487.0302003
Bolívar-González A, Valdez-Melara M, Gatica-Arias A (2018) Responses of Arabica coffee *(Coffea arabica* L. var. *Catuaí)* cell suspensions to chemically induced mutagenesis and salinity stress under *in vitro* culture conditions. In Vitro Cell Dev Biol Plant 54:576–589. https://doi.org/10.1007/s11627-018-9918-x
Cabral PGC, Maciel-Zambolim E, Oliveira SAS, Caixeta ET, Zambolim L (2016) Genetic diversity and structure of *Hemileia vastatrix* populations on *Coffea* spp. Plant Pathol 65(2):196–204. https://doi.org/10.1111/ppa.12411
Dhumal K, Bolbhat S (2012) Induction of genetic variability with gamma radiation and its applications in improvement of horsegram. In: Adrovic F (ed) Gamma radiation. IntechOpen, pp 207–228. https://doi.org/10.5772/37885
Eskes AB (1982) The use of leaf disk inoculations in assessing resistance to coffee leaf rust (*Hemileia vastatrix*). Neth J Plant Pathol 88(4):127–141. https://doi.org/10.1007/BF01977270
Khan S, Al-Qurainy F, Anwar F (2010) Sodium azide: a chemical mutagen for enhancement of agronomic traits of crop plants. Environ We Int J Sci Technol 4:1–21
Laskar RA, Chaudhary C, Khan S, Chandra A (2018) Induction of mutagenized tomato populations for investigation on agronomic traits and mutant phenotyping. J Saudi Soc Agric Sci 17(1):51–60. https://doi.org/10.1016/J.JSSAS.2016.01.002

Leguizamón J, Orozco L, Gómez L (1998) Períodos de incubación (pi) y de latencia (pl) de la roya del cafeto en la zona cafetera central de Colombia. Cenicafé 49(56)

Lima JD, Maigret B, Fernandez D, Decloquement J, Pinho D, Albuquerque EVVS, Rodrigues MO, Martins NF (2020) Searching *in silico* novel targets for specific coffee rust disease control. In: Kowada L, de Oliveira D (eds) Advances in bioinformatics and computational biology. BSB 2019. Lecture notes in computer science, vol 11347. Springer, Cham. https://doi.org/10.1007/978-3-030-46417-2_10

Mago R, Till B, Periyannan S, Yu G, Wulff BBH, Lagudah E (2017) Generation of loss-of-function mutants for wheat rust disease resistance gene cloning. In: Periyannan S (ed) Wheat rust diseases: methods in molecular biology, vol 1659. Humana Press, New York. https://doi.org/10.1007/978-1-4939-7249-4_17

McCook S (2006) Global rust belt: *Hemileia vastatrix* and the ecological integration of world coffee production since 1850. J Glob Hist 1(2):177–195. https://doi.org/10.1017/S174002280600012X

Melese Ashebre K (2016) The role of biotechnology on coffee plant propagation: a current topics paper. J Biol Agric Healthc 6(5). www.iiste.org

Mishra MK, Slater A (2012) Recent advances in the genetic transformation of coffee. Biotechnol Res Int 2012:1–17. https://doi.org/10.1155/2012/580857

Rozo-Peña YI, Cristancho-Ardila M (2011) Evaluación de la susceptibilidad de *Hemileia vastatrix* Berk and Br a fungicidas del grupo de los triazoles. Cenicafé 61(4)

Talhinhas P, Batista D, Diniz I, Vieira A, Silva D, Loureiro A, do Céu Silva M (2017) The coffee leaf rust pathogen *Hemileia vastatrix*: one and a half centuries around the tropics. Mol Plant Pathol 18(8):1039–1051. https://doi.org/10.1111/mpp.12512

Vargas-Segura C, López-Gamboa E, Araya-Valverde E, Valdez-Melara, M., Gatica-Arias A (2019) Sensitivity of seeds to chemical mutagens, detection of DNA polymorphisms and agro-metrical traits in M1 generation of coffee (*Coffea arabica* L.). J Crop Sci Biotechnol 22(5):451–464. https://doi.org/10.1007/S12892-019-0175-0

Waller JM, Bigger M, Hillocks RJ (2007) Coffee pests, diseases and their management, pp 1–434. https://doi.org/10.1079/9781845931292.0000

Zambolim L (2016) Current status and management of coffee leaf rust in Brazil. Trop Plant Pathol 41(1):1–8. https://doi.org/10.1007/S40858-016-0065-9

A PCR-Based Assay for Early Diagnosis of the Coffee Leaf Rust Pathogen *Hemileia vastatrix*

Weihuai Wu, Le Li, Kexian Yi, Chunping He, Yanqiong Liang, Xing Huang, Ying Lu, Shibei Tan, Jinlong Zheng, and Rui Li

Abstract Early detection and identification of plant pathogens is one of the most important strategies for sustainable plant disease management. Fast, sensitive, and accurate methods that are cost-effective are crucial for plant disease control decision-making processes. Coffee leaf rust (CLR) caused by *Hemileia vastatrix* is a devastating worldwide fungal disease which causes serious yield losses of coffee, especially relevant for *Coffea arabica*. A rapid PCR assay for detecting and characterizing *H. vastatrix* with high specificity, high sensitivity and simple operation has been developed based on specific amplification of the Internal Transcribed Spacer (ITS) region of ribosomal genes. The specificity of the primers was determined using isolates DNA of *H. vastatrix, Coleosporium plumeriae,* and other fungal species that infect coffee plants and are common in coffee leaves, such as *Lecanicillium* sp (the *H. vastatrix* hyperparasite fungi)*, Cercospora coffeicola, Colletotrichum gloeosporioides,* amongst others. Results showed specific amplification of a 396-bp band from *H. vastatrix* DNA with a detection limit of 10 pg/µl of pure genomic DNA of the pathogen. The PCR assay described in the current chapter allows to detect *H. vastatrix* rapidly and reliably in naturally infected coffee tissues, vital for the early detection and diagnostics of *H. vastatrix* and CLR epidemiology.

1 Introduction

Accurate identification and diagnosis of plant diseases are vital for prevention of the spread of invasive pathogens (Balodi et al. 2017). So far, advances in the development of molecular methods have provided diagnostic laboratories with powerful tools for the detection and identification of phytopathogens, among which polymerase chain reaction (PCR) and other DNA-based techniques proved to be rapid and highly suitable approaches to improve the accuracy and efficiency of plant pathogen detection and characterization (Lévesque et al. 1998; Haudenshield et al. 2017). Detection

W. Wu · L. Li · K. Yi (✉) · C. He · Y. Liang · X. Huang · Y. Lu · S. Tan · J. Zheng · R. Li
Environment and Plant Protection Institute, Chinese Academy of Tropical Agricultural Sciences, Haikou 571101, Hainan, China
e-mail: yikexian@126.com

© The Author(s) 2023
I. L. W. Ingelbrecht et al. (eds.), *Mutation Breeding in Coffee with Special Reference to Leaf Rust*, https://doi.org/10.1007/978-3-662-67273-0_18

protocols used for the diagnosis or quarantine measures should be reproducible and cost effective, time saving and simple in procedure (Elnifro et al. 2000; Hayden et al. 2008; Tomkowiak et al. 2019). In addition, sensitivity to pathogen concentration, and specificity to genetic variability within a target pathogen population are also high priorities for molecular detection (Balodi et al. 2017).

The Internal Transcribed Spacer (ITS) of the ribosomal DNA show high inter-species variability and intra-species stability and conservation, and hence is considered a reliable DNA marker to identify and classify the pathogenic fungi (Glynn et al. 2010). PCR assays based on the ITS region have been widely used for the detection of fungal pathogens in different crops such as sunflower, tobacco, soybean, cedar trees, miscanthus and others (Guglielmo et al. 2007; Chen et al. 2008; Torres-Calzada et al. 2011; Capote et al. 2012), relating to the pathogens of *Phytophthora* (Grünwald et al. 2012; Patel et al. 2016), *Puccinia* (Guo et al. 2016), *Verticillium* spp. (Nazar et al. 1991), *Pleurotus* spp. (Ma and Luo 2002), *Pyricularia and anthracnose* (Sugawara et al. 2009), *Saccharomyces saccharum* (Anggraini et al. 2019), *Podosphaera xanthii* (Tsay et al. 2011) and *Golovinomyces cichoracearum* (Troisi et al. 2010). This technique was applied to differentiate two pathotypes of *Verticillium alboatrum* infecting hop, to distinguish 11 taxons of wood decay fungi infecting hardwood trees, and to differentiate multiple *Phytophthora* species from plant material and environmental samples (Shamim et al. 2017; Belete and Boyraz 2019).

Coffee leaf rust (CLR), a major disease of Arabica coffee (*Coffea arabica* L.), is caused by the obligate biotrophic fungus *Hemileia vastatrix Berkeley and Broome* (Talhinhas et al. 2017). The infection of coffee leaves by *H. vastatrix* starts with urediniospore germination, appressorium formation over stomata, penetration, and inter- and intracellular colonization without any visible symptoms in the early stages of the infection in the field conditions < 10 days (Talhinhas et al. 2017; Silva et al. 2018). In field conditions, the visible rust spores can be observed about 20 days after the first infection of *H. vastatrix* (Schieber 1972). So far, the traditional method for detecting and characterizing CLR was time-consuming and laborious, and relied on conventional morphological examination requiring professional taxonomic knowledge and extensive experience (McCartney et al. 2003; Silva et al. 2012). Hence, rapid and high-throughput identification and detection methods for *H. vastatrix* are required to recognize the infection as early as possible before the appearance and spread of CLR spores in the leaf surface. Early detection methods can facilitate implementing proper management approaches to prevent the development and spread of the coffee leaf rust pathogen (Sankaran et al. 2010).

The present study was undertaken with the objective of early detection of *H. vastatrix* based on the PCR amplification of a specific ITS region in the rDNA of *H. vastatrix*. A simple, accurate and rapid PCR-based assay for CLR is presented as a reliable technique to monitor *H. vastatrix* in the early stages of the infection, as well as to provide scientific basis for the prevention and control of CLR.

2 Materials

1. ddH$_2$O.
2. 1 X TE buffer (pH 8.0).
3. CTAB.
4. KAc.
5. Chloroform.
6. Isoamyl alcohol.
7. Isopropanol.
8. 75% ethanol.
9. Anhydrous ethanol.
10. Phenol.
11. Na$_2$Ac.
12. *rTaq* (Dalian TaKaRa Co., Ltd., 5 U/μl).
13. 10X PCR Buffer (Mg^{2+} plus).
14. dNTPs (2.5 mM).
15. Biowest regular agarose G-10 (CB005-100G).
16. Tris/borate electrophoresis buffer.
17. Microwave.
18. GoldView II Nuclear Staining Dyes (5,000×) (Solarbio® LIFE SCIENCES).
19. Electrophoresis tank.
20. DL 2000 Marker (Dalian TaKaRa Co., Ltd.).
21. RNAse A solution (Solarbio® LIFE SCIENCES, 10 mg/ml).
22. Water bath.
23. Specific primers (*see* Fig. 1).
24. Genomic DNA of the pathogen (*see* Fig. 2).
25. Ice.
26. Ice machine.
27. Autoclave.
28. Mortar.
29. Measuring cylinder (100 ml).
30. Scissors.
31. Liquid nitrogen.
32. Micropipette (1,000, 200, 10, 2.5 μl).
33. Centrifuge tube (1.5, 2 ml).
34. NanoDrop 2000c Spectrophotometer (Thermo Scientific, USA).
35. PTC-100™ Programmable Thermal Controller (MJ Research Inc, USA).
36. BIO-RAD GelDoc 2000 GelDoc 2000™.
37. Power/PAC300.
38. PCR tubes (0.2 ml).
39. Tips (1,000, 200, 10 μl).
40. Absolute alcohol.
41. Refrigerated Centrifuge Sigma 3k15.
42. SCILOGEX_D2012_Centrifuge.
43. Refrigerator.

```
  1 TCCGTAGGTG AACCTGCGGA AGGATCATTA AAAAATTAAG AGTGCACTTA
 51 ATTGTGGCTT GAAATTTTAC TTATTTACAC CCAACGTCTT CGGGACACTG
101 CGGCAATTTA TTGCTTAGCG AATTTAACCC CTGCGGTTAG GCATATATAA
151 TTCTCTCTGA GGGTTGTATG TGTTCTAATC TTTGTTTTTT TTATTTTCAA
201 CCACAAATTT ATACATATGT ATATATGTAT TATTTACTAT CAAGTAAATA
251 AATATAAAAC TTTTAACAAT GGATCTCTTG GCTCTCACAT CGATGAAGAA
301 CACAGTGAAA TGTGATAAGT AATGTGAATT GCAGAATTCA GTGAATCATC
351 GAATTTTTGA ACGCATATTG CGCCTTTTGG CTATTCCAAA AGGTACACCT
401 GTTTGAGAGT ATGAAAGGTC AGGGTGTTGA GAGAGTTATT AAAAAAAAGA
451 AAAAGGCAAA GTAACACTTT AAGTGTTATT TTGTCTTTGA TTTTTTTTTT
501 TCTCTTAGCA TCTTGGATAT TGGGTGCTTG CCATTATTAA GTTTGATGGC
551 TCACCTTAAA TTTATAAGTT GTTTTTTATT AAGGATGAAA AGTCTTTTGA
601 TGGCTTGATG TTATTGATAT ATGAAATGTC ATTCATCAAG AAATCAGGGG
651 GGTGACTAAC CTTGATGAGA AATGTTGACT TTATAAACAC ATGTTAAAAG
701 ATAAATAAAA AGTAAAAGAT AAAAAAAATT AAAAATTAAA AATTAAAGGG
751 AAAGAAGTAA AAAGAATGAG AGGTATGACA TATTTTGAAA ATATGTCCCT
801 TGTTCTCTTT TTATTTTGAT TATCTTTATT TTTTTATTTT TAATTTTAAT
851 TTTTATGTTT TAAGTTTTAT TTTATTTTTT ACTTCTTTAT AATGAATCTC
901 AAATCAGGTG GGACTACCCG CTGAACTTAA GCATATCAAT AAGCGGAGGA
```

Fig. 1 Primers Hv-ITS-F/R designed for *H. vastatrix* PCR assay based on rDNA-ITS sequences

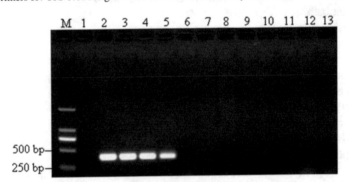

Fig. 2 Example of specificity test of Hv-ITS-F/R primer sets. The DNA of 4 strains of *H. vastatrix* (lanes 2–5), 8 other fungi (lanes 6–13) (*see* Note 5) and sterilized ddH$_2$O as the negative control (lane 1) were amplified by PCR using Hv-ITS-F/R primers. Primers for Hv-ITS-F/R amplify a 396-bp specific band from the DNA of *H. vastatrix*, while no bands were observed from the DNA of other fungi. **M:** DL 2000 DNA marker; **1:** ddH$_2$O control; **2–5:** *H. vastatrix*; **6:** *Colletotrichum gloeosporioides*; **7:** *Lecanicillium* sp.; **8:** *Cercospora coffeicola*; **9:** *Coleosporium plumeriae*; **10:** *Colletotrichum falcatum*; **11:** *Ustilago scitaminea*; **12:** *Leptosphaeria sacchari*; **13:** *Aspergillus niger*

3 Methods

3.1 Designing the Specific Primers for Hemileia vastatrix

1. The primers Hv-ITS-F/R were designed to specifically amplify the ITS2 region of *H. vastatrix*. The sequence of the forward primer Hv-ITS-F is 5'-GGTACACCTGTTTGAGAGTATG-3', and the sequence of the reverse primer is Hv-ITS-R is 5'-CAAAATATGTCATACCTCTCATTCT-3 (*see* Fig. 1).
2. Primer sequences of Hv-ITS-F and Hv-ITS-R were used as inputs for a BLAST search against the NCBI database to confirm the specificity. The primers were synthesized by Invitrogen Biotechnology (Shanghai) Co., Ltd.
3. Upon delivery, dilute lyophilized primers to the concentration of 10 μM by adding 0.1 X TE buffer. Store at − 20 °C for later use.

3.2 Total DNA Extraction from Suspected Diseased Leaves or Typical Diseased Samples

The CTAB method (Siegel et al. 2017) was used to extract DNA from diseased leaves.

1. Preheat the CTAB extraction buffer to 65 °C in a water bath.
2. Grind approximately 1 g of diseased leaf tissue into a fine powder in a mortar using liquid nitrogen (*see* Note 1).
3. Add 15 ml of pre-heated CTAB buffer into each tube. Mix well and incubate at 65 °C for 30 min. Turn the tubes upside down every 10 min to resuspend the samples in the buffer (*see* Note 2).
4. Add 3 ml of 5 M KAc to the tube containing the lysate and let it stand on ice for 20 min.
5. Add the same volume of a chloroform:iso-amyl alcohol (24:1) mixture to the tube, mix well and centrifuge at 12,000 rpm at 4 °C for 15 min.
6. Repeat step 4.
7. After centrifugation, transfer supernatant into a new tube.
8. Add 12 ml of a pre-cooled isopropanol, mix by inverting and put at −20 °C to fully precipitate the DNA.
9. Centrifuge the tube at 10,000 rpm for 15 min to pellet the DNA.
10. Rinse the pellet twice with 75% ethanol, and once with anhydrous ethanol. Air-dry the DNA pellet and dissolve in 10 ml TE buffer.
11. Treat the DNA samples with 1 μl RNase (10 mg/ml) at room temperature for 1–2 h.
12. Add the same volume of phenol: chloroform: isoamyl alcohol (25:24:1), mix well and then centrifuge at 12,000 rpm at 4 °C for 15 min.
13. Transfer the supernatant to a new tube, mix with 1 ml ice-cold 3 M Na_2Ac, and 20 ml of anhydrous ethanol, and place at − 20 °C overnight.
14. Centrifuge at 12,000 rpm for 30 min at 4 °C.

15. Discard the supernatant, rinse the DNA pellet with 75% ethanol and dissolve in 1 ml TE after drying.
16. Determine the DNA concentration by e.g., a NanoDrop 2000c.
17. Store the DNA at − 20 °C until further use (*see* Note 3).

3.3 Preparation of the PCR Reaction Mixture and PCR Amplification

1. Prepare a 20 μl PCR reaction mix as follows (*see* Note 4):

10X PCR Buffer	2 μl
dNTPs (2.5 mM)	1.6 μl
Forward Primer (10 μM)	1 μl
Reverse Primer (10 μM)	1 μl
rTaq (5 U/μl)	0.1 μl
DNA template	1 μl
ddH$_2$O	14.2 μl

2. Mix all components, spin briefly and immediately place in a thermocycler (here a gradient Mastercycler was used).
3. Set the thermocycler conditions as following: initial denaturation at 94 °C for 3 min, denaturation at 94 °C for 30 s, annealing at 62 °C for 30 s, extension at 72 °C for 1 min, 35 cycles; final extension time at 72 °C for 5 min.
4. Upon termination store samples at 15 °C.

3.4 Gel Electrophoresis

1. Prepare a 1% agarose gel by mixing 1 g of agarose and 100 ml of TBE buffer (pH 8.0).
2. Melt thoroughly in a microwave.
3. Allow the mixture to cool down to 40 °C, add 1 μl GoldView DNA dye solution (1 μl/100 ml gel) and mix. Pour the gel and allow to solidify.
4. Load 10 μl of PCR products and run at 120 V for 20 min.
5. View the gel under the UV light. The *H. vastatrix* positive samples are defined as the ones that show a specific single band of 396-bp (*see* Figs. 2 and 3).

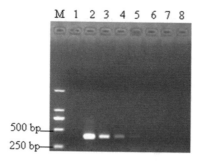

Fig. 3 Example sensitivity test of primer sets Hv-ITS-F/R. Prepare a series of DNA concentrations to determine the sensitivity of the detection system. The initial genomic DNA concentration of *H. vastatrix* was adjusted to 10 ng/μL, with serial tenfold dilutions to reach 10^{-5} ng/μl. The results showed that samples with DNA concentration of 10 pg/μL or higher yielded a clearly visible 396-bp band while samples with a lower concentration were negative. **M:** DL 2 000 DNA marker; **1:** ddH$_2$O control; **2:** 10 ng/μl; **3:** 1 ng/μl; **4:** 10^{-1} ng/μl; **5:** 10^{-2} ng/μl; **6:** 10^{-3} ng/μl; **7:** 10^{-4} ng/μl; **8:** 10^{-5} ng/μl

4 Notes

1. The leaf samples should be fully ground into a fine powder. To prevent sample cross-contamination, change gloves after finishing each sample.
2. All the tubes, tips and utensils should be sterilized prior to use.
3. To prevent cross contamination, the pipette tips must be used once after contact with samples.
4. The PCR reaction mix is prepared on ice in a clean environment.
5. The urediniospores of *Coleosporium plumeriae* and the other fungal isolates were extracted using a Fungal DNA kit (E.Z.N.A.TM Fungal DNA Kit, Omega, Bio-tek, USA) according to the manufacturer's protocol.

Acknowledgements Funding for this work was provided by the National Key R&D Program of China (2018YFD0201100), the IAEA Collaborative Research Project D22005 (No. 20380), the International Exchange and Cooperation Project funded by the Agricultural Ministry 'Construction of Tropical Agriculture Foreign Cooperation Test Station and Training of Foreign Managers in Agricultural Going-Out Enterprises' (SYZ2019-08) and the Central Public-interest Scientific Institution Basal Research Fund for Chinese Academy of Tropical Agricultural Sciences (No. 1630042017021).

References

Anggraini I, Ferniah RS, Kusdiyantini DE (2019) Isolasi khamir fermentatif dari batang tanaman tebu (*Saccharum officinarum*. L) dan hasil identifikasinya berdasarkan sekuens internal transcribed spacer. Berkala bioteknologi 2(2):12–22

Balodi R, Bisht S, Ghatak A, Rao KH (2017) Plant disease diagnosis: technological advancements and challenges. Indian Phytopathol 70(3):275–281

Belete T, Boyraz N (2019) Biotechnological tools for detection, identification and management of plant diseases. Afr J Biotechnol 18(29):797–807

Capote N, Pastrana AM, Aguado A, Sánchez-Torres P (2012) Molecular tools for detection of plant pathogenic fungi and fungicide resistance. Plant Pathol 151–202

Chen RS, Huang CC, Li JC, Tsay JG (2008) First report of *Simplicillium lanosoniveum* causing brown spot on *Salvinia auriculata* and *S. molesta* in Taiwan. Plant Dis 92(11):1589–1589

Elnifro EM, Ashshi AM, Cooper RJ, Klapper PE (2000) Multiplex PCR: optimization and application in diagnostic Virology. Clin Microbiol Rev 13(4):559–570

Glynn NC, Dixon LJ, Castlebury LA, Szaboc LJ, Comstock JC (2010) PCR assays for the sugarcane rust pathogens *Puccinia kuehnii* and *P. Melanocephala* and detection of a SNP associated with geographical distribution in *P. kuehnii*. Plant Pathol 59(4):703–711

Grünwald NJ, Werres S, Goss EM, Taylor CR, Fieland VJ (2012) *Phytophthora obscura* sp. nov., a new species of the novel *Phytophthora* subclade 8d. Plant Pathol 61:610–622

Guglielmo F, Bergemann SE, Gonthier P, Nicolotti G (2007) A multiplex PCR-based method for the detection and early identification of wood rotting fungi in standing trees. J Appl Microbiol 103(5):1490–1507

Guo DD, Zhao SH, Zhang HZ, Zhao J, Jing L (2016) Fungicides screening for management of sunflower rust caused by *Puccinia henlianthi*. Chin J Oil Crop Sci 38(3):388–394

Haudenshield JS, Song JY, Hartman GL (2017) A novel, multiplexed, probe-based quantitative PCR assay for the soybean root- and stem-rot pathogen, *Phytophthora sojae*, utilizes its transposable element. PLoS ONE 12(4):e0176567

Hayden MJ, Nguyen T, Waterman A, Chalmers KJ (2008) Multiplex-ready PCR: a new method for multiplexed SSR and SNP genotyping. BMC Genomics 9(1):80–80

Lévesque CA, Harlton CE, de Cock AWAM (1998) Identification of some oomycetes by reverse dot blot hybridization. Phytopathology 88:213–222

Ma FY, Luo XC (2002) Phylogeny of Pleurotus inferred from PCR-RFLP of 28S ribosomal DNA. J Huazhong (Cent China) Agric Univ 21(3):201–205

McCartney HA, Foster SJ, Fraaije BA, Ward E (2003) Molecular diagnostics for fungal plant pathogens. Pest Manag Sci 59(2):129–142

Nazar RN, Hu X, Schmidt J, Culham D (1991) Potential use of PCR-amplified ribosomal intergenic sequences in the detection and differentiation of *verticillium wilt* pathogens. Physiol Mol Plant Pathol 39(1):1–11

Patel JS, Vitoreli A, Palmateer AJ, El-Sayed A, Noman DJ, Goss EM, Brennan MS, Ali GS (2016) Characterization of *Phytophthora* spp. Isolated from ornamental plants in Florida. Plant Dis 100:500–509

Sankaran S, Mishra A, Ehsani R, Davis C (2010) A review of advanced techniques for detecting plant diseases. Comput Electron Agric 72(1):1–13

Schieber E (1972) Leaf blight incited by *Ascochyta coffeae* on coffee in Guatemala. Plant Dis Reporter 56(9):753–754

Shamim M, Kumar P, Kumar RR, Kumar M, Kumar RR, Singh KN (2017) Assessing fungal biodiversity using molecular markers. Mol Markers Mycol 15:305–333. https://doi.org/10.1007/978-3-319-34106-4

Siegel CS, Stevenson FO, Zimmer EA (2017) Evaluation and comparison of FTA card and CTAB DNA extraction methods for non-agricultural taxa. Appl Plant Sci 5(2):1600109

Silva AVC, Santos ARF, Lédo AS, Feitosa RB, Almeida CS, Silva GM, Rangel MSA (2012) Moringa genetic diversity from germplasm bank using RAPD Markers. Trop Subtrop Agroecosyst 15:31–39

Silva DN, Varzea V, Paulo OS, Batista D (2018) Population genomic footprints of host adaptation, introgression and recombination in coffee leaf rust. Mol Plant Pathol 19(7):1742–1753

Sugawara K, Matsudate A, Ito Y, Nanai T (2009) Anthracnose of Christmas rose caused by *Colletotrichum* sp. J Gen Plant Pathol 75(2):163–166

Talhinhas P, Batieta D, Diniz I, Vieira A, Sliva DN, Loureiro A, Tavares S, Pereira AP, Azinheira HG, Guerra-Guimarães L, Várzea V, Silva MDC (2017) The coffee leaf rust pathogen *Hemileia vastatrix*: one and a half centuries around tropics. Mol Plant Pathol 18(8):1039–1051

Tomkowiak A, Bocianowski J, Radzikowska D, Kowalczewski PL (2019) Selection of parental material to maximize heterosis using SNP and SilicoDarT markers in maize. Plants 8:349

Torres-Calzada C, Tapia-Tussell R, Quijano-Ramayo A, Martin-Mex R, Rojas-Herrera R, Higuera-Ciapara I, Perez-Brito D (2011) A Species-specific polymerase chain reaction assay for rapid and sensitive detection of *Colletotrichum capsici*. Mol Biotechnol 49(1):48–55

Troisi M, Bertetti D, Garibaldi A, Gullino ML (2010) First report of powdery mildew caused by *Golovinomyces cichoracearum* on Gerbera (*Gerbera jamesonii*) in Italy. Plant Dis 94(1):130–130

Tsay JG, Chen RS, Wang HL, Wang WL, Weng BC (2011) First report of powdery mildew caused by *Erysiphe diffusa*, *Oidium neolycopersici*, and *Podosphaera xanthii* on papaya in Taiwan. Plant Dis 95:1188

Molecular Characterisation of Induced Mutations in Coffee

Targeted Sequencing in Coffee with the Daicel Arbor Biosciences Exome Capture Kit

Norman Warthmann

Abstract Exome Capture is a molecular biology technique that, in combination with Next Generation DNA sequencing technologies (NGS), allows for selectively sequencing the predicted genes of an organism. Such capture sequencing provides a compromise between genome coverage and sequencing cost. The capture reaction is an additional step in an otherwise standard sequencing protocol and exome capture effectively enriches the sequencing library for DNA molecules that overlap with predicted genes (the exome). This enables genome-wide assessments while focusing on the gene space. Capture sequencing is particularly attractive in species with large genomes, where whole genome sequencing in larger numbers of samples would be cost-prohibitive at present prices. Plant Breeding and Genetics Laboratory (PBGL) developed an Exome Capture Kit for *Coffea arabica* in collaboration with Daicel Arbor Biosciences (Ann Arbor, MI, USA). Use of the kit achieves eightfold enrichment, and hence approx. eightfold reduction in sequencing cost for a whole genome assessment of *Coffee arabica* plants. The kit is available as a regular product from Daicel Arbor Biosciences and this protocol describes the kit and gives detailed instructions on how to perform the capture reaction.

1 Introduction

With today's cheap Next Generation DNA sequencing (NGS) virtually all DNA variation in genomes can be readily identified, including new mutations. Such knowledge makes the breeding process more efficient. Being able to comprehensively catalogue genome-wide DNA variation at the population-scale opens the door for genomic prediction as well as for tracking genetic variation through the breeding process.

Despite the low prices, sequencing cost is currently still of concern when applying whole genome approaches on a large number of samples, particularly when high sequencing depth is required. An example is mutation detection in mutant M_1 populations, where induced mutations are in hemizygous, and often in chimeric state.

N. Warthmann (✉)
Plant Breeding and Genetics Laboratory (PBGL), Joint FAO/IAEA Centre for Nuclear Applications in Food and Agriculture, International Atomic Energy Agency, Seibersdorf, Austria
e-mail: n.warthmann@iaea.org; norman@warthmann.com

© The Author(s) 2023
I. L. W. Ingelbrecht et al. (eds.), *Mutation Breeding in Coffee with Special Reference to Leaf Rust*, https://doi.org/10.1007/978-3-662-67273-0_19

For many genomics-supported breeding applications it is sufficient to sequence only a representative subset of the genome. This can save cost. There exist several approaches to achieve such complexity reduction. One of them is 'target sequence capture', a molecular biology procedure that enriches for predefined regions of the genome (targets) prior to sequencing. Probes complementary to target DNA sequences are designed at large scale and used to effectively capture, i.e., pull-out, the desired molecules from sequencing libraries, thereby enriching for target molecules. The so enriched libraries are then subjected to Next Generation Sequencing (NGS) and the resulting sequencing data mostly consist of sequences representing the target regions. In the case of exome capture, those target regions are the predicted genes, the exome.

Applying target capture requires an up-front investment: It needs prior knowledge of the DNA sequence of target regions and the production of probes. In case of exome capture, which intends to enrich for (all) genes, the selection of target regions is based on a suitable reference genome and a genome annotation, which has to be available or generated. The number of exons in a eukaryotic genome is large, and the necessary number of probes can be in the hundreds of thousands. In human medical applications, including diagnostics, exome capture sequencing is standard procedure for more than a decade (Choi et al. 2009). Exome capture sequencing has gained traction in plant breeding for important food crops with very large genomes such as wheat (Dong et al. 2020; Gardiner et al. 2019) and barley (Mascher et al. 2013; Russell et al. 2016), with several commercial suppliers offering competing exome capture panels and kits.

To enable cost-effective whole genome approaches in coffee breeding, we developed and provide an Exome Capture Kit for *Coffea arabica*. This is in collaboration with Daicel Arbor Biosciences (Ann Arbor, MI, USA), hereafter "Arbor". *Coffea arabica* is an allotetraploid and the genome is the result of a merger of *C. eugenoides* and *C. canephora* (Scalabrin et al. 2020). The design is based on a public *C. arabica* genome assembly and annotation (Cara_1.0, NCBI accession number GCF_003713225.1, derived from cultivar 'Caturra red', isolate CCC135-36), which we augmented with a public *C. arabica* chloroplast sequence (NCBI accession number: NC_008535.1).

This chapter details the design of Daicel Arbor Biosciences' Exome Capture Kit, provides a step-by-step protocol for its use, and describes a validation experiment of exome capture sequencing of 41 indexed samples in a single capture experiment.

2   Materials

Main inputs to the exome capture procedure are a whole-genome DNA sequencing library outfitted with respective adaptors and Arbor's Exome Capture Kit. Additional requirements for equipment, consumables, and reagents are listed below. Most of these should already be at hand as they will have been used when preparing the NGS library. For post-capture library amplification, Arbor recommends KAPA HiFi DNA polymerase.

2.1   The Exome Capture Kit

Main component of the Exome Capture Kit are thousands of probes that are complementary to the thousands of target regions. They function as baits to fish their complementary targets from an NGS library in solution. In case of this Arbor kit the baits are biotinylated RNA molecules, and the target is the exome extracted from a publicly available coffee reference genome and annotation (NCBI).

2.1.1   Exome Capture Kit Design Details

Initial target intervals for probe design included a *C. arabica* Chloroplast (NC_008535.1) in its entirety and all annotations containing the string "exon" found in the *C. arabica* genome assembly GCF_003713225.1 (https://www.ncbi.nlm.nih.gov/assembly/GCF_003713225.1).

The exonic intervals of the genome assembly were merged into non-overlapping regions representing 94.5 Mbp total exome space. The regions were padded with 50 nt on either side (i.e., 5'- and 3'-ends) and new overlaps re-merged, which resulted in 121.0 Mbp sequence space for initial probe design. Regions were divided into non-overlapping 100nt intervals and the best 80nt candidate probe hybridization site was chosen using Arbor's proprietary algorithm. Candidate probe sequences with strong predicted affinity to regions outside of the target regions were removed. The final predicted retrievable space of the filtered probe set was estimated by aligning the remaining probes back to the genome (megablastn, BLAST + version 2.6.0 +, default parameters) and padding each probe hit with 200nt on either side. Merging these regions results in 151.8 Mbp total genome space (represented in file DAB_CoffeeExomeV1_capspace.bed.gz), of which 87.2 Mbp overlap with the original exon region intervals (overlap represented in file: DAB_CoffeeExomeV1_exonspace.bed.gz). These files can be downloaded from the kit's dedicated section on the Arbor website https://arborbiosci.com/genomics/targeted-sequencing/mybaits/mybaits-custom-predesigned-community-panels/plants-and-fungi/.

The probes were synthesized in four distinct sets: Subgenome "C" (=*canephora*), Subgenome "E" (=*eugenoides*), Subgenome "O" ("other" = unassigned contigs), and "Chlor" (=chloroplast). The probe sets can be used separately or combined as the user sees fit depending on the application. To generate a pool of all nuclear genome probes, the "C" sub-genome module should comprise 47.4% of the pool by volume, the "E" sub-genome module 49.1%, and the "other" sub-genome module 3.5%. If the user aims to enrich the chloroplast as well, that module can comprise a final 0.1% of the final pool, though optimization for the tissue type might be required.

2.1.2   Availability of the Exome Capture Kit

The Coffee Exome V1 kit is available from Daicel Arbor Biosciences as part of their Community Panels series (https://arborbiosci.com/genomics/targeted-sequencing/ mybaits/mybaits-custom-predesigned-community-panels/plants-and-fungi/).   The design ID is D10496CFEXM. Order inquiries should be directed to sales@arbor. daicel.com.

2.2   NGS Library Requirements

In principle, libraries prepared for Illumina short-read as well 3rd-generation long-read sequencing technologies can be used. This protocol describes the exome capture reaction for Illumina sequencing libraries with dual-index-barcoded Nextera-type adaptors. For different adaptors, such as 'TruSeq', the protocol is the same, but different blockers and universal amplification primers will be required. Please consult the respective manual from Arbor.

Input requirement	
100–500 ng dsDNA in 7 µl	Nextera/Illumina short-read sequencing library

2.3   Equipment

1. Heat Block for 1.5 ml microfuge tubes.
2. Thermal cycler (PCR machine) with heated lid suitable for desired vessel size.
3. Qubit Instrument or equivalent for fluorescence-based dsDNA quantification.
4. Optional: Fragment analyser to establish DNA fragment size distribution.

2.4   Consumables and Reagents (Non-standard)

1. Coffee Exome V1 Capture Kit, Daicel Arbor Biosciences, Community Panel design ID D10496CFEXM
2. Magnet for 1.5 ml Eppendorf tubes (e.g., DynaMag™-2, Invitrogen™, ThermoFisher #12321D)
3. Magnet for PCR-strips/tubes (e.g., DynaMag™-96 Side Magnet (Invitrogen™ ThermoFisher #12331D)
4. KAPA HiFi HotStart ReadyMix (Roche #KK2601)
5. Resuspension Buffer (self-prepared): 10 mM TrisCl, 0.05% Tween-20, pH 8.0–8.5
6. Protein LoBind® Tube, 1.5 ml (Eppendorf #0030108116)
7. Agencourt Ampure XP beads (Beckman Coulter, Agencourt #A63881)
8. Qubit™ dsDNA HS Reagent (Invitrogen™, ThermoFisher #Q32851)
9. Optional (when using manufacturer's deprecated protocol version 4): xGen Universal Blockers-NXT Mix, Integrated DNA Technologies Inc. (IDT): Catalogue No. 1079584.

2.5 PCR Primers

Universal amplification primers post-capture amplification of the NGS library must match the respective NGS library type. This protocol uses Nextera-type/Illumina libraries.

Name	Alias	Sequence[a]
Seib_275	Nextera libraries-universal-FWD	A*ATGATACGGCGACCACCGAGA
Seib_276	Nextera libraries-universal-REV	C*AAGCAGAAGACGGCATACGAGA

[a]the star (*) denotes a PTO-binding

3 Methods

Figure 1 provides an overview of the subsequent steps, their approximate duration, and required consumables and equipment.

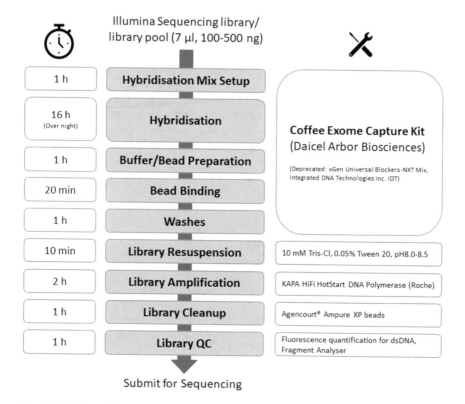

Fig. 1 Workflow of the exome capture procedure with time estimates and required consumables

Sequencing libraries are combined with various blockers (=Hybridisation Mix Setup) and then incubated with the baits/probes at 65 °C for the actual capture (=Hybridisation). The hybridisation is usually performed overnight. The next day, buffer and beads for the binding and washes are prepared and the bait/target hybrid molecules are captured with beads (=Bead Binding). A total of 4 washes at 65 °C remove unbound and unspecific DNA molecules (=Washes). The target molecule library is then recovered from the beads and amplified to desired amount (Library Resuspension, Library Amplification) and bead cleaned for sequencing (=Library Cleanup, Library QC).

All reagents required for the actual capture and wash reactions are included in the Daicel Arbor Biosciences Kit. Reagents for resuspension, amplification, final bead clean-up, and QC will have to be provided by the user.

3.1 Hybridisation Mix Setup

The following describes the preparation of the baits, the setup of the hybridization mix. All consumables for the hybridisation are contained in the Exome Capture Kit.

3.1.1 Combining Baits

Pool the different sub-genome probe sets in representative ratios (*see* Note 1). Below table gives the necessary amounts for one capture reaction, scale if required.

Bait	Amount	Ratio in final pool (%)
C. canephora ("C")	2.61 µl	47.4
C. eugenoides ("E")	2.70 µl	49.1
Other ("O")	0.2 µl	3.5
Chloroplast ("Chlo")	1 µl of a 1:1000 dilution	≤ 0.1
	6.5 µl total	

3.1.2 Set Up the Hybridisation Mix

Component	Amount
Hyb N	9.25 µl
Hyb D	3.5 µl
Hyb S	0.5 µl
Hyb R	1.25 µl
Baits	5.5 µl
	20 µl total

3.1.3 Set Up the Blockers Mix

The Blockers Mix has changed between Arbor myBaits kit manuals versions v4 and
v5. Version v5 should be used. Version v4 is given for backwards compatibility only.

1. Set up the Blockers Mix.
   (Amounts are given for one capture reaction, scale as appropriate)

Component	Blockers Mix v5	Blockers Mix v4 (deprecated)
Block X	0.5 µl	–
IDT blocker (*see* Note 2)	–	2 µl
Block O	2.5 µl	2.5 µl
H_2O (*see* Note 3)	2.5 µl	2.5 µl
	5.5 µl total	7 µl total

3.2 Hybridisation

During hybridization the binding of the probes/baits to the complimentary molecules
in the NGS library occurs. Hybridisation is performed at 65 °C after denaturation
at 95 °C. Use PCR tubes/or strips and perform the Incubation program in a thermal
cycler. Use a heated lid to minimize condensation. The hybridisation is a 2-step
process, where blockers and library are denatured at 95, and the Hybridisation mix
is added after, the library has been cooled down to 65 °C. (Amounts given are per
capture reaction).

1. Create the incubation program in a thermal cycler.

Incubation program	
95 °C	5 min
65 °C	5 min
65 °C	Forever

2. For hybridisation, combine components and incubate as per table below.

Component	Amount
Blockers mix	5 μl
Sequencing library (100–500 ng dsDNA)	7 μl (mix by pipetting)
• Denare in thermal cycler (95 °C, 5 min) • Let the cycler reach hybridization temperature (65 °C) • Equilibrate Hybridisation mix in thermal cycler (65 °C, 5 min)	
Add hybridisation mix to library/blocker, mix by pipetting, ~ 5 ×	18.5 μl
	30.5 μl total
• Incubate as 65 °C for 16 + h (in practice: overnight)	

3.3 Bead Binding and Washes

During binding, the bait-target hybrids are collected with streptavidin coated magnetic beads and subsequently washed with warm buffer (65 °C) to remove non-target DNA. 'Wash buffer X' and beads and need to be prepared before use.

3.3.1 Prepare 'Wash Buffer X'

Amounts given are per capture reaction. Scale up if you have more than one.

Component	Amount
Hyb S	6.25 μl
H2O	618 μl
Wash Buffer	156 μl
	780.25 μl total

3.3.2 Prepare Beads

1. Aliquot 30 μl beads in a 1.5 ml protein low-bind Eppendorf tube.
2. Pellet the beads on a magnet for 2 min.
3. Discard supernatant.
4. Conduct 3 washes:

   • Add 200 μl Binding Buffer and thoroughly resuspend the beads,
   • Pellet the beads on the magnet for 2 min,
   • Remove and discard the supernatant.

5. Resuspend beads in 70 μl binding buffer.
6. Transfer to PCR tube/strip.

3.3.3   Bead Binding Reaction

At this point the hybridization reaction should have been in the thermal cycler for the past 16 + hours and still be in the cycler at 65 °C. In the below we will add the prepared magnetic beads to our hybridisation reaction. Those beads will then bind the baits.

1. For bead-binding the baits, combine components and incubate as per table below.

Component	Amount (in μl)
Prepared Beads in PCR tube	70 μl
• Equilibrate bead aliquots in thermal cycler at 65 °C for 2 min) (place them alongside the hybridization reaction in the thermal cycler)	
Transfer capture reaction(s) to the bead aliquot(s)	30.5 μl
• mix by pipetting, ~ 5 × • replace the lids	
	100 μl total
• Incubate in thermal cycler at 65 °C for 5 min (Flick/spin the tubes after 2.5 min to keep beads suspended)	

2. Take out from the thermal cycler.
3. Pellet the beads on a magnet until the solution is clear, discard supernatant.
4. Immediately perform 4 subsequent washes with pre-warmed 'Wash buffer X' (see next step: 3.3.4 Bead Washing).

3.3.4   Bead Washing

Repeat the below steps 4 times for a total of 4 washes. After the last wash, remove all wash buffer and proceed without delay to 3.4 Library Resuspension.

4 ×	• Add 180 μl warmed wash buffer X to the beads, mix by pipetting
	• Incubate in thermal cycler (65 °C, 5 min). Flick/spin the tubes after 2.5 min to keep beads suspended
	• Pellet the beads on a magnet until the solution is clear, discard supernatant

Proceed without delay with the next step: 3.4 Library Resuspension.

3.4   Library Resuspension

Add 30 μl of 10 mM Tris-Cl, 0.05% Tween-20 (pH 8.0–8.5) to the washed beads and resuspend the 'enriched library' by pipetting.

3.5   Library Amplification

Set up the PCR reaction mix as per below with universal primers suitable for your library type. The resuspended, 'enriched library' is of sufficient volume to conduct two PCRs as per Arbor protocol. Overamplification of the library should be avoided. Pooling of independent PCRs can reduce error.

3.5.1   PCR Primers

Amplification primers for Nextera libraries		
Universal forward primer [i5]	AATGATACGGCGACCACCGAGA	Tm = 66.2
Universal reverse primer [i7]	CAAGCAGAAGACGGCATACGAGA	Tm = 64.4

3.5.2   PCR Reaction Mix

Component	Amount
H2O	5 μl
KAPA HiFi HotStart ReadyMix (2 ×)	25 μl
Universal forward primer [i5], 10 μM	2.5 μl
Universal reverse primer [i7], 10 μM	2.5 μl
Enriched library (on beads)	15 μl
	50 μl total

3.5.3   PCR Program

Step	Temperature (°C)	Time	
1	98	2 min	
2	98	20 s	8–14 cycles
3	60	30 s	
4	72	Length-dependent[a]	
5	72	5 min	
6	15	Forever	

[a]Recommended elongation times (by average insert size): 500 bp: 30 s, 500–700 bp: 45 s, > 700 bp: 1 min

3.6 Library Clean-Up

1. Optional: Pool several PCRs.
2. Perform least two rounds of bead clean-ups: 1× bead clean-up, followed by a 0.7× bead clean-up. Initial clean-up and volume reduction can be more cost-effective using a column-based PCR clean-up kit (e.g., Qiagen).

3.7 Library QC and Quantification

Sequencing service providers will have minimum requirements with respect to DNA amount and quality and often require a minimum 'molarity', which can be calculated from average fragment size and weight. The size distribution should be determined with a Fragment Analyzer and the amount of dsDNA in ng by fluorescence-based DNA quantification. Molarity can then be calculated using the formula below:

$$\frac{concentration\left(\frac{ng}{\mu l}\right) * 10^6}{660 * Average\ fragment\ length} = Molarity\left(\frac{nmol}{l}\right)$$

The formula was copied from https://bitesizebio.com/23105/quantifying-your-ngs-libraries/. Illumina has published a technical note on the quantification of Nextera Libraries of similar content: https://www.illumina.com/documents/products/techno tes/technote_nextera_library_validation.pdf.

4 Performance of the Exome Capture Kit—Example Project

To test the performance of the PBGL/Daicel Arbor Biosciences Exome Capture Kit, we performed exome capture and sequencing on an Illumina/Nextera NGS library pool of 41 DNA samples, aligned the resulting sequencing reads to the reference genome and assessed the fraction of reads that matched the exome and the coverage. We used the same reference genome and annotation that had been used to design the kit.

4.1 Example Project: Sequencing a Mutant Population (M_1V_1)

The work was performed at the PBG Laboratory, Seibersdorf, Austria and entailed individual DNA isolations from 41 leaf samples derived from *Coffea arabica* plants that had been grown in tissue culture, sequencing library construction for each sample (Nextera), pooling of all samples, performing the exome capture reaction on the pool of 41 samples, and submitting the library pool to a service provider for Illumina short-read sequencing (PE150). During library preparation, each sample received an individual molecular barcode (index), so the sequencing reads could be associated to the respective samples after DNA sequencing. We aligned the raw reads (fastq files) to the *Coffea arabica* reference genome Cara_1.0 (NCBI assembly GCF_003713225.1) with software bwa mem (Li and Durbin 2009). From these alignments (bam files) we evaluated the quality of the capture and enrichment with the R-package TEQC (Hummel et al. 2011, 2020).

4.1.1   Input NGS Library

An Illumina DNA sequencing library pool with 41 individually-indexed coffee samples was prepared following a transposase-mediated protocol (Nextera-type) as detailed in the IAEA-PBGL protocol: Library Preparation for Medium- to High-throughput DNA Sequencing on the Illumina Sequencing Platform, A Laboratory Protocol (IAEA 2022a). The library pool was size selected with Ampure XP beads (one-sided, $0.7\times$) to an average insert size of ~ 540 bp and a lower size limit of above 300 bp (Fig. 2). Seven microliter (7 µl) containing 300 ng of this Illumina/Nextera sequencing library pool was the input for the exome capture reactions.

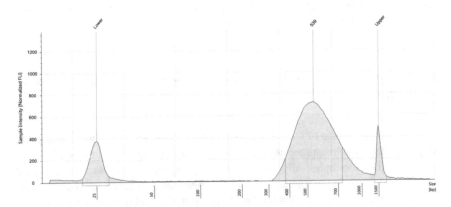

Fig. 2 Size distribution of the input DNA sequencing library pool of 41 individually indexed coffee samples (Illumina/Nextera), assessed with ©Agilent Technologies, Inc. TapeStation, high sensitivity D1000 ScreenTape®

4.1.2   Exome Capture

One capture reaction was performed on this pool of 41 samples following Arbor protocol version 4: Baits (5.5 µl) were combined with the hybridisation components to 20 µl Hybridisation Mix. Blockers (2 µl IDT Blocker, 2.5 µl Block O) were added to 7.5 µl of the Illumina library resulting in 12 µl total. 18.5 µl of Hybridisation mix were combined with the 12 µl library/blocker mix and hybridization was allowed to occur in a PCR machine for 16 h at 65 °C. The bait/library hybrids were captured (with streptavidin-coated beads) and washed with 1× Buffer X (618 µl H2O, 156 µl wash buffer, 6.25 µl Hyb S). Beads were resuspended in 30 µl 10 mM TrisCl, 0.05% TWEEN-20, pH 8.0, and two independent enrichment PCRs (50 µl, KAPA HiFi) were performed, each with 15 µl of the bead suspension as template, 13 PCR cycles with 45 s extension time. Both PCRs were pooled (100 µl total) and subjected to PCR purification (Qiagen MinElute) and two subsequent bead-cleanups for size selection (1 and 0.7× with Ampure XP beads). Final DNA amount was assessed by fluorescence measurement (Qubit). A one in four dilution was assessed for size distribution on the Agilent TapeStation.

4.1.3   Output Exome Enriched NGS Library

DNA amount of the exome enriched library was assessed by fluorescence measurement (Qubit). A one in four dilution was assessed for size distribution on the Agilent TapeStation (Fig. 3). Average fragment size of the library was ~ 570 bp, which corresponds to an average insert size of ~ 460 bp, adaptors subtracted.

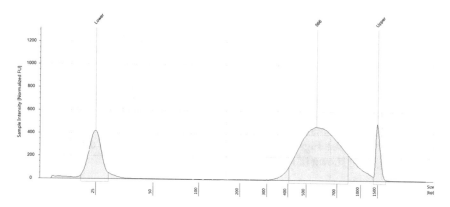

Fig. 3 Size distribution of the Exome Captured library as shipped to the sequencing service provider, 1/4 dilution assessed with ©Agilent Technologies, Inc. TapeStation, high sensitivity D1000 ScreenTape®

Fig. 4 Requesting 400 Gbp raw data output resulted in 3.2 billion sequencing reads with fairly even distribution across the 41 samples. Median is 75 Mio reads

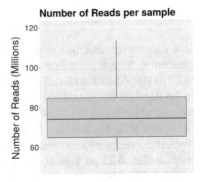

4.1.4   DNA Sequencing

The exome-enriched library along with the list of sample indices was submitted to a sequencing service provider for Illumina DNA sequencing PE150 (paired-end reads with 150 bp read length). We shipped 200 ng (50 µl, 4 ng/µl) and requested 400 Gbp raw data output. We received a total of 3.2 billion reads. They were fairly well distributed across the 41 samples (Fig. 4), with between 58 and 113 Mio reads per sample (Median: 75 Mio).

4.1.5   Analysis and Results

We aligned all 3.2 billion sequencing reads to the coffee reference genome; the same annotated reference assembly that had been used to derive the targets (Cara_1.0, NCBI accession number: GCA_003713225.1). The reads were aligned with software bwa mem (Li and Durbin 2009) as part of our automated analysis workflow: A Software Workflow for Automated Analysis of Genome (Re-) Sequencing Projects, A Laboratory Protocol (IAEA 2022b). Software and documentation are available on PBGL's github page (https://github.com/pbgl).

The on-target enrichment for each individual sample was assessed from the alignments to the reference, represented in per sample.bam files, with the R-Bioconductor package TEQC (Hummel et al. 2011, 2020). Target definitions were the actual exons of the annotation (*see* Figs. 5 and 6 for results). As an example, a representative genomic region is shown in Fig. 7.

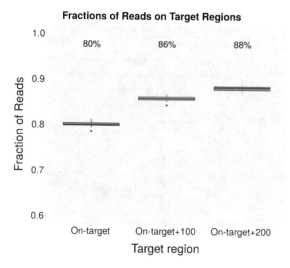

Fig. 5 For each individual sample we assessed what fraction of the sequencing reads that align to the genome match annotated genes. Counting strictly the region annotated as exons we reach 80% with a very little variation between samples. When extending the target space by 100 or 200 bp to either side this fraction increases. This is expected, because the probes are fishing molecules from a library with an average insert size of 460 bp (Fig. 3). We can conclude that close to 90% of the sequencing reads are matching the target space

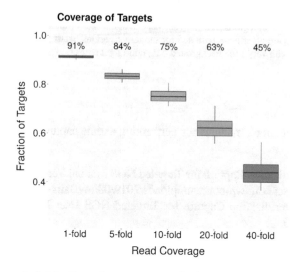

Fig. 6 We assessed what fractions of genes are covered at least 1, 5, 10, 20 or 40-fold. More than 90% of annotated gene is covered at least one-fold and ¾ of the genes are covered more than tenfold

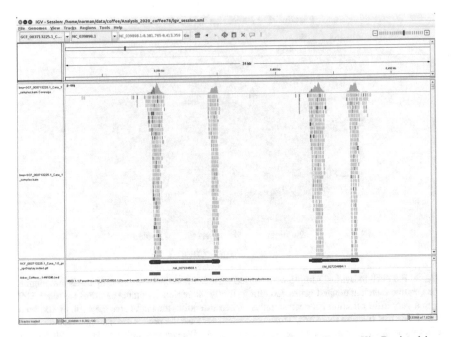

Fig. 7 Visualization of successful target enrichment by the Exome Capture Kit. Depicted is a representative genomic region (screenshot of the Integrative Genomics Viewer, IGV, *see* Note 4), showing the alignments of sequencing reads (bam file) of 41 coffee samples on the Coffee arabica reference genome. Target regions (red bars) correspond to the exons (thick blue bars) of genes (blue bars). The libraries are effectively enriched for the target regions, reads (grey bars) pile on target regions (red bars) with very little background, i.e., non-target reads

5   Manuals

1. The manufacturer's manuals for performing exome capture reactions with this kit

myBaits, Hybridization Capture for Targeted NGS Manual Version 4.01 April 2018, https://arborbiosci.com/wp-content/uploads/2019/08/myBaits-Manual-v4.pdf.

myBaits, Hybridization Capture for Targeted NGS User Manual Version 5.00 September 2020,
https://arborbiosci.com/wp-content/uploads/2020/08/myBaits_v5.0_Manual. pdf.

2. Sequencing library preparation

The custom-indexed Nextera NGS libraries for Illumina Sequencing were prepared following the PBGL protocol: *Library Preparation for Medium- to High-throughput DNA Sequencing on the Illumina Sequencing Platform, A Laboratory Protocol* (IAEA 2022a).

3. Sequence read mapping

Read mapping with software bwa mem (*see* Note 5) (Li and Durbin 2009) was performed as part of PBGL's automated software workflow: *A Software Workflow for Automated Analysis of Genome (Re-) Sequencing Projects, A Laboratory Protocol* (IAEA 2022b).

4. Quality assessment of the capture reactions

TEQC: Quality control for target capture experiments, Hummel et al. (2020). DOI:10.18129/B9.bioc.TEQC, TEQC, R package version 4.18.0. https://bioconductor.org/packages/release/bioc/html/TEQC.html (Hummel et al. 2011).

6 Notes

1. *Coffee arabica* is an allotetraploid of *Coffea eugenoides* and *Coffea canephora*. We developed separate probe sets for the different sub genomes, so that they can be used independently, if desired. For use in *Coffea arabica* they need to be pooled in representative ratios. If the user aims to enrich the chloroplast as well, that module can comprise a final 0.1% of the final pool, though optimization for the tissue type might be required.
2. xGen® Universal Blockers-NXT Mix, Catalog no. 1079584, purchased from Integrated DNA Technologies Inc. (IDT, www.idtdna.com).
3. If the amount of DNA in library is limiting, then H_2O can be replaced with additional sequencing library.
4. https://software.broadinstitute.org/software/igv/.
5. https://bio-bwa.sourceforge.net/bwa.shtml.

Acknowledgements This work was funded by FAO/IAEA. The Exome Capture Kit design was a contribution by Daicel Arbor Biosciences (Ann Arbor, Michigan, USA). Mr Florian Goessnitzer (IAEA) provided tissue from in vitro coffee plantlets.

References

Choi M, Scholl UI, Ji W, Liu T, Tikhonova IR, Zumbo P, Nayir A, Bakkaloğlu A, Ozen S, Sanjad S, Nelson-Williams C, Farhi A, Mane S, Lifton RP (2009) Genetic diagnosis by whole exome capture and massively parallel DNA sequencing. Proc Natl Acad Sci U S A 106:19096–19101. https://doi.org/10.1073/pnas.0910672106

Dong C, Zhang L, Chen Z, Xia C, Gu Y, Wang J, Li D, Xie Z, Zhang Q, Zhang X, Gui L, Liu X, Kong X (2020) Combining a new exome capture panel with an effective varBScore algorithm accelerates BSA-based gene cloning in wheat. Front Plant Sci 11:1249. https://doi.org/10.3389/fpls.2020.01249

Gardiner L-J, Brabbs T, Akhunov A, Jordan K, Budak H, Richmond T, Singh S, Catchpole L, Akhunov E, Hall A (2019) Integrating genomic resources to present full gene and putative promoter capture probe sets for bread wheat. GigaScience 8. https://doi.org/10.1093/gigascience/giz018

Hummel M, Bonnin S, Lowy E, Roma G (2011) TEQC: an R package for quality control in target capture experiments. Bioinformatics 27:1316–1317. Oxford, England. https://doi.org/10.1093/bioinformatics/btr122

Hummel M, Bonnin S, Lowy E, Roma G (2020). TEQC: Quality control for target capture experiments. R package version 4.18.0.

Li H, Durbin R (2009) Fast and accurate short read alignment with Burrows-Wheeler transform. Bioinformatics 25:1754–1760. Oxford, England. https://doi.org/10.1093/bioinformatics/btp324

Mascher M, Richmond TA, Gerhardt DJ, Himmelbach A, Clissold L, Sampath D, Ayling S, Steuernagel B, Pfeifer M, D'Ascenzo M, Akhunov ED, Hedley PE, Gonzales AM, Morrell PL, Kilian B, Blattner FR, Scholz U, Mayer KFX, Flavell AJ, Muehlbauer GJ, Waugh R, Jeddeloh JA, Stein N (2013) Barley whole exome capture: a tool for genomic research in the genus Hordeum and beyond. Plant J Cell Mol Biol. 76:494–505. https://doi.org/10.1111/tpj.12294

IAEA (2022a) Library preparation for medium- to high-throughput DNA sequencing on the illumina sequencing platform. A laboratory protocol, https://www.iaea.org/resources/manual/library-preparation-for-medium-to-high-throughput-dna-sequencing-on-the-illumina-sequencing-platform-a-laboratory-protocol

IAEA (2022b) A software workflow for automated initial analysis of high-throughput DNA sequencing project. A laboratory protocol, https://www.iaea.org/resources/manual/a-software-workflow-for-automated-initial-analysis-of-high-throughput-dna-sequencing-project-a-laboratory-protocol

Russell J, Mascher M, Dawson IK, Kyriakidis S, Calixto C, Freund F, Bayer M, Milne I, Marshall-Griffiths T, Heinen S, Hofstad A, Sharma R, Himmelbach A, Knauft M, van Zonneveld M, Brown JWS, Schmid K, Kilian B, Muehlbauer GJ, Stein N, Waugh R (2016) Exome sequencing of geographically diverse barley landraces and wild relatives gives insights into environmental adaptation. Nat Genet 48:1024–1030. https://doi.org/10.1038/ng.3612

Scalabrin S, Toniutti L, Di Gaspero G, Scaglione D, Magris G, Vidotto M, Pinosio S, Cattonaro F, Magni F, Jurman I, Cerutti M, Suggi Liverani F, Navarini L, Del Terra L, Pellegrino G, Ruosi MR, Vitulo N, Valle G, Pallavicini A, Graziosi G, Klein PE, Bentley N, Murray S, Solano W, Al Hakimi A, Schilling T, Montagnon C, Morgante M, Bertrand B (2020) A single polyploidization event at the origin of the tetraploid genome of Coffea arabica is responsible for the extremely low genetic variation in wild and cultivated germplasm. Sci Rep 10:4642. https://doi.org/10.1038/s41598-020-61216-7

High Resolution Melt (HRM) Genotyping for Detection of Induced Mutations in Coffee (*Coffea arabica* L. var. Catuaí)

Andrés Gatica-Arias, Alejandro Bolívar-González, Elodia Sánchez-Barrantes, Emanuel Araya-Valverde, and Ramón Molina-Bravo

Abstract Arabica coffee (*C. arabica* L.) is a highly valued agricultural commodity on the world market. Tons of products are traded internationally, and it has become an extremely valuable resource. However, the species is threatened by the alarmingly low genetic diversity present among its wild populations and agronomic varieties. It is highly relevant to exploit different mechanisms to increase genetic variability in coffee. One of such methods is the induction of variability through chemical or physical mutagenesis. In this work, a population of 320 coffee plants (*Coffea arabica* L. var. Catuaí) originated from chemically mutagenized embryogenic callus was analysed. Here we describe a protocol for detection of induced mutations using High Resolution Melting (HRM) on a Real Time PCR machine with HRM capabilities. The protocol allows to detect mutations in pooled DNA samples of up to four M_2 mutant plants. The procedures and example data are presented for mutation detection in the CaWRKY1 gene. This procedure can be applied for mutation detection in other genes of interest to coffee breeders and scientists.

1 Introduction

The genetic improvement of crops depends on the selection of genotypes with the desired, novel agronomic characteristics. Genetic variation provides the main resource to develop varieties adapted to different scenarios. Arabica coffee (*Coffea arabica* L.) is an allopolyploid species ($2n = 4x = 44$) that resulted from hybridization

A. Gatica-Arias (✉) · A. Bolívar-González · E. Sánchez-Barrantes
Laboratorio Biotecnología de Plantas, Escuela de Biología, Universidad de Costa Rica, 2060 San Pedro, Costa Rica
e-mail: andres.gatica@ucr.ac.cr

R. Molina-Bravo
Laboratorio de Cultivo de Tejidos y Células Vegetales, y Laboratorio de Biología Molecular, Universidad Nacional, Heredia, Costa Rica

E. Araya-Valverde
Centro Nacional de Innovaciones Biotecnológicas, San José, Costa Rica

© The Author(s) 2023
I. L. W. Ingelbrecht et al. (eds.), *Mutation Breeding in Coffee with Special Reference to Leaf Rust*, https://doi.org/10.1007/978-3-662-67273-0_20

between two species extremely close to *C. eugenioides* and *C. canephora* (Scalabrin et al. 2020). The genetic variation present in Arabica coffee doesn't represent the entire possible conglomerate of spontaneous mutations. Rather, they result from the recombination of genotypes within populations and their continuous interaction with both biotic and abiotic environmental elements (Oladosu et al. 2016). Therefore, the availability of genotypes to introduce into a breeding program is limited.

Through induced mutagenesis it is possible to generate heritable changes in the genome of an organism, without the need for genetic segregation or recombination (Oladosu et al. 2016). These changes can be generated in genes that regulate characteristics of interest and finally allow their improvement or functional analysis. Mutation induction has been carried out in different tissue types through irradiation and exposure to chemical agents (Serrat et al. 2014). One of the most widely used chemical agents, is ethyl methanesulfonate (EMS), which mainly induces C-T substitutions that result in C/G to A/T transitions (Kim et al. 2006).

A fundamental aspect of the genetic improvement through induced mutagenesis is the process of identifying those plants with the mutations of interest in their genome. This can be done in two phases: 1) screening or detection of the mutants, and 2) confirmation of the mutation (Forster and Shu 2012). To achieve this, it is important to have an efficient and scalable detection strategy to increase the probability of detecting new genetic variants within mutant populations. The High-Resolution Melting (HRM) analysis can facilitate the detection of variants in genes. This technique does not involve any enzymes, but rather requires the presence of saturating fluorochromes that interact with the double-stranded DNA. In this way, a heteroduplex structure, with less stability, is denatured at lower temperatures than the DNA copies, a process that is monitored by the decrease in fluorescence emission (Szurman-Zubrzycka et al. 2016). When trying to detect mutations in large populations, it is convenient to pool the DNA of individuals, thus reducing the number of samples to be analyzed and consequently the cost. However, this clustering decreases the sensitivity and makes it difficult to detect low frequency mutations (Simko 2016). COLD-PCR (lower denaturation temperature co-amplification) can be applied to increase the sensitivity of HRM analysis by preferentially amplifying mismatched DNA. This is a modification of PCR where the reaction is carried out at a denaturation temperature at which the heteroduplex DNA is denatured in a greater proportion than the other DNA types (Chen and Wilde 2011). This chapter describes the PCR-HRM based detection of variants in genomic sequences of Arabica coffee plants var. Catuaí developed via chemical mutagenesis.

Fig. 1 Coffee (*Coffea arabica* L. var. Catuaí) M_2 mutant population. **a** M_2 mutant plants in the experimental field, **b** fresh material brought from the field to the lab for DNA extraction, **c** young and disease-free leaves used for DNA extraction

2 Materials

2.1 Plant Material

1. M_2 mutant coffee population (*e.g., Coffea arabica* L. var. Catuaí) (*see* Fig. 1a *see* Note 1).

2.2 Reagents

1. 1 Kb DNA ladder (*e.g.,* Thermo Scientific Cat Nr.: SM0311).
2. 18S primers (18S_F: 5′-AGGTAGTGACAATAAATAACAA-3′ and 18S_R: 5′-TTTCGCAGTTGTTCGTCTTTC-3′) (*see* Note 2).
3. 6X loading dye (*e.g.,* Thermo Scientific Cat Nr.: R0611).
4. Agarose (*e.g.,* Phytotechnology Cat Nr.: A110).
5. Chloroform (*e.g.,* Sigma Cat Nr.: 288306) (*see* Note 3).
6. dNTPs (25 mM each) (*e.g.,* Thermo Scientific Cat Nr.: R1121).
7. EDTA (*e.g.,* Phytotechnology Cat Nr.: E582).
8. Ethanol (*e.g.,* Sigma Cat Nr.: 459844).
9. GelRed™ (*e.g.,* Gold Biotechnology, Inc Cat Nr.: G-720-500).
10. Hexadecyltrimethylammonium bromide (CTAB) (*e.g.,* Bio Basic Inc. Cat Nr.: CB0108).
11. Isopropanol (*e.g.* Sigma Cat Nr.: I9516).
12. MeltDoctor™ HRM Master Mix (*e.g.,* Thermo Scientific Cat Nr.: 4415440).
13. 2-Mercaptoethanol (*e.g.,* Phytotechnology Cat Nr.: M649).
14. Mix for real-time PCR with HRM (*e.g.,* Melt Doctor™, Thermo Scientific Cat Nr.: 4415440).
15. NaCl (*e.g.,* Sigma Cat Nr.: S9888).
16. Phenol (*e.g.,* Sigma Cat Nr.: P1037).
17. Polyvinylpyrrolidone (PVP-40) (*e.g.,* Phytotechnology Cat Nr.: P728).
18. RNAase (10 mg/mL) (*e.g.,* Thermo Scientific Cat Nr.: EN0531).

19. Taq DNA polymerase recombinant (5 U/μl) (*e.g.*, Thermo Scientific Cat Nr.: EP0402).
20. TBE buffer (5X) (*e.g.*, Phytotechnology Cat Nr.: T773).
21. Tris-HCl (pH 8.0) (*e.g.*, Phytotechnology Cat Nr.: T764).

2.3 Equipment

1. Analytical balance.
2. Autoclave.
3. High-resolution real-time PCR instrument (*e.g.*, CFX96 real-time PCR system, Bio-Rad Laboratories, Hercules, CA, USA).
4. Electrophoresis apparatus (electrophoresis tank and power supply).
5. Freezer (− 20 °C).
6. Gel imaging documentation system.
7. High-resolution real-time PCR instrumentation.
8. Hot plate shaker.
9. Microwave.
10. PCR workstation with UV light.
11. pH meter.
12. Refrigerated centrifuge.
13. Spectrophotometer (*e.g.*, NanoDrop 2000, Thermo Scientific, Osterode am Harz, Germany).
14. Thermal cycler.
15. Thermomixer block.
16. Vortex stirrer.
17. Water bath.

2.4 Software

1. Precision Melt Analysis™ Software (CFX96 real-time PCR system, Bio-Rad Laboratories, Hercules, CA, USA).
2. Molecular Evolutionary Genetics Analysis (MEGA) (https://www.megasoftw are.net/).

3   Methods

3.1   Preparation of Stock Solutions

1. 100 mL extraction buffer: add 10.0 mL 1 M Tris/HCl (pH 8.0), 28.0 mL 5 M NaCl, 4.0 mL 0.5 M EDTA (pH 8.0), 2 g CTAB, 2 g PVP and dissolve by adding 58.0 mL of molecular grade water (*see* Note 4). Once completely dissolved, add 40 μL of 2-mercaptoethanol for each 20 mL of extraction buffer.
2. 1X TE [10 mM Tris/HCl (pH 8.0) containing 1 mM EDTA (pH 8.0)]: add 1.0 mL Tris/HCl (1 M, pH 8.0) and 0.2 mL EDTA (0.5 M, pH 8.0) and molecular grade water to a final volume of 100 mL.
3. Chloroform:phenol (24:1): in a ducted chemical fume hood, combine 48 mL chloroform with 2 ml phenol (pH 8.0) in a 100 mL glass bottle with a lid.

3.2   DNA Extraction

The following protocol describes the procedure for the extraction of genomic DNA from young and disease-free coffee leaves. It has been optimized to eliminate or reduce oxidation during the extraction. Although it can be used on dry material, the recommendation is to use fresh tissue, which yields better quality DNA. To transfer the fresh material from the field to the lab, it is recommended to cut the branch and place it in a bag with water until it reaches the laboratory (*see* Fig. 1b).

1. Place approx. 50 mg fresh weight plant material in a 2 mL reaction tube (*see* Fig. 1c).
2. Add 600 μL of extraction buffer, macerate by hand using mortar and pestle until the samples are homogeneous and mix with the vortex.
3. Incubate the sample at 65°C for 12 min, every 6 min invert the tubes about 5 times.
4. Add 600 μL of chloroform:phenol (24:1) in the ducted chemical fume hood (*see* Note 5).
5. Mix by inversion 15 times. Do not use vortex at this stage.
6. Centrifuge the sample for 5 min at 13,000 rpm, at 4 °C.
7. Transfer 350 μL of the supernatant to a new 1.5 mL reaction tube. Be careful not to contaminate the tip with the organic phase (bottom phase).
8. Add 350 μL of a cold isopropanol, shake 20 times by inversion and incubate 15 min at − 20 °C.
9. Centrifuge the sample for 7 min at 13,000 rpm, at 4 °C.
10. Discard the supernatant by decanting. Be careful not to lose the pellet.
11. Add 500 μl of cold 70% v/v ethanol.
12. Centrifuge the sample for 2 min at 13,000 rpm, at 4 °C.
13. Carefully remove the ethanol by decanting. Be careful not to lose the pellet.
14. Dry the pellet at 45°C until there are no ethanol residues.

15. Resuspend the pellet in 50 μl of 1X TE.
16. Add 1 μL of RNase (conc. stock solution 10 mg/mL).
17. Incubate for 30 min at 37 °C.
18. Store at − 20 °C.

3.3 Determination of DNA Integrity

1. Weigh 0.8 g agarose.
2. Mix agarose with 100 mL 1X TBE in a flask.
3. Microwave for 1–2 min until the agarose is completely dissolved, avoid boiling the solution.
4. Let the agarose solution cool down for about 5 min.
5. Add 2 μl of GelRed™ solution per 100 mL gel.
6. Pour the gel slowly avoiding any air bubbles.
7. Once solidified, place the agarose gel into an electrophoresis tank filled with 1XTBE.
8. Add 6X loading buffer to each DNA sample at a final concentration of 1X (*e.g.*, 2 μL 6X loading dye + 5 μL DNA + 5 μL molecular grade water).
9. Load 2–3 μL of a molecular weight ladder (*e.g.*, 1 Kb DNA Ladder) in the first and last lane of the gel.
10. Load your samples into the remaining wells.
11. Run the gel at 100 V for about 45–60 min or until the dye has migrated approximately 75–80% in the gel.
12. Visualize the DNA fragments using a gel imaging device (*see* Fig. 2a).
13. Quantify the concentration of the DNA using the NanoDrop 2000 spectrophotometer (*see* Fig. 2b and Note 6).

3.4 PCR Amplification of the 18S Gene

1. Thaw reagents on ice, vortex and centrifuge prior use.
2. Prepare a MastexMix by pipetting on ice the following components in a final volume of 25 μl: 1X Taq PCR buffer, 0.25 mM of each dNTPs, 0.2 μM of each primer, 1.5 mM of $MgCl_2$, 0.5 U Taq polymerase, and 1 μl DNA (100 ng/μl). Include a negative control (no-DNA template) (*see* Notes 7 and 8).
3. Place the tubes in the thermal cycler with the following program: 95 °C for 10 min followed by 30 cycles at 94 °C for 30 s, 53 °C for 30 s, 72 °C for 90 s and a final step of 72 °C for 10 min.
4. Prepare a 1.5% m/v agarose gel in 1X TBE as described previously (*see* Sect. 3.3).
5. Load 15 μl of the PCR product on the gel.
6. Run at 100 V for 1 h.
7. A PCR product of approximately 481 bp should be visible (*see* Fig. 2c).
8. Record the amplification of the 18S gene.

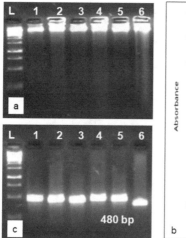
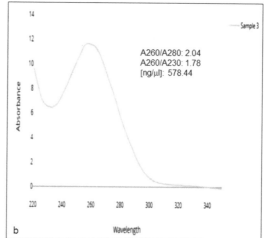

Fig. 2 Quantification and determination of DNA integrity, **a** DNA isolated from young and disease-free leaves. **L:** 1 Kb DNA Ladder, **1–6:** samples, **b** quantification and DNA purity determined using a NanoDrop 2000 (Thermo Scientific) spectrophotometer, **c** PCR amplification of a 480 bp fragment of the 18S gene **L:** 100 bp DNA Ladder, **1–6:** samples

3.5 Literature Mining and Selection of Candidate Genes for Mutation Screening

1. Go to PubMed database (http://www.ncbi.nlm.nih.gov/pubmed/) (*see* Fig. 3, **step 1**).
2. Enter the desired word combination associated with a particular topic, *e.g.*, "*coffee and rust resistance genes*" (*see* Fig. 3, **step 2**).
3. Click on "Search" (*see* Fig. 3, **step 3**).
4. Filter the results in terms of article type, text availability, publication date or investigated species (*see* Fig. 3, **step 4**).
5. Click on the "Abstract" of the desired article (*see* Fig. 3, **step 5**).
6. Select 'Related information' menu within abstract content to link to other related NCBI databases for the selected record (*e.g.*, gene and protein sequence) (*see* Fig. 3, **step 6**).
7. Retrieve gene and protein sequence (*see* Fig. 3, **step 7**).

3.6 Primer Design

1. Go to *Primer3* primer design tool (version 4.1.0) (*Primer3 Input*) (*see* Fig. 4, **step 1**).
2. Paste a raw nucleotide sequence (5′–3′) of your gene target in the box, *e.g.*, *CaWRKY1* (GenBank: DQ335599.1) (*see* Fig. 4, **step 2** and Note 9).

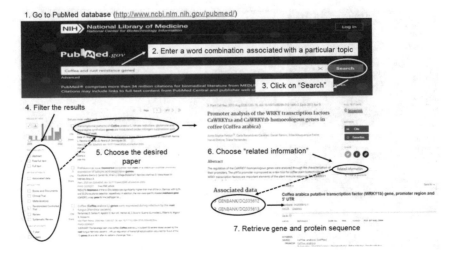

Fig. 3 Literature mining and selection of candidate genes for mutation screening. For details *see* Sect. 3.5

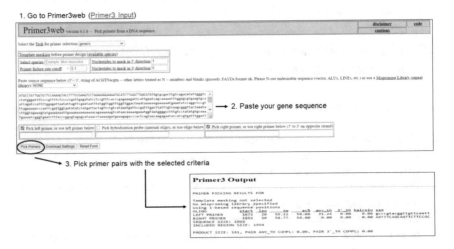

Fig. 4 Designing primers for the amplification of a specific gene fragment using Primer3. For details *see* Sect. 3.6

3. Design primers for HRM analysis with the following criteria:

- A length of 18–25 nucleotides.
- The melting temperature (Tm) between 55 and 65 °C, and not more than 3 °C difference of each other.
- The GC content between 40 and 60%, with the 3′ of a primer ending in C or G to promote binding.
- Balanced distribution of GC-rich and AT-rich domains.

- Lack of runs of four or more of one base or dinucleotide repeats.
- Lack of intra-primer homology (more than three bases that complement within the primer) or inter-primer homology (forward and reverse primers having complementary sequences).

4. Click 'Pick Primers' command to obtain primer pairs which best match the selected criteria (*see* Fig. 4, **step 3**).
5. Check the absence of secondary structures and properties of designed primers using appropriate software, *e.g.*, the Oligo Analyser™ tool (https://www.idtdna.com/pages/tools/oligoanalyzer).
6. For each primer pairs, complete the information shown in Table 1.
7. Order primer pairs through a specialized company (e.g., Macrogen, South Korea).

3.7 *In Silico Analysis of Primer Specificity*

1. Go to National Centre for Biotechnology Information (NCBI) database (https://www.ncbi.nlm.nih.gov/) (*see* Fig. 5, **step 1**).
2. Click "BLAST" button (*see* Fig. 5, **step 2**).
3. Choose the type of BLAST based on your goal. In our case, we chose "Nucleotide BLAST" (*see* Fig. 5, **step 3**).
4. Paste the primer sequences in the box, *e.g.*, *CaWRKY1_F:* 5-TGAGTATGTTTCCGGCCACC-3 (*see* Fig. 5, **step 4**).
5. Select "Somewhat similar sequences (blastn)" (*see* Fig. 5, **step 5**).
6. Click "BLAST" button (*see* Fig. 5, **step 6**).
7. Check the E-value (*see* Note 10), percentage identity (*see* Note 11) and query cover (*see* Note 12) of the designed primer pairs (*see* Fig. 5, **step 7**).

3.8 *In Silico PCR*

1. Access Primer BLAST at NCBI (https://www.ncbi.nlm.nih.gov/tools/primer-blast/index.cgi) (*see* Fig. 6, **step 1**).
2. Paste the gene sequence in FASTA format and your designed primer pairs (*see* Fig. 6, **step 2**).
3. Scroll down and enter an organism name (*Coffea arabica* (taxid:13443) (*see* Fig. 6, **step 3**).
4. Click "Get primers" button (*see* Fig. 6, **step 4**).
5. Check product length, Tm, GC%, and self-complementarity (*see* Fig. 6, **step 5**).

Table 1 Properties of designed primers. Example primers designed for the CaWRKY1 gene

Gene	Orientation	Sequence (5'–3')	Length (bp)	Position in the gene	%GC	Tm (ºC)	Product (pb)
CaWRKY1	Forward	TGAGTATGTTTCCGGCCACC	20	116	55	60	896
	Reverse	CACTAGGGCCCAAGTCCAAG	20	1011	60	60	

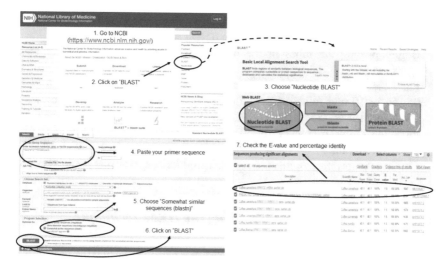

Fig. 5 Primer specificity in silico analysis performed at NCBI. For details *see* Sect. 3.7

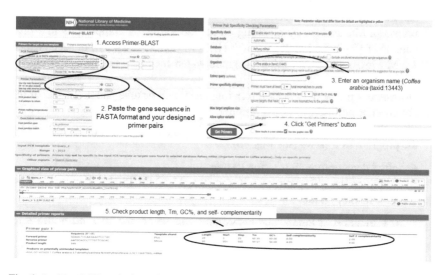

Fig. 6 In silico PCR analysis performed at NCBI. For details *see* Sect. 3.8

3.9 Nested PCR

1. Prepare fourfold DNA pools that will serve as the template for HRM-PCR reactions. DNA from each of four M_2 individuals should be mixed in equal amounts to obtain a final concentration of 30 ng/μl.

2. Prepare a Master Mix by pipetting on ice the following components in a final volume of 25 μl: 1X Taq PCR buffer, 0.25 mM of each dNTPs, 0.2 μM of each

primer, 1.5 mM of $MgCl_2$, 1.5 U Taq polymerase, and 1 µl DNA (90 ng/µl). Include a negative control, as well as a no-DNA template.

3. Before placing the tubes in the instrument, briefly spin for 10 s at room temperature.

4. Place the tubes in the thermal cycler with the following touch down program: 95 °C for 10 min followed by a touch down phase consisting of 13 cycles at 95 °C for 60 s, 65–53 °C for 60 s, 72 °C for 120 s and a final amplification phase involving 12 cycles at 95 °C for 60 s, 53 °C for 60 s, 72 °C for 120 s and a final 72 °C for 10 min.

5. Confirm the specific amplification of each product by agarose gel electrophoresis.

3.10   Mutation Identification Using the HRM Technique

The mutation identification using PCR-HRM was performed using the nested PCR methodology.

1. Design the plate following the steps defined by the software (*e.g.*, HRM Software, Applied Biosystems ™) (*see* Fig. 7a).

2. Prepare a Master Mix by pipetting on ice the following components in a final volume of 10 µl: 5 µl Melt Doctor™ HRM Master Mix, 0.4 µM of each primer, 2 mM $MgCl_2$ and PCR molecular grade water. Include a negative control (no-DNA template).

3. Distribute 9 µL of the reaction mix into each well or tube and add 1 µL of the nested PCR product (1:1000 dilution).

4. Place the tubes or plate in the high-resolution real-time PCR instrument and set the program as follows (*see* Fig. 7b):

   - Initial denaturation: 95 °C for 10 min.
   - Amplification (30 cycles): 95 °C for 30 s, 57 °C for 15 s, and 72 °C for 20 s
   - High-resolution melting:

     – Formation of homo- and heteroduplexes: 95 °C for 30 s, 40 °C for 60 s with continuous fluorescence acquisition (60–95 °C).

5. Close wells/tubes. Spin down to ensure that the entire volume is at the bottom of the tubes.

6. Place the reaction tubes in the thermal cycler and start the run.

7. After finishing the real-time PCR, open the precision melt analysis software and save the generated melt file (*see* Fig. 8).

8. Open the melt file and analyse the results (*see* Note 13).

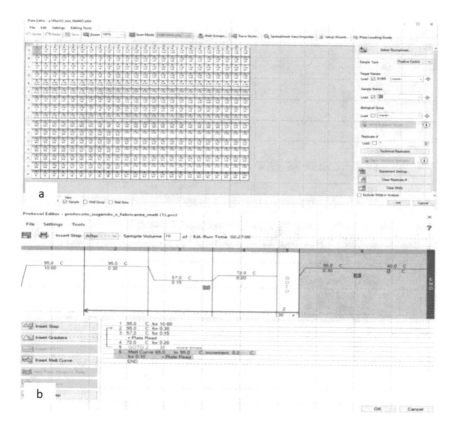

Fig. 7 Mutation identification using the HRM technique, **a** example of an HRM-PCR plate, **b** example of an HRM-PCR program of temperatures with melt curve protocol adjusted for posterior analysis

3.11 DNA Sanger Sequencing for HRM Validation

1. Confirm a potential mutation by Sanger sequencing the fragment being analysed for the identified M_2 individuals.
2. Check the quality of electropherograms using for example MEGA software.
3. Analyse the sequencing results using the NovoSNP 3.0.1 program (Weckx et al. 2005).

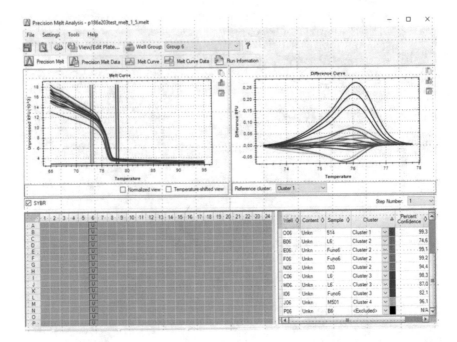

Fig. 8 Example of a melt file generated with the precision melt analysis, showing the normalized melt curve, the difference curve, the plate, and the classification of species by cluster

4 Notes

1. This protocol has been established using a M_2 mutant population obtained from mutagenized M_0 coffee seeds (*Coffea arabica* L. var. Catuaí). It can be used as a reference for other coffee varieties, nevertheless, it is recommended to optimize the HRM-PCR parameter for each variety used.

2. PCR amplification of the 18S endogenous gene is used as internal control for PCR evaluation of the quality and integrity of extracted plant DNA.

3. Read the Materials Safety Data Sheet (MSDS) of the reagents being used and follow the recommendation of the manufacturer. It is very important to wear personal protective equipment (gloves, safety glasses with side shields or chemical goggles; lab coat, closed-toe shoes, and full-length pants).

4. Prepare all solutions using ultrapure molecular grade water (deionized water) and analytical grade reagents.

5. DNA extraction should be performed in a dedicated molecular biology laboratory equipped with a ducted fume hood, toxic waste disposal and decontamination procedures.

6. The ratio of absorbance at 260 and 280 nm is used to assess the purity of DNA. A ratio of ~ 1.8 is generally accepted as "pure" for DNA. Moreover, the ratio 260/230 is used as a secondary measure of nucleic acid purity; 260/230 values in the

range of 2.0–2.2 indicate the absence of contaminants (such as carbohydrates and phenol).

7. Use molecular biology grade consumables (*e.g.*, tips, reaction tubes, PCR tubes, real-time PCR strips and caps) for DNA extraction and PCR analysis (sterile, DNase and RNase free). Other materials and consumables can be purchased in non-sterile conditions and autoclaved (121 °C, 15 min) before use.

8. To avoid cross contamination, it is highly recommended to work in a PCR work-station, especially when performing all tasks associated with the preparation of PCR or HRM-PCR mixes.

9. The genomic regions of the *CaKO, CaPOP, CaWRKY1, CCoAOMT1, ITS2, LOX1, SUS2, ABC*, and *FLC* genes were selected based on the potential effect that a variation could have on certain phenotypes of *C. arabica* L. plants.

10. The Expected value (E value) is a parameter that describes the number of hits one can "expect" to see by chance when searching a database of a particular size. The lower the E-value, or the closer it is to zero, the more "significant" the match is.

11. The percent identity is a number that describes how similar the query sequence is to the target sequence (how many characters in each sequence are identical). The higher the percent identity is, the more significant the match.

12. The query cover is a number that describes how much of the query sequence is covered by the target sequence. If the target sequence in the database spans the whole query sequence, then the query cover is 100%. This tells us how long the sequences are, relative to each other.

13. The Precision Melt Analysis™ Software uses predefined automated parameters, which might be adjusted according to the type of analysis that is being processed. See the user guide for more details (https://www.bio-rad.com/sites/default/files/webroot/web/pdf/lsr/literature/10000080911.pdf).

Acknowledgements Funding for this work was provided by the University of Costa Rica, the Ministerio de Ciencia, Tecnología y Telecomunicaciones (MICITT), the Consejo Nacional para Investigaciones Científicas y Tecnologicas (CONICIT) (project No. 111-B5-140; FI-030B-14) and a fellowship granted to Alejandro Bolivar-González by Centro Nacional de Alta Tecnología (CeNAT). A. Gatica-Arias acknowledged the Cátedra Humboldt 2023 of the University of Costa Rica for supporting the dissemination of biotechnology for the conservation and sustainable use of biodiversity.

References

Chen Y, Wilde HD (2011) Mutation scanning of peach floral genes. BMC Plant Biol 11:1–8

Forster BP, Shu QY (2012) Plant mutagenesis in crop improvement: basic terms and applications. In: Shu QY, Forster BP, Nakagawa H (eds) Plant mutation breeding and biotechnology. CABI, Wallingford, pp 9–20

Kim Y, Schumaker KS, Zhu JK (2006) EMS mutagenesis of Arabidopsis. Methods Mol Biol 323:101–103

Oladosu Y, Rafii MY, Abdullah N, Hussin G, Ramli A, Rahim HA, Usman M (2016) Principle and application of plant mutagenesis in crop improvement: a review. Biotechnol Biotechnol Equip 30:1–16

Scalabrin S, Toniutti L, Di Gaspero G, Scaglione D, Magris G, Vidotto M, Bertrand B (2020) A single polyploidization event at the origin of the tetraploid genome of *Coffea arabica* is responsible for the extremely low genetic variation in wild and cultivated germplasm. Sci Rep 10:1–13

Serrat X, Esteban R, Guibourt N, Moysset L, Nogués S, Lalanne E (2014) EMS mutagenesis in mature seed-derived rice calli as a new method for rapidly obtaining TILLING mutant populations. Plant Methods 10. https://doi.org/10.1186/1746-4811-10-5

Simko I (2016) High-resolution DNA melting analysis in plant research. Trends Plant Sci 21:528–536

Szurman-Zubrzycka M, Chmielewska B, Gajewska P, Szarejko I (2016) Mutation detection by analysis of DNA heteroduplexes in TILLING populations of diploid species. In: Jankowicz-Cieslak J, Tai T, Kumlehn J, Till B (eds) Biotechnologies for plant mutation breeding. International Atomic Energy Agency. Springer Open, pp 292–303

Weckx S, Del-Favero J, Rademakers R, Claes L, Cruts M, De Jonghe P, Van Broeckhoven C, De Rijk P (2005) novoSNP, a novel computational tool for sequence variation discovery. Genome Res 15(3):436–442. https://doi.org/10.1101/gr.2754005

Protocols for Chromosome Preparations: Molecular Cytogenetics and Studying Genome Organization in Coffee

Le Li, Trude Schwarzacher, Paulina Tomaszewska, Qing Liu, Xiaoyu Zoe Li, Kexian Yi, Weihuai Wu, and J. S. Pat Heslop-Harrison

Abstract Cytological preparations from cell nuclei are required to count the number of chromosomes (including determining ploidy or aneuploidy), to investigate their morphology and organization. The results are valuable for genetic and evolutionary studies, and in breeding programs to understand species relationships, polyploidy, and potential introgression of chromosomes in hybrids between different species. Preparation of good chromosome spreads with well-separated metaphase chromosomes is the foundation of cytogenetic research including chromosomal mapping based on FISH (fluorescence in situ hybridization). FISH combined with specific locus probes correlated with molecular markers to specific chromosomes for integrating physical and linkage maps as well as studying the genetic evolution of allopolyploidization, has rarely been applied in *Coffea* spp. despite being a global high-value crop. Cytogenetic studies of *Coffea* are limited by the small size and similar morphology of the chromosomes, but FISH can help to map sequences to

L. Li
NHC Key Laboratory of Tropical Disease Control/Key Laboratory of Tropical Translational Medicine of Ministry of Education, School of Tropical Medicine, Hainan Medical University, NO.3 of Xueyuan Road, Haikou, Hainan 571199, P. R. China

T. Schwarzacher (✉) · P. Tomaszewska · J. S. Pat Heslop-Harrison (✉)
Department of Genetics and Genome Biology, University of Leicester, University Road, Leicester LE1 7RH, UK
e-mail: ts32@le.ac.uk

J. S. Pat Heslop-Harrison
e-mail: phh4@le.ac.uk; phh@molcyt.com

T. Schwarzacher · Q. Liu · X. Z. Li · J. S. Pat Heslop-Harrison
Key Laboratory of Plant Resources Conservation and Sustainable Utilization/Guangdong Provincial Key Laboratory of Applied Botany, South China Botanical Garden, Chinese Academy of Sciences, Tianhe District, Xingke Road 723, Guangzhou 510650, P. R. China

P. Tomaszewska
Department of Genetics and Cell Physiology, Faculty of Biological Sciences, University of Wrocław, 50-328 Wrocław, Poland

K. Yi · W. Wu
Environment and Plant Protection Institute, Chinese Academy of Tropical Agricultural Science (CATAS), Longhua District, Xueyuan Road 4, Haikou, Hainan 571101, P. R. China

© The Author(s) 2023
I. L. W. Ingelbrecht et al. (eds.), *Mutation Breeding in Coffee with Special Reference to Leaf Rust*, https://doi.org/10.1007/978-3-662-67273-0_21

chromosome arms and identify individual chromosomes. This chapter presents protocols for germinating seeds and growing coffee plants involving pre-treatment and fixation of root-tips where the meristems of actively growing roots have many divisions. Mitotic metaphase chromosome preparation on microscope slides is described, as well as preparing probes of 5S and 18S rDNA to be used for FISH. The FISH experiments involve a two-step protocol with pre-treatments and setting up the hybridization on day 1 and the detection of probe sites on day 2 after overnight hybridization. A final section gives advice about visualization using a fluorescent microscope and capturing images.

1   Introduction

The vast majority of commercial coffee comes from two closely related species, *Coffea canephora* and *Coffea arabica* (Melese and Kolech 2021). *C. canephora* (known as robusta; 35% of world production) is diploid with two sets of chromosomes, while *C. arabica* (65% of production) is tetraploid, with four chromosome sets from two ancestral species (see other chapters in this volume). Thus, the genome of *C. arabica* has four sets of chromosomes ($2n = 4x = 44$), and four copies of most genes (compared to the two in *C. canephora* ($2n = 2x = 22$) (see Hamon et al. 2009). At meiosis, the tetraploid *C. arabica* behaves as a diploid and chromosomes pair and recombine. The genus *Coffea* includes in total over 90 species, with wild species of coffee including both diploids and polyploids (see Lashermes et al. 1997; Melese and Kolech 2021). The use of this germplasm through hybridization can increase the genetic base of the crop, and may be used to transfer useful genes from wild relatives into a crop variety.

The study of chromosome numbers in an accession of a species is important to give the ploidy-level (Tomaszewska et al. 2021). Identification of individual chromosomes by morphological analysis or by in situ hybridization with DNA probes, can be used to link the genetic map to physical chromosomes (Paesold et al. 2012), and track chromosomes in breeding programmes involving hybridization and recombination (eg, in cereals, Patokar et al. 2016 or Ali et al. 2016). In crosses involving wide species or polyploid species, cytogenetic study of the chromosome numbers and morphology is particularly valuable to define which crosses may be most easy to make, and to determine new combinations of chromosomes in hybrids and backcross derivatives. In some cases, recombination between chromosomes of different species is required to introgress useful agronomic characteristics without undesirable characters. Aneuploidy, involving the loss or gain of one or more chromosomes (e.g., Niemelä et al. 2012 in *Brassica*; Tomaszewska et al. 2023 in *Urochloa*), is found occasionally, along with other types of chromosome rearrangements such as inversions, deletions or translocations (eg, Forsström et al. 2002; Liu et al. 2019; Tomaszewska and Kosina 2021), particularly in irradiated material and after some tissue culture protocols.

The key method for molecular cytogenetics is fluorescent in situ hybridization (FISH) that allows visualizing the location of DNA sequences to be determined along chromosomes, providing cytogenetic maps of chromosomes (Schwarzacher 2003; Heslop-Harrison and Schwarzacher 2011; Bačovský et al. 2018). Many repetitive sequences can be used to provide chromosomal landmarks to identify chromosomes, show aspects of genome organization and evolution, and track chromosome presence or rearrangement through evolution and crossing programmes. High-resolution FISH mapping on mitotic chromosomes is a powerful technique to help integrate physical and genetic maps and also to evaluate genome assembly quality (Szinay et al. 2008). Although now less used than in the past, chromosome banding can also offer chromosome differentiation and identification (e.g. Schwarzacher 2003, Kumar et al. 2021 and in coffee Pierozzi et al. 1999). Meiotic analysis can show the pairing of chromosomes and reveals any translocations between chromosomes, either during evolution or following breakage and rejoining (e.g. Lashermes et al. 2000).

Preparation of high quality metaphase, with well-spread chromosomes free of cytoplasm and other cellular material, containing high number of divisions, is a prerequisite for cytogenetic studies such as chromosome counting, morphological analysis and mapping, chromosome banding procedures, and in situ hybridization. In this chapter, we describe the basic methods of chromosome preparations and fluorescent in situ hybridization (FISH) in coffee species, including protocols to obtain root tips with abundant metaphases from seedlings, root tip sampling and fixation, mitotic chromosome spread preparation and the basic steps of FISH for repetitive DNA using rDNA probes. Schwarzacher and Heslop-Harrison (2000) and Schwarzacher (2016) give more details about chromosome preparation and in situ hybridization for many species.

Cytogenetic maps provide an efficient tool for gene localization, validation of contig order from sequence analysis, characterization of gene regions or physical genetic distances, and the presence of chromosomal rearrangements such as inversions or translocations (Heslop-Harrison and Schwarzacher 2011). In the protocol below, we show the use of the repetitive rDNA as probes to identify coffee chromosomes carrying the 35S/45S and 5S rDNA loci, which provide robust and useful landmarks on chromosomes (e.g., Široký et al. 2001; Ali et al. 2016; and for coffee Hamon et al. 2009). Many other repetitive probes can be used to provide landmarks. Generally, short lengths of probes from single-copy genes or groups of genes (single-copy FISH) do not work reliably in plants. In most cases, repetitive sequences—simple sequence repeats, tandem repeats from satellite DNA, or rDNA genes are used as probes to provide chromosomal landmarks (e.g. Liu et al 2019; Agrawal et al. 2020; Rathore et al. 2022). Large insert clones, in particular BACs, or large pools of synthetic oligonucleotide probes (typically 20,000 bp or more) may also be used (see Niemelä et al. 2012; Zaki et al. 2021), particularly when any repetitive sequences within the probes have been removed. Mitotic metaphase chromosome FISH can locate probes with a longitudinal resolution of about 2–20 Mb. Meiotic chromosome preparations can also be used, and in some circumstances may be readily available. In particular, pachytene chromosomes from meiotic prophase may give resolution of in situ signal of 50–100 kb (e.g. Szinay et al 2008, Mandáková et al.

2019). Other systems with statistical analysis of hybridization sites within interphase nuclei, stretched chromosomes (digested with proteinase K), and fibre FISH to DNA fibres from the nucleus extended to nearly their full molecular length can be used to give higher resolution of a few kb (see Schwarzacher and Heslop-Harrison 2000).

The application of FISH to coffee mitotic chromosomes has provided opportunities for identifying chromosomes and mapping genes and sequences of interest in several *Coffea* species. On several cultivated and wild species of coffee, mitotic chromosomes, as well as meiotic pachytene have been used to map using repetitive sequences of 45S rDNA and 5S rDNA or BACs linked to resistance genes as probes for in situ hybridization (Lombello and Pinto-Maglio 2004a, b, c; Pinto-Maglio 2006; Herrera et al. 2007; Hamon et al. 2009; Iacia and Pinto-Maglio 2013). Coffee mitotic chromosomes in metaphase are small (1–3 µm) and have similar morphologies, making their individual identification difficult (Krug 1934, 1937; Mendes 1957). On these small chromosomes, the exact location of a small repetitive sequence by in situ hybridization is also difficult to determine (Lombello and Pinto-Maglio 2003; Herrera et al. 2007), although the repetitive probes and BACs can identify clearly chromosome arms and domains such as terminal, intercalary or centromeric.

Most chromosome analyses rely on mitotic metaphases, so dividing tissue is essential and is best obtained from healthy, disease free, and rapidly growing plants (Schwarzacher and Heslop-Harrison 2000; Schwarzacher 2016). Among plant tissues containing actively dividing cells, root-tip meristems are one of the most commonly used, but other plant tissues such as meristematic cells from young shoots, leaves or emerging buds as well as hairy root cell culture lines or liquid tissue culture cells (Anamthawat-Jónsson and Thórsson 2003; Bačovský et al. 2018). Calli or protoplasts (Nishibayashi et al. 1989) from tissue culture can be also used for chromosome preparation but it is difficult to obtain many metaphases, and to spread chromosomes sufficiently that they can be counted. The best source to obtain fresh coffee root tips for chromosome preparation is from seedlings or young plants. When growing tropical species in temperate climates, it is important to understand temperature and light requirements and we give suggestions to achieve best results.

Although several methods for chromosome preparation are available, we recommend to use the squashing method for coffee. Mitotic chromosomes are released with cells from fixed rapidly dividing root tips, and spread onto a microscopic slide into a single layer by gentle pressure on the digested tissue through squashing. Classically, this preparation method was used in combination with acetocarmine staining to analyse the number and shape of metaphase chromosomes. Not discussed below, but preparations for chromosome counting can be made by acid digestion of root-tips, staining with Feulgen's reagent, before spreading of cells and chromosomes on a microscope slide. The squashing method does not require costly equipment and usually gives high quality metaphase spreads. The protocols for labelling probes for rDNA is based on PCR amplification and the two step FISH experiments follow our previous published protocols (Schwarzacher and Heslop-Harrison 2000) and although modifications and optimizations for coffee chromosomes are included below.

2 Materials

It is assumed that a well-equipped laboratory with molecular biology and microscopy facilities and consumables are available; these include microcentrifuge (Eppendorf) tubes, automatic pipettes, tips, micro centrifuge, balance, stirrer and mixer. Only more specialized or essential equipment is described in the following list of materials.

2.1 Seed Germination and Plant Cultivation

1. Plant material: Coffee seeds or young plants growing in pots. Here, we used, diploid coffee ($2n = 2x = 22$) including *C. canophera* 'Roburst' and tetraploid *C. arabica* ($2n = 4x = 44$) cultivars.
2. Trays and plant pots (8–20 cm diameter depending on size of plants).
3. Nutrient potting soil and fine sand mixtures as required.
4. Sand bed: fill small trays with about 8 cm deep sand.
5. Plastic bags (new or recycled as long as they are clean and disease free).
6. Plastic film.
7. Water (*see* Note 1).
8. Appropriately shaded greenhouse or growth chamber at 25–28 °C and 14 h day/ 10 h night light cycle.

2.2 Fresh Root Sampling and Fixation

1. α-bromo-naphthalene (saturated 0.05% aqueous solution): prepare by mixing a few drops of α-bromo-naphthalene liquid in 500 ml distilled water, shake vigorously to make a saturated solution and allow to settle (a small amount of the α-bromo-naphthalene should remain at the bottom of the vessel). Keep at room temperature up to six months (*see* Note 2).
2. Fixative: prepare fresh for each experiment by mixing 96 or 100% (v/v) ethanol and glacial acetic acid 3:1; do not keep fixative for more than 30 mins before use.
3. Small tubes with tight caps (e.g. microcentrifuge tubes, Bijou tubes or freezer vials; should hold about 2–10 ml).
4. Fine forceps and scissors.

2.3 Chromosome Preparation

1. Enzyme buffer (10 mM, pH 4.6): for a 100 mM stock solution, mix 100 mM citric acid and 100 mM tri-sodium citrate in a ratio of 2:3 and autoclave (once opened can be stored at 4 °C for a few days to 2 weeks). Before use dilute buffer with distilled water to 10 mM.

2. Enzyme solution: 20 U/ml cellulase (e.g. Sigma C1184), 10 U/ml 'Onozuka' RS cellulase and 20 U/ml pectinase (e.g. Sigma P4716 from *Aspergillus niger*; solution in 40% glycerol) in 10 mM enzyme buffer. Store in 2–5 ml aliquots at − 20 °C. The enzyme solution can be re-used a few times (*see* Note 3).
3. 0.01 M HCl.
4. 60% (v/v) acetic acid.
5. Dry ice or liquid nitrogen (*see* Note 4).
6. Fine forceps, razor blades and dissecting needles.
7. Microscope glass slides: specified for microscopy, pre-cleaned and if available specially treated for better adhesion (e.g. Thermofisher, UK, Superfrost or Superfrost plus; Citotest, PR China) (*see* Note 5).
8. Coverslips: 18 × 18 mm No. 1 (*see* Note 6).
9. Petri dishes (9 cm diameter or smaller).
10. Filter paper (approx. 9 cm diameter).
11. Autoclave or masking tape.
12. Diamond pen to scratch glass.
13. Plastic slide box holding 20 or 50 slides.
14. Dissecting microscope.
15. Phase contrast microscope.
16. Spirit lamp (alcohol) for flaming.

2.4 Probe Labelling

1. DNA template PCR amplification of probe DNA. Here we give the rDNA probes that we used on coffee chromosomes, but can also be used on any plant species.

   a. 5S rDNA: DNA of clone containing the 5S rDNA repeat, pTa794 from wheat, *Triticum aestivum* (Gerlach and Dyer 1980), insert length 410 bp.
   b. 18S-5.8S-26S rDNA (35S or 45S rDNA): total genomic DNA of wheat or rice for amplification (coffee DNA would probably work too).

2. Primers for PCR amplification of probe DNA

   a. For clones M13 sequencing primers; eg. M13 forward (GTA AAA CGA CGG CCA GT) and M13 reverse (GGA AAC AGC TAT GAC CAT G); this will amplify the insert plus about 30–50 bp on each side depending on the cloning site.
   b. 35S/45S rDNA: Primers based on 18S rDNA sequence of rice (Chang et al. 2010) rice_18S_P1 forward (CGA ACT GTG AAA CTG CGA ATG GC) and rice_18S_P2 reverse (TAG GAG CGA CGG GCG GTG TG); the product will be about 2.7 kb depending on species of the template DNA. (*see* Note 7).

3. Standard PCR reagents, nucleotides and TAQ polymerase.

4. Gel electrophoresis system, agarose, running buffer, DNA ladder and loading buffer.

5. Labelling kit: based on random priming or nick translation (e.g. Invitrogen, Roche).

6. Labelled nucleotides if they are not included in the labelling kit: digoxigenin-11-dUTP and/or biotin-16-dUTP (e.g. Roche or other suppliers).

7. Purification tubes to clean PCR product or labelled probe to remove unincorporated nucleotides, unwanted enzymes, and salts (*see* Note 8).

2.5 FISH Day 1: Pre-treatments and Hybridization

1. 20X SSC (Sodium citrate buffer): 3 M NaCl and 0.3 M sodium citrate, adjust to pH 7, and autoclave before storage at room temperature or in small aliquots at − 20 °C. Before use dilute with distilled water to 2X SSC.

2. RNase solution: 100 μg/ml in 2X SSC; prepare 250 μl per slide from a 10 mg/ml stock solution of DNAse-free RNAse in 10 mM Tris–HCl, pH 8; store small aliquots at − 20 °C.

3. 0.01 M HCl

4. Pepsin solution (optional, *see* Note 9): 1–10 μg/ml in 0.01 M HCl; prepare from a 1 mg/ml (ca. 4000 U/mg) stock solution; store stock solution in aliquots at − 20 °C.

5. 4% paraformaldehyde solution: ready made in PBS (e.g. Thermo Scientific) (*see* Note 10).

6. Acetic acid 100% (v/v).

7. Ethanol series: 96% (v/v), 85% (v/v) and 70% (v/v) in water

8. Hybridization mixture (*see* Table 1); all solutions can be stored in 0.5–1 ml aliquots at − 20 °C if not indicated otherwise. Before use, slowly defrost ingredients and keep on ice.

   a. Formamide (molecular grade).

   b. Dextran sulfate: 50% (w/v) solution in water, heat to dissolve and sterilize by forcing through a 0.22 μm filter.

   c. SDS solution: 10% (w/v) sodium dodecyl sulfate (also called sodium lauryl sulfate) in water, filter sterilize; store at room temperature.

   d. Salmon sperm DNA: 4 μg/μl sonicated or autoclaved DNA (also suitable are herring sperm or *E. coli* DNAs).

   e. 20X SSC (sterile): see above Sect. 2.1.

   f. Optional EDTA (Ethylene-diamine-tetra-acetic acid): 100 mM, pH 8.

   g. Probe DNA from protocol 3.4.

   h. Molecular grade distilled water.

9. Micro-centrifuge tubes (1.5 and 0.2 ml).

10. Plastic coverslips: 25 × 25 mm pieces from autoclavable plastic bags.

11. Plastic or glass Coplin jars holding eight slides and 80–100 ml solution.

12. Humid chamber; a plastic or metal box with preferably curved lid to avoid condensation to drop on the slides; line with filter paper or tissue and moisten with water or 2X SSC.
13. Hybridization oven or platform (optional) (*see* Note 11).
14. Shaking platform (optional).
15. 37 °C incubator.
16. Water bath or PCR machine.

2.6 FISH Day 2: Detection of Hybridization Sites and Mounting of Slides

1. Post-hybridization wash solution: dilute from 20X SSC to 2X SSC and 0.1X SSC as required.
2. Detection buffer: 4X SSC containing 0.2% (v/v) Tween-20.
3. Blocking solution: 5% BSA (Bovine serum albumin, heat shock fraction, pH7): 5% (w/v) in detection buffer. Prepare about 1 ml for 8 slides; weigh out approximate amount of BSA in a weighing boat or small tube and add corresponding volume of detection buffer (200 μl for each 0.01 g BSA).
4. Detection solution: dilute antibodies about 1:200–1:500 in blocking solution (final concentration is 1–6 μg protein /ml); if two labels are used, antibodies can be combined; prepare 50 μl per slide. For the detection of biotin and digoxigenin we recommend to use the following detection reagents:

    a. streptavidin (e.g. Alexa Fluor 594 conjugated; 1mg dilute with distilled water to 200 μg/ml; Molecular Probes, Invitrogen).
    b. anti-digoxigen antibody (e.g. FITC conjugated FAB fragment, 200 μg/ml; Roche).

5. DAPI antifade solution: mix 5 μl of DAPI stock (100 μg/ml in water) and 245 μl of antifade solution (Vectashield or Citifluor) (*see* Note 12).
6. Filter paper.
7. Large coverslips (No. 0, 24 × 30 or 40 mm) (*see* Note 6).
8. Plastic coverslips: 25 × 25 mm pieces from autoclavable plastic bags.
9. Coplin jars: holding eight slides and 80–100 ml solution.
10. Humid chamber, see 2.6.12.
11. 37 °C incubator.
12. 45 °C water bath.
13. Shaking platform or water bath.

2.7 Microscopy and Image Analysis

1. Epifluorescent microscope with multiple wavelength light illumination, appropriate filter blocks for the fluorochromes to be imaged (*see* Table 2), 20, 40 and 100X lenses suitable for UV fluorescence and digital high resolution low light sensitive B&W or colour camera (*see* Note 13).
2. Immersion oil suitable for UV fluorescence microscopy (e.g. Nikon Immersion Oil F or Zeiss immersion oil 518F).
3. Image analyzing program, e.g. Adobe Photoshop, ImageJ (formerly NIH image; https://imagej.nih.gov/ij/) or microscope software (e.g. Nikon NIS).

3 Methods

3.1 Seed Germination and Plant Cultivation

Coffee seeds are able to germinate with water when isolated from the yellowish-green fruits at around 225d after anthesis (Eira et al. 2006; Bytof et al. 2007). Coffee embryos are very sensitive to low temperature and are damaged when seeds are kept at temperatures below 25 °C. However, seed storage for medium periods of a few weeks to months at 25 °C is possible if environmental relative humidity is maintained around 50%, while for conservation at freezing temperatures, a lower moisture content of coffee seeds and hermetic conditions are required (Eira et al. 1999, 2006; Patui et al. 2014).

Once seeds have germinated and seedlings are transferred to soil, favorable temperatures and high air humidity near saturation are the most important conditions required for good growth. *C. arabica* is well adapted to cooler temperatures with the optimum growth at mean annual temperature in natural conditions ranging from 18 to 22 °C, while *C. canephora* is better adapted to higher temperatures, with the optimum growth at annual mean temperature ranging from 22 to 30 °C (Pohlan and Janssens 2010).

In a greenhouse (warm tropical or temperate conditions, 25–28°C, 14h day) or growth chamber, the following method yields coffee plants for molecular biology and cytology experiments and analysis. In order to maintain high humidity of 80–90% needed for good growth, individual plants are covered with plastic bags.

1. Fresh seeds are best collected as ripe berries from disease-free plants, growing areas or plantations; pulp them and, after removing the mucilage by fermentation and washing in water, dry the seeds in the shade.
2. Germinate fresh seeds, less than 2 months old, as they tend to lose viability thereafter (*see* Note 14).

Use either of the following methods in the green house or growth chamber.

a. Spread the seeds on a sand bed and cover with a layer of sand (about 2 cm). Water well and cover with a plastic film/bag or moist organic materials. Keep covered and moist until the radicles emerge and the seedlings reach 10–20 cm height.

b. Pre-germinate the seeds layered between moist gauze. As soon as the radicles emerge, transfer to pots or trays with a soil and sand mixture and cover with plastic film or bag. Keep covered until seedlings reach 20–40 cm.

3. Transfer plants to individual 15 cm diameter pots with 8:2 soil sand mixture and cover with plastic bags to keep moist at 80–90% humidity.

4. Keep *C. arabica* at around 25 °C, and *C. canephora* at about 28 °C, at 14 h day length with natural or artificial light avoiding exposure to direct sun light).

5. Grow plants for several days to weeks until roots develop and reach the side of the pots (*see* Note 15). For growth of root tips with many divisions, make sure the soil is not too wet and waterlogged, but does not dry out either.

3.2 Fresh Root Sampling and Fixation

In order to maximise the number of metaphases, the best period to sample fresh root tips is 4–4.5 h after sunrise or lights coming on. Only collect from healthy and well growing plants and only take roots with white ends indicative of new growth (Fig. 1). Steps are carried out at room temperature unless otherwise stated; use about 3X as much solution as plant material and make sure that roots are well covered by each solution.

1. Forceps and scissors should be washed in distilled water before sampling (*see* Note 16).

2. Carefully remove plants from the pot and collect roots from the edge or by rinsing away soil (Fig. 1).

3. Cut roots about 1–2 cm from the tip with a clean forceps or scissors and immediately transfer to a small tube with 1.5–2 ml α-bromo-naphthalene solution.

4. Incubate at room temperature for 2 h, followed with 4 °C for another 2 h.

5. Quickly blot roots dry on a filter paper and transfer roots to freshly prepared fixative.

6. Keep at room temperature for 2 h and then transfer to 4 °C for a minimum of 2 days. For long term storage of several months keep fixed roots at − 20 °C (*see* Note 17).

7. Fixed root-tips can be transported between laboratories and are 'dead' so there is no risk of transfer of pathogens. To obtain optimum chromosome preparations, they should be transported under cold conditions.

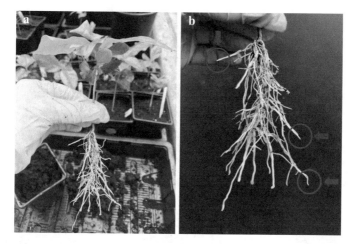

Fig. 1 Coffee seedlings and fresh roots, **a** seedlings in good health have many freshly growing roots to sample. **b** The arrows and the circles mark the fresh roots which are the best choices for chromosome preparation. The sampled roots should be less than 1 cm suitable for the small tubes

3.3 Chromosome Preparation

Having good preparations with plenty well spread metaphase chromosomes that are free of cytoplasm is the most crucial prerequisite of successful chromosome banding or in situ hybridization. Here we describe the method using proteolytic enzymes including pectinase and cellulase to remove cell walls and squashing dissected meristematic tissue in acetic acid between glass slide and cover slip. The method is modified from Schwarzacher et al. (1980) and Schwarzacher and Heslop-Harrison (2000). Steps are carried out at room temperature in a Petri dish if not otherwise stated.

1. Wash fixed root tips in distilled water in a Petri dish for 30 min.
2. Remove any dirt from the root, discard and cut away any unwanted material. Roots should just be 0.5–1 cm long and should have a clear white tip.
3. Incubate root tips in 0.01 M HCl for 10 min.
4. Wash the root tips in 1X enzyme buffer 3 times for 5 min each.
5. Place the root tips one by one on a clean slide, place in a Petri dish and fix with tape (Fig. 2).
6. Apply enzyme solution unto the root tips, 150–200 µl for 10 root tips, close the Petri dish and incubate at 37 °C for 35 min.
7. Remove the enzyme solution using a pipette, and then wash the root tips with enzyme buffer 3 times for 5 min (*see* Note 18).
8. Put a drop of water on the treated root tips to keep moist. (*see* Note 14).
9. Make sure that root tips are kept moist and do not dry out during the following steps of chromosome preparation.

Fig. 2 Enzyme treatment of the root tips

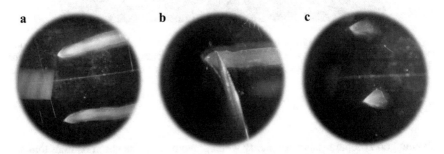

Fig. 3 Example of roots with pointed root caps and meristematic cells just behind. Cutting away non-dividing tissue is essential for preparations with high metaphase index

10. Dissect the root meristem (Fig. 3).

    a. Put a drop of 60% acetic acid on a clean slide and then place one or two treated root tips in it. Leave for 1–3 min.
    b. Under a dissection microscope, separate the root cap (<0.1 cm) from the root tip using a clean needle or forceps and discard.
    c. Dissect the root meristem that contains the dividing cells in the acetic acid and separate individual cells by tapping or squeezing with a fine forceps; remove non-meristematic tissues and mix cells evenly.

11. Place a small cover slip on the preparation. Cover one corner with a tissue or filter paper and hold with one finger to prevent sliding. Then vertically squeeze the slide gently using another finger. Tap the slide vertically with an eraser or needle until bubbles disappear between the cover slip and the slide.

12. Put the slide with the cover slip in a folded filter paper and press with the thumb or palm of your hand on the coverslip area; in order to avoid shearing of cells and chromosomes slowly and carefully increase the pressure.

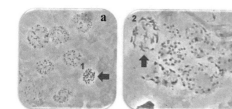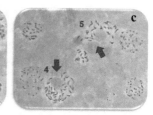

Fig. 4 Examples of the chromosomes under phase contrast microscope. Arrows mark several typical features of the chromosome preparation. **a** Nuclei are at good density; arrow1 shows the chromosomes from a single cell, but not well spreads. **b** Cells are too dense and nuclei overlap; arrow2 shows pro-metaphase chromosomes, that are not well spread. Arrow3 shows metaphase chromosomes, but they are squashed to hard, so some chromosomes are distorted or destroyed. **c** Well separated cells and metaphase chromosomes. Arrows 4 and 5 are good examples of metaphase chromosome preparation, well spreads and very little cytoplasm. The round opaque object is the nucleolus

13. Check the slide under phase-contrast microscopy to assess morphology and number of metaphase and chromosomes. Metaphases should be frequent (about 5–10% of cells) and metaphase chromosomes should be free of cytoplasm and dirt, well spread and with little overlaps, but should not be distorted (Fig. 4). Chromosomes and nuclei appear light grey with little contrast when flat and squashed well, but are either black or very bright if not squashed enough.

14. If necessary to clear cytoplasm or spread chromosomes more, a drop of acetic acid can be applied to the edge of the cover slip and the slide heated for a few seconds (not higher than 60 °C) with an alcohol flame and squashed again.

15. When preparations are satisfactory, freeze slide on dry ice for 5–10 min or dip into liquid nitrogen for a few seconds; then remove the cover slip with a razor blade and let the slide air dry.

16. Store slides with chromosomes in the dark at 4 °C in a small plastic slide box. For long term storage, keep at − 20 °C.

3.4 Probe Labelling

An important factor for the success of FISH experiments is the choice of probes. The amount of target sequences are critical and low copy sequences present at less than 10–15 kb at one site within the chromosomes are not suitable whereas repetitive sequences in large arrays such as the rDNAs or tandem repeats are ideal targets. Template DNAs to be used for generating probes can be inserts of clones, PCR products or total genomic DNA. Probes after labelling should be 100–300 bp long to allow for sufficient penetration to the DNA within the chromosomes, but shorter probes of 30–100 bp are also suitable, while probes longer than 500 bp are not recommended. Many different labelling kits are available commercially and use DNA polymerases in random priming or nick-translation that automatically generate

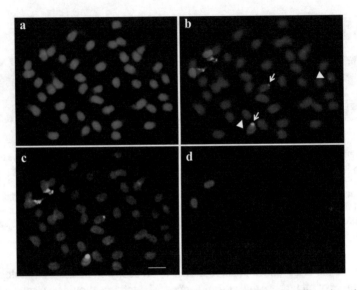

Fig. 5 Fluorescent in situ hybridization of a root tip metaphase of *C. arabica* (2n = 4x − 44). **a** The chromosomes are stained with DAPI (**a**) showing centromeres as small gaps or slightly brighter bands (**b**). Overlay of DAPI image with 18S rDNA signal in green (**c**) and 5S rDNA signal in red (**d**). One pair of chromosomes has a major 5S rDNA site near the centromere and a terminal 18S rDNA site of the small chromosome arm. Additionally, there is one pair of minor 18S rDNA sites (arrows) and one pair of minor 5S rDNA sites (arrowhead). Bar = 5 μm

probes of suitable lengths from larger templates. However, template DNAs longer than 2 kb do not label efficiently and will need cutting with enzymes, sonication or heat (Schwarzacher and Heslop-Harrison 2000; Salvo-Garrido et al. 2001).

Either labelled dUTP or dCTP are used and dependent on the labelled attached different ratios to unlabelled dTTP or dCTP nucleotides are recommended to allow efficient incorporation by the DNA polymerase. Manufacturers give detailed instructions of the procedure and recommendation for amounts of reagents to be used, but we have found that often the amount of expensive labelled nucleotides can be reduced when they are fresh and have not undergone several freeze-thaw cycles. We recommend to use biotin and digoxigenin as labels and here we give the rDNA probes that we used on coffee chromosomes (Fig. 5), but they can also be used on any plant species. Similarly, any cloned DNA or amplified PCR product that represents the repeats to be visualized in the species of interest are also suitable.

1. Amplify the probe DNA by PCR.

   a. For 5S rDNA: use miniprep DNA of clone pTa794 with M13 primers to amplify the insert by PCR using an annealing temperature of 56 °C. Expected product insert size plus about 80 bp (Fig. 6).

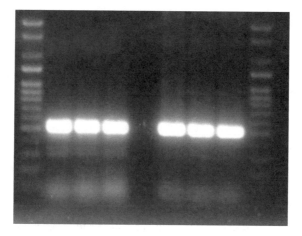

Fig. 6 PCR amplification of pTa794 insert using M13 primers. The image shows the following lanes from left to right: 100 bp ladder (with 100, 200, 300, 400, 500, 600, 700, 800, 900 bp, 1, 1.25, 1.5 and 2 kb band), 3 PCR replicas of the same miniprep DNA, empty lane, 3 PCR replicas of a different miniprep DNA, 100 bp ladder. The correct product is a very strong band at about 500 bp as the primers used are outside the cloning site and add 81 bp to the insert length of 410 bp. Some background smear and bands are also visible. It is therefore best to cut the band and purify before labelling

    b. For 35S/45S rDNA: use total genomic DNA from wheat or rice with the rice_ 18S primers to amplify the 18S rDNA sequence by PCR using an annealing temperature of 68 °C.

2. Check the PCR product on a 1.2% agarose gel.

    a. If there is a single sharp band of the expected size, then the entire PCR product can be used for labelling after purification.

    b. If there are several bands, or a smear, then cut out the band of the expected size, extract DNA from the band and purify.

3. Label probe DNA following the instructions of the labelling kit.
4. Clean the labelled probe using purification tubes (*see* Note 8) and resuspend in 20–30 μl distilled water.
5. Store probes at − 20 °C until use avoiding freeze thaw cycles (*see* Note 19).

3.5   *FISH Day 1: Pre-treatment and Hybridization*

For in situ hybridization, the protocol of Schwarzacher et al. 1989 and Schwarzacher and Heslop-Harrison (2000) is used with some adaptations to coffee chromosomes reported by Pinto-Maglio (2006) and several optimizations proposed here. Steps are carried out at room temperature unless otherwise indicated. Washing steps and incubation in buffers are carried out in Coplin jars (holding 8 slides and 80–100 ml solution); specific reagents are applied in small volumes of 200–300 μl per slide and covered with a plastic cover slip and incubated in a humid chamber.

1. Take slides from protocol 3.3. (Step 15), warm to room temperature if they were kept in the fridge or freezer for storage and check again for quality. Mark the area of chromosome preparation with a diamond pen and number slides for easy identification during the following steps.
2. Post-fixation of slides.

   a. Incubate slides in fixative (ethanol/acetic acid 3:1) for 10–30 min.
   b. Wash with 100% ethanol 2 times for 5 min each.
   c. Air-dry.

3. RNase treatment.

   a. Apply 200 μl RNase solution to each slide and cover with a plastic cover slip.
   b. Incubate for 1 h at 37 °C in a humid chamber.
   c. Wash 2 times in 2X SSC for 5 min each.

4. Pepsin treatment (optional; slides that are not used for this step should be kept in 2X SSC at room temperature).

   a. Incubate slides in 0.01 M HCl for 2 min.
   b. Shake of excess solution and apply 200 μl Pepsin solution to each slide and cover with a plastic cover slip.
   c. Incubate at 37 °C for 10 min in a humid chamber.
   d. Rinse in distilled water for 1 min.
   e. Wash in 2X SSC for 5 min.

5. Post-fixation with paraformaldehyde.

   a. In the fume hood, incubate slides in 4% paraformaldehyde solution for 10 min.
   b. Wash in 2X SSC for 5 min.

6. Dehydrate slides through an ethanol series and air dry.
7. Prepare hybridization mixture.

   a. Decide on probes (from **Protocol 3.4**) and amounts to be used, normally the final concentration of the probes should be 1–3 ng/μl in a hybridization mixture (*see* Table 1). Each slide can be probed with two different probes (e.g. 5S and 35/45S rDNA), but each probe needs to be labelled with a different hapten (e.g. biotin and digoxigenin) so it can be detected with a different antibody linked to a different fluorochrome (*see* **Protocol 3.6** and Fig. 5).
   b. Calculate and make master mix for all slides plus one following Table 1. Mix well and keep on ice.
   c. Prepare the hybridization mixture for each slide in a separate tube by adding master mix, probe and water following Table 1. Mix gently but thoroughly.
   d. Denature hybridization mixtures at 75 °C for 10 min and stabilize on ice for 10 min.

8. Hybridization.

    a. Apply hybridization mixture on each slide and cover with a plastic coverslip.

    b. Denature chromosomes and hybridization mixture in hybridization oven. This step is critical, and time and temperature will need adjusting even if the same species or variety is used. It is influenced by n the way chromosome preparations are made, how plants were grown, how old fixations were when used for preparation and how long slides were stored before FISH. As a guide, use 72–75 °C for 5–8 min.

    c. Hybridize slides in the hybridization oven or a humid chamber at 37 °C overnight (about 16 h).

3.6 FISH Day 2: Detection of Hybridization Sites and Mounting of Slides

Original methods for FISH used 20 or 50% formamide for washing steps (Schwarzacher and Heslop-Harrison 2000), but to avoid using this toxic chemical, we now routinely use low salt conditions for stringency washes as this also reduces the background created by formamide. Take care that the slides do not dry out during all steps of the protocol. Washes are carried out in a shaking waterbath if available, otherwise gentle shaking by hand is recommended once every 30–60 s. We describe here the use of two probes labelled with digoxigenin and biotin and they must be detected with two different colours, we recommend to use FITC for digoxigenin detection and Alexa594 for biotin detection (*see* Sect. 2.6 step 4), but other fluorochromes can be used too (Schwarzacher and Heslop-Harrison 2000). For visualization of chromosomes, slides are stained with DAPI (4', 6-diamino-2-phenylindole) and mounted in antifade solution.

1. Post-hybridization washes.

    a. Prepare post-hybridization wash solutions and heat in 45 °C waterbath.

    b. Collect slides from hybridization oven, carefully examine for bubbles, extra water or dried out patches and note down if there are any irregularities.

    c. Put slides from hybridization in 2X SSC at 35–40 °C to float off coverslips.

    d. Wash slides in 2X SSC at 42 °C for 2 min.

    e. Wash twice in 0.1X SSC at 40–45 °C for 5 min; record temperature.

    f. Wash in 2X SSC for 5 min. Allow to cool to room temperature.

2. Detection.

    a. Transfer slides to detection buffer

    b. Shake of excess solution and apply 200 µl of blocking solution to each slide and cover with a plastic cover slip. Incubate at RT or 37 °C for 10 min.

    c. Remove coverslip, drain slides and apply 40–50 µl of appropriate detection solution to each slide (*see* Sect. 2.6 and Table 2).

    d. Replace the coverslip and incubate at 37 °C for 1 h.

    e. Wash slides in detection buffer at 40–42 °C 3 times for 5 min each.

3. Drain slides and add one drop (20–30 µl) of DAPI anti-fade solution.
4. Put a large cover slip on each slide, cover with a tissue-paper and squash gently but firmly.
5. Keep slides in dark at 4 °C until observations (*see* Note 20).

3.7 Microscopy and Image Analysis

For visualization of probe hybridization and chromosome staining, an epi-fluorescence microscope equipped with suitable filters for the fluorochromes used in the detection step. A selection of filters are given in Table 2. Apart from a 20 or 40X lens for scanning the slides, you will need a top of the range 63 or 100X lens for image capture that all need to be specified for UV fluorescence. Also make sure you use immersion oil that is specified for fluorescence analysis (*see* Notes 21 and 22). The microscope should be located on a stable surface in a completely dark room with a comfortable adjustable chair and ideally a small lamp with a dimmer switch, so operation for several hours is possible.

1. Before you start, make sure that the system is set up correctly, the illumination lamp is centered, and no stray light can enter the lenses. Digital camera systems are now universal to capture images, and indeed allow capture of weak signal that is not visible to the eye, but no matter how expensive, they cannot make up for deficiencies in the set-up of the microscope.
2. Familiarize yourself with the camera control programme, the illumination and various microscope buttons and levers so you are able to operate the system in the dark and quickly as fluorochromes and FISH signals may be very weak and fade rapidly even when viewed with good antifade mountants.
3. Scan slide under DAPI with a low power lens. This will not fade your FISH signal, but close lamp shutter when not viewing mainly to avoid destruction of microscope lenses and filters by UV light.
4. When you find suitable nuclei or a metaphase, change the filter to green or red fluorescence for the FISH signal and then change to the 100X lens. This avoids fading the signal with UV, but might make refocusing difficult. Only view DAPI when absolutely essential and then keep time to a minimum.
5. Capture the image by sequential exposure with the different filter sets. Start with the FISH signal; the focus of the red and green fluorescence is interchangeable so use the stronger image for focusing. Generally, take the red or weaker image first, but it is advisable to do a few the other way round (*see* Note 23). Capture the DAPI image last, making sure that you focused the image again.
6. Save raw images before image adjustment are carried out.
7. To overlay the individual images and for image analysis, use either the built-in camera or microscope software or export as 'tif' files to use with Adobe Photoshop or NIH Image. Use only those features that are applied to all pixels of the image. Do not save with lossy formats such as JPG. Figure 5 gives some examples of coffee chromosomes after hybridization with 5S and 18S rDNA.

4   Notes

1. We recommend using bottled drinking water for seed germination and root growth as it is not contaminated with chlorine, heavy metal ions, or other water purification media or toxins. Distilled water can be used, but it is less favorable as it does not contain some salts or minerals.

2. Alternate arresting agents for many plant species is 2 mM 8-Hydroxyquionoline, incubate at temperature of plant growth for 30–90 min and then at 4 °C for 1–2 h or overnight.

3. Concentrations of enzymes might need adjusting if a new batch or different sources of enzymes are used. Addition of pectolyase (0.1–1% solution) or viscozyme (0.1–0.5%) can be considered. Enzyme mixtures can be reused several times: after use, centrifuge in a micro-centrifuge, transfer the supernatant to a new tube, mark for reuse and freeze (not recommended for screening lines of similar material since cells may remain in the solution). Increase digestion time slightly after each round of use.

4. If dry ice or liquid nitrogen are not available, slides can be frozen on a metal plate in a − 80 °C freezer.

5. If pre-treated slides are not available scrub slides in detergent and incubate in 96% ethanol with a few drops of HCl for better adhesion of cells and chromosomes.

6. Use a small coverslip (e.g. 18 × 18 mm, medium thick, No. 1) for making preparations (Sect. 3.3), and as best nuclei and metaphases are often near the periphery of the preparation use larger coverslips (e.g. 24 × 30 or 40 mm; thin No. 0 is essential for oil immersion microscopy) for mounting and observation (Sect. 3.6).

7. The primers were originally designed to rice but do work with wheat DNA as template; if genomic DNA from other species is used, primers might not bind well and no, or only weak amplification might result. Then the primers will need re-designing using sequence information from your or a related species. Once a probe has been made, we found that due to high similarity of all 18S sequences, FISH is successful in most cases.

8. If purifications columns are not available, ethanol precipitation using sodium acetate or lithium chloride can be used (see standard molecular biology protocols).

9. Pepsin is an endopeptidase that breaks down proteins and can be used to remove cytoplasm if chromosomes are not free and clearly visible. Pepsin is most active at low pH and therefore is made up in HCl. Adjust concentration and time according to amount of cytoplasm present.

10. If readymade paraformaldehyde solution is not available this can be made from powder; in the fume hood add 4 g paraformaldehyde to 80 ml water. Heat to 60 °C for 10 min, add a few drops of 4 M NaOH to clear the solution. Cool down to room temperature. Adjust pH to 7 with H_2SO_4. Make up to final volume of 100 ml with water. **n.b.** One drop of NaOH leaves the solution at approximately the correct pH.

11. If a hybridization platform is not available a humid chamber can be used either floating in a water bath or placed in an incubator.

12. If commercial antifade solution is not available, glycerine can be mixed with 4X SSC 1:1.

13. Top of the range fluorescence microscopes with more or less automation are available from the Leitz, Nikon, Olympus and Zeiss. We have used them all and recommend choosing the manufacturer that gives you a good prize, has a competent salesman and offers a reliable after sale service. We prefer to spend our money on lenses and filter sets rather than automation. For years, mercury vapour lamps were used for illumination, but recently powerful LED lamps have become available and are easy to use, low in energy consumption and safe.

14. If seeds are old or do not germinate, try to increase the germination rate by treating seeds in a water bath at 38 °C.

15. To stimulate root growth from older coffee plants, repot in the same or larger pot with 3 cm new soil at the bottom.

16. Fixatives or toxic chemicals on glassware, tubes or tools used for root collection, on your hands or in the atmosphere greatly reduce metaphase index. Hence it is important to have clean tools and containers with airtight lids. Collect roots for pre-treatment first, before handling fixative.

17. Fixed material can be stored for several weeks to months before making chromosome preparations as long as it does not get warm. But even when stored cold, roots tend to get hard, and it becomes difficult to remove cytoplasm. Storing or shipping ready-made chromosome slide preparations (Sect. 3.5) may sometimes be the better option when FISH has to be done later or in another remote lab.

18. Roots can be kept at 4 °C for up to 24 h if making chromosome preparations is not possible in the same day.

19. Incorporation of label into the probe DNA can be checked with a simple test blot. Pipet a small drop on a Southern hybridization membrane and follow protocols for colorimetric detection of biotin and digoxigenin labelled DNA using alkaline phosphatase linked antibodies (*see* e.g. Roche diagnostics, Eisel et al. 2008);

20. For long-term storage or transport, cover slips can be sealed with gum or nail varnish.

21. The optics and immersion oil used for visualization must be specified for fluorescent applications including UV. Immersion oil must be kept in the dark at room temperature: heating in the sun in a salesman's car, mixing oils, or absorption of water will make oils auto-fluorescent or UV opaque.

22. https://www.olympus-lifescience.com/en/microscope-resource/primer/techniques/fluorescence/filters/ or https://www.nikon.com/products/microscope-solutions/explore/microscope-abc/learn-more-microscope/filters/index.htm or https://www.semrock.com/introduction-to-fluorescence-filters.aspx)

23. Microscope filter sets are highly specific, but due to the broad spectrum of excitation and emission of fluorochromes, it is unavoidable that bleed-through of signal occurs particularly when one probe is very strong. Particularly after excitation with the correct wavelength bleed through can be stronger, hence photographing colours in reverse order can help identifying problems.

Table 1 FISH hybridization mixture

	Component	Function	Final concentration in hybridization mix	Amount for one slide (μl)
MASTER MIX	100% Formamide[a]	Reduces stability of DNA duplex	50%	20
	20X SSC[a]	Sodium ions increase stability of DNA duplex	2X	4
	50% Dextran sulphate[a]	Inert medium to increase solution volume	10%	8
	10% SDS[a]	Detergent to improve probe penetration/distribution	0.125–0.5%	0.5–2
	100 mM EDTA (optional)[a]	Hinders DNase activity	1.25 mM	0.5
	Salmon sperm DNA 4 μg/μl[a]	Blocks sites that attract unspecific binding of probe	2–4 μg	0.5–1
Probe 1		Labelled DNA to bind to specified target; probes need to have different labels to allow separate detection	20–100 ng	1–2
Probe 2			20–100 ng	1–2
Water		To make up to full volume		x
Total				40

[a] Components of master mix. Make this for the number of slides plus one extra

Table 2 Common fluorochromes and microscope filter sets used for FISH analysis. When using DAPI and two FISH probes, images can be conveniently displayed in RGB mode with each captured image in a separate channel. All microscope and filter manufacturers have useful descriptions, graphics and often active visualizations of filter/fluorochrome combinations (*see* Note 22)

	Colour	Display colour	Example filter characteristics		
			Exciter	Dichroic mirror	Emitter
DAPI	Cyan	Blue	365/10 × 10X (360–370 nm)	400DCLP	BA400 (> 400 nm)
DAPI	Blue	Blue	D350/50 × 50X (325–375 nm)	400DCLP	D460/50 m (435–485 nm)
FITC, Alexa 488	Green	Green	D480/30 × 30X (465–490 nm)	505DCLP	D535/40 m (515–555)
TRITC, Alexa594, CY3, Texas Red	Orange	Red	D540/25 × 25X (527–552 nm)	565DCLP	D605/55 m (638–693 nm)

Acknowledgements The work was carried out in the framework of the IAEA Coordinated Research Programme CRP22005 "Efficient Screening Techniques to Identify Mutants with Disease Resistance for Coffee and Banana". P.T. has received support from the European Union's Horizon 2020 research and innovation programme under the Marie Sklodowska-Curie grant agreements No 844564 and No 101006417 for analysis of polyploid chromosomal evolution. We thank John Bailey and Adel Sepsi for sharing useful information about some of the reagents and details described in the protocols.

References

Agrawal N, Gupta M, Banga SS, Heslop-Harrison JS (2020) Identification of chromosomes and chromosome rearrangements in crop Brassicas and *Raphanus sativus*: a cytogenetic toolkit using synthesized massive oligonucleotide libraries. Front Plant Sci 11:598039. https://doi.org/10.3389/fpls.2020.598039

Ali N, Heslop-Harrison JP, Ahmad H, Graybosch RA, Hein GL, Schwarzacher T (2016) Introgression of chromosome segments from multiple alien species in wheat breeding lines with wheat streak mosaic virus resistance. Heredity 117(2):114–123

Anamthawat-Jónsson K, Thórsson AT (2003) Natural hybridisation in birch: triploid hybrids between *Betula nana* and *B. pubescens*. Plant Cell Tissue Organ Cult 75(2):99–107

Bačovský V, Hobza R, Vyskot B (2018) Technical review: cytogenetic tools for studying mitotic chromosome. In: Plant chromatin dynamics. Humana Press, New York, pp 509–535

Bytof G, Knopp SE, Kramer D et al (2007) Transient occurrence of seed germination processes during coffee post-harvest treatment. Ann Bot 100(1):61–66

Chang K-D, Fang S-A, Chang F-C, Chung M-C (2010) Chromosomal conservation and sequence diversity of ribosomal RNA genes of two distant Oryza species. Genomics 96:181–190

Eira MTS, Silva EA, De Castro RD et al (2006) Coffee seed physiology. Braz J Plant Physiol 18(1):149–163

Eira MTS, Walters C, Caldas LS (1999) Water sorption properties in *Coffea* spp. seeds and embryos. Seed Sci Res 9(4):321–330

Eisel D, Seth O, Grünewald-Janho S, Kruchen B (2008) DIG application manual for non-radioactive in situ hybridization, 4th edn. Roche Diagnostics GmbH, Mannheim

Forsström PO, Merker A, Schwarzacher T (2002) Characterisation of mildew resistant wheatrye substitution lines and identification of an inverted chromosome by fluorescent in situ hybridisation. Heredity 88(5):349–355

Gerlach WL, Dyer TA (1980) Sequence organization of the repeating units in the nucleus of wheat which contain 5S rRNA genes. Nucleic Acids Res 8:4851–4865

Hamon P, Siljak-Yakovlev S, Srisuwan S, Robin O, Poncet V, Hamon S, De Kochko A (2009) Physical mapping of rDNA and heterochromatin in chromosomes of 16 *Coffea* species: a revised view of species differentiation. Chromosome Res 17(3):291–304

Herrera JC, D'Hont A, Lashermes P (2007) Use of fluorescence in situ hybridization as a tool for introgression analysis and chromosome identification in coffee (*Coffea arabica* L.). Genome 50(7):619–626

Heslop-Harrison JS, Schwarzacher T (2011) Organization of the plant genome in chromosomes. Plant J 66:18–33

Iacia AAS, Pinto-Maglio CAF (2013) Mapping pachytene chromosomes of coffee using a modified protocol for fluorescence in situ hybridization. AoB Plants 2013:5

Krug CA (1934) Contribuição para o estudo da citologia do gênero *Coffea*. Boletim Técnico 11:1–10. Instituto Agronômico, Campinas

Krug CA (1937) Observaçõe scitológicas em *Coffea*. III. Boletim Técnico 37:1–19. InstitutoAgronômico, Campinas

Kumar S, Kiso A, Kithan NA (2021) Chromosome banding and mechanism of chromosome aberrations. In: Larramendy ML, Soloneski S (eds) Cytogenetics-classical and molecular strategies for analysing heredity material. IntechOpen, London, pp 45–60

Lashermes P, Combes MC, Trouslot P, Charrier A (1997) Phylogenetic relationships of coffee-tree species (*Coffea* L.) as inferred from ITS sequences of nuclear ribosomal DNA. Theor Appl Genet 94(6–7):947–55

Lashermes P, Paczek V, Trouslot P, Combes MC, Couturon E, Charrier A (2000) Brief communication. Single-locus inheritance in the allotetraploid *Coffea arabica* L. and interspecific hybrid *C. arabica* X *C. canephora*. J Hered 91(1):81–5

Liu Q, Li X, Zhou X, Li M, Zhang F, Schwarzacher T, Heslop-Harrison JS (2019) The repetitive DNA landscape in *Avena* (Poaceae): chromosome and genome evolution defined by major repeat classes in whole-genome sequence reads. BMC Plant Biol 19:226

Lombello RA, Pinto-Maglio CAF (2003) Cytogenetic studies in *Psilanthuse bracteolatus* Hiern., a wild diploid coffee species. Cytologia 68(4):425–429

Lombello RA, Pinto-Maglio CAF (2004a) Cytogenetic studies in *Coffea* L. and *Psilanthus* Hook.f. using CMA/DAPI and FISH. Cytologia 69(1):85–91

Lombello RA, Pinto-Maglio CAF (2004b) Heterochromatin and rDNA sites in *Coffea* L. chromosomes revealed by FISH and CMA/DAPI. I: *C. humilis, C. kapakata, C.* sp. Moloundou and *C. stenophylla*. Caryologia 57(1):11–17

Lombello RA, Pinto-Maglio CAF (2004c) Heterochromatin and rDNA sites in Coffea L. chromosomes revealed by FISH and CMA/DAPI II: *C. canephora* cv. Apoatã, *C. salvatrix* and *C. sessiliflora*. Caryologia 57(2):138–143

Mandáková T, Pouch M, Brock JR, Al-Shehbaz IA, Lysak MA (2019) Origin and evolution of diploid and allopolyploid camelina genomes was accompanied by chromosome shattering. Plant Cell 31:2596–2612

Melese YY, Kolech SA (2021) Coffee (*Coffea arabica* L.): methods, objectives, and future strategies of breeding in Ethiopia. Sustainability 13(19):10814

Mendes AJT (1957) Citologia das espécies de *Coffea*: sua importância para o melhoramento do cafeeiro. In: Instituto Agronômico. I Curso de Cafeicultura, 3rd edn. IAC, Campinas, pp 37–45

Niemelä T, Seppänen M, Badakshi F, Rokka VM, Heslop-Harrison JP (2012) Size and location of radish chromosome regions carrying the fertility restorer Rfk1 gene in spring turnip rape. Chromosome Res 20(3):353–361

Nishibayashi S, Hayashi Y, Kyozuka J et al. (1989) Chromosome variations in protoplast-derived calli and in plants regenerated from the calli of cultivated rice (*Oryza sativa* L.). Japan J Genet 64(5):355–361

Paesold S, Borchardt D, Schmidt T, Dechyeva D (2012) A sugar beet (*Beta vulgaris* L.) reference FISH karyotype for chromosome and chromosome-arm identification, integration of genetic linkage groups and analysis of major repeat family distribution. Plant J 72(4):600–611

Patokar C, Sepsi A, Schwarzacher T, Kishii M, Heslop-Harrison JS (2016) Molecular cytogenetic characterization of novel wheat-*Thinopyrum bessarabicum* recombinant lines carrying intercalary translocations. Chromosoma 125(1):163–172

Patui S, Clincon L, Peresson C et al (2014) Lipase activity and antioxidant capacity in coffee (*Coffea arabica* L.) seeds during germination. Plant Sci 219:19–25

Pierozzi NI, Pinto-Maglio CA, Cruz ND (1999) Characterization of somatic chromosomes of two diploid species of *Coffea* L. with acetic orcein and C-band techniques. Caryologia 52(1–2):1–8

Pinto-Maglio CAF (2006) Cytogenetics of coffee. Braz J Plant Physiol 18(1):37–44

Pohlan HAJ, Janssens MJ (2010) Growth and production of coffee. Soils Plant Growth Crop Prod 3:101

Rathore P, Schwarzacher T, Heslop-Harrison JS, Bhat V, Tomaszewska P (2022) The repetitive DNA sequence landscape and DNA methylation in chromosomes of an apomictic tropical forage grass, *Cenchrus ciliaris*. Front Plant Sci

Salvo-Garrido H, Travella S, Schwarzacher T, Harwood WA, Snape JW (2001) An efficient method for the physical mapping of transgenes in barley using in situ hybridization. Genome 44:104–110

Schwarzacher T (2016) Preparation and fluorescent analysis of plant metaphase chromosomes. In: Caillaud M-C (ed) Plant cell division: methods and protocols, methods in molecular biology, vol 1370, pp 87–103

Schwarzacher T, Heslop-Harrison JS (2000) Practical in situ hybridization. Bios Scientific, Oxford, pp 203. ISBN: 9781859961384

Schwarzacher T (2003) DNA, chromosomes, and in situ hybridization. Genome 46(6):953–962

Schwarzacher T, Ambros P, Schweizer D (1980) Application of Giemsa banding to orchid karyotype analysis. Plant Syst Evol 134:293–297

Schwarzacher T, Leitch AR, Bennett MD, Heslop-Harrison JS (1989) In situ localization of parental genomes in a wide hybrid. Ann Bot 64:315–324

Široký J, Lysák MA, Doležel J et al (2001) Heterogeneity of rDNA distribution and genome size in *Silene* spp. Chromosome Res 9(5):387–393

Szinay D, Chang SB, Khrustaleva L, Peters S, Schijlen E, Bai Y, Stiekema WJ, Van Ham RCHJ, de Jong H, Klein Lankhorst RM (2008) High-resolution chromosome mapping of BACs using multi-colour FISH and pooled-BAC FISH as a backbone for sequencing tomato chromosome 6. Plant J 56:627–637

Tomaszewska P, Pellny TK, Hernández LM, Mitchell RAC, Castiblanco V, De Vega JJ, Schwarzacher T, Heslop-Harrison JS (2021) Flow cytometry-based determination of ploidy from dried leaf specimens in genomically complex collections of the tropical forage grass *Urochloa* s.l. Genes 12:957

Tomaszewska P, Vorontsova MS, Renvoize SA, Ficinski SZ, Tohme J, Schwarzacher T, Castiblanco V, de Vega JJ, Mitchell RAC, Heslop-Harrison JS (2023) Complex polyploid and hybrid species in an apomictic and sexual tropical forage grass group: genomic composition and evolution in *Urochloa* (*Brachiaria*) species. Ann Bot. https://doi.org/10.1093/aob/mcab147

Tomaszewska P, Kosina R (2021) Cytogenetic events in the endosperm of amphiploid *Avena magna* × *A. longiglumis*. J Plant Res 134:1047–1060

Zaki NM, Schwarzacher T, Singh R, Madon M, Wischmeyer C, Hanim Mohd Nor N, Zulkifli MA, Heslop-Harrison JS (2021) Chromosome identification in oil palm (*Elaeis guineensis*) using in situ hybridization with massive pools of single copy oligonucleotides and transferability across Arecaceae species. Chromosome Res 29:373–390